Advances in Intelligent Systems and Computing

Volume 1101

The series "Advances in Intelligent Systems and Computing" contains publications on theory, applications, and design methods of Intelligent Systems and Intelligent Computing. Virtually all disciplines such as engineering, natural sciences, computer and information science, ICT, economics, business, e-commerce, environment, healthcare, life science are covered. The list of topics spans all the areas of modern intelligent systems and computing such as: computational intelligence, soft computing including neural networks, fuzzy systems, evolutionary computing and the fusion of these paradigms, social intelligence, ambient intelligence, computational neuroscience, artificial life, virtual worlds and society, cognitive science and systems, Perception and Vision, DNA and immune based systems, self-organizing and adaptive systems, e-Learning and teaching, human-centered and human-centric computing, recommender systems, intelligent control, robotics and mechatronics including human-machine teaming, knowledge-based paradigms, learning paradigms, machine ethics, intelligent data analysis, knowledge management, intelligent agents, intelligent decision making and support, intelligent network security, trust management, interactive entertainment, Web intelligence and multimedia.

The publications within "Advances in Intelligent Systems and Computing" are primarily proceedings of important conferences, symposia and congresses. They cover significant recent developments in the field, both of a foundational and applicable character. An important characteristic feature of the series is the short publication time and world-wide distribution. This permits a rapid and broad dissemination of research results.

** Indexing: The books of this series are submitted to ISI Proceedings, EI-Compendex, DBLP, SCOPUS, Google Scholar and Springerlink **

More information about this series at http://www.springer.com/series/11156

Debabala Swain · Prasant Kumar Pattnaik ·
Pradeep K. Gupta
Editors

Machine Learning and Information Processing

Proceedings of ICMLIP 2019

 Springer

Editors
Debabala Swain
Department of Computer Science
Rama Devi Women's University
Bhubaneswar, Odisha, India

Prasant Kumar Pattnaik
School of Computer Engineering
Kalinga Institute of Industrial Technology
Deemed to be University
Bhubaneswar, Odisha, India

Pradeep K. Gupta
Department of Computer Science
and Engineering
Jaypee University of Information
Technology
Waknaghat, Himachal Pradesh, India

ISSN 2194-5357   ISSN 2194-5365   (electronic)
Advances in Intelligent Systems and Computing
ISBN 978-981-15-1883-6   ISBN 978-981-15-1884-3   (eBook)
https://doi.org/10.1007/978-981-15-1884-3

This Springer imprint is published by the registered company Springer Nature Singapore Pte Ltd.
The registered company address is: 152 Beach Road, #21-01/04 Gateway East, Singapore 189721, Singapore

Organization

Chief Patron

Pramod Kumar, President, ISB&M School of Technology, Pune, India

Patron

Pankaj Kumar Srivastav, Principal, ISB&M School of Technology, Pune, India

Steering Committee

Jayasri Santhappan, NorthBay Solution, TX, USA
Maheshkumar H. Kolekar, IIT Patna, India
Raghvendra Kumar, LNCT Group of Colleges, Jabalpur, India

General Chairs

Debabala Swain, Rama Devi Women's University, Bhubaneswar, India
Prasant Kumar Patnaik, KIIT Deemed University, Bhubaneswar, India

Technical Program Committee Chair

Pradeep K. Gupta, Jaypee University of Information Technology, Solan, India

Convener

Vivek Srivastav, ISB&M School of Technology, Pune, India

Organizing Chairs

Chanakya Jha, ISB&M School of Technology, Pune, India
M. P. Yadav, ISB&M School of Technology, Pune, India

Organizing Secretaries

Pallavi Jha, ISB&M School of Technology, Pune, India
Anirudha Bulbule, ISB&M School of Technology, Pune, India

Organizing Committee

Yogendra Jain, ISB&M School of Technology, Pune, India
Vivek Doke, ISB&M School of Technology, Pune, India
Ashwin Somkuwar, ISB&M School of Technology, Pune, India
Prakash Gaikwad, ISB&M School of Technology, Pune, India

Finance Chair

Bijay Ku Paikaray, Centurion University of Technology and Management, Odisha, India

Technical Program Committee

Manas Ranjan Patra, Berhampur University, India
Priyadarshini Adyasha, Telecom SudParis, France
Mohit Mittal, Kyoto Sangyo University, Japan
Saurabh Bilgaiyan, KIIT Deemed University, India
Banchhanidhi Dash, KIIT Deemed University, India
Prashant Gadakh, I2IT, Pune, India
Ramakrishna Gandi, AITS, Tirupati, India
Monalisa Jena, Rama Devi Women's University, India
Ashok Misra, Centurion University of Technology and Management, India
Maulika Patel, G H Patel College of Engineering and Technology, India
Jayanta Mondal, Mody University, Jaipur, India
Annannaidu Paidi, Centurion University of Technology and Management, India
Sony Snigdha Sahoo, Utkal University, India
Santwana Sagnika, KIIT Deemed University, India
Puran Singh, Vishwakarma Institute of Technology, India
Debabrata Swain, Vishwakarma Institute of Technology, India
Shanmuk Srinivas Amiripalli, GITAM University, India

Preface

In the current era of computing, technologies like machine learning, information processing, Internet of things, and data analytics have their own significance. Numerous studies and researches are continuously proposed and implemented using these techniques. This volume contains papers presented at the First International Conference on Machine Learning and Information Processing (ICMLIP 2019) held during December 27–28, 2019, at ISB&M School of Technology, Pune, India. The main objective of organizing this conference was to support and promote innovative research works of the students, researchers, academics, scientists, and industry persons of next generation into a common platform for mutual benefits and knowledge sharing.

The Program Committee of ICMLIP 2019 is very much appreciative to the authors who have shown immense interest in the form of paper submissions and huge publicity throughout India and abroad. So that, a total of 154 papers were received, out of which 48 papers were accepted for presentation and publication, after going through a rigorous peer review process in Springer AISC series. We are very much thankful to our reviewers for their sincere timely efforts in filtering the high-quality papers.

The conference could not be completed without the distinguished personalities including Dr. Jayasri Santhappan, NorthBay Solution, TX, USA; Prof. Maheshkumar H. Kolekar, IIT Patna, India; Prof. Anu Gokhle, Illinois State University, USA; Dr. Gopichand Agnihotram, Wipro Technologies Pvt. Ltd, Bangalore; Dr. Sarita Supakar, RDWU, Bhubaneswar; and Dr. Niranjan Acharya, RDWU, Bhubaneswar, for their support and guidance.

We are very grateful for the participation of all speakers in making this conference a memorable event. The Program Committee of ICMLIP 2019 is obliged to Dr. Pankaj Kumar Srivastav, Principal, ISB&M, Pune, Prof. Vivek Srivastav, and all faculty members of the Applied Science Department, ISB&M, Pune, for their valuable backing throughout the conference to make it a grand success.

Bhubaneswar, India   Debabala Swain
Bhubaneswar, India   Prasant Kumar Pattnaik
Waknaghat, Himachal Pradesh   Pradeep K. Gupta

Contents

About the Editors

Debabala Swain is working as Associate Professor in Department of Computer Science, Rama Devi Women's University, Bhubaneswar, India. She has more than a decade of teaching and research experience. Dr. Swain has published number of Research papers in peer-reviewed International Journals, Conferences and book chapters. Her area of research interest includes High performance Computing, Information Security, Machine Learning, IoT.

Prasant Kumar Pattnaik Ph.D (Computer Science), Fellow IETE, Senior Member IEEE is a Professor at the School of Computer Engineering, KIIT Deemed University, Bhubaneswar. He has more than a decade of teaching and research experience. Dr. Pattnaik has published numbers of Research Papers in peer-reviewed International Journals and Conferences. He also published many edited book volumes in Springer and IGI Global Publication. His areas of interest include Mobile Computing, Cloud Computing, Cyber Security, Intelligent Systems and Brain Computer Interface. He is one of the Associate Editor of Journal of Intelligent & Fuzzy Systems, IOS Press and Intelligent Systems Book Series Editor of CRC Press, Taylor Francis Group.

Dr. Pradeep K. Gupta is working as Associate Professor in Dept. of Computer Science and Engineering in Jaypee University of Information Technology, Solan, India. He has more than a decade of teaching and research experience. Dr. Gupta has published number of Research Papers in peer-reviewed International Journals and Conferences. His areas of interest include Machine Learning, Cloud Computing, IoT.

Mobile Device Transmission Security Policy Decision Making Using PROMETHEE

Sambhrant Samantraj, Satwik Dash and Prasant Kumar Patnaik

Abstract This paper focuses on the brand of mobile devices having the better quality transmission security policy available in the market as per the need of the customers. However, the criteria for each brand of the mobile device have its own functions. So, to choose the reasonable one among accessible options is a challenge and leads to decision-making issues. These issues might be tended to by multiple-criteria decision-making (MCDM) approach. PROMETHEE is one of the decision-making processes that encourage clients to choose the appropriate option depending on their own observation and the criteria they take into consideration. The experiment has been performed based on the feedback provided and the outcome is recorded.

Keywords Multiple-criteria decision making (MCDM) · PROMETHEE · Mobile transmission

1 Introduction

The increasing demand of mobile phones among all types of users brings a significant change in the market scenario. But mobile phones are not free from the perspective of security threats [1]. The quantity of aggressor performing program assault is expanding ongoing year, whose target is running various sorts of applications on cell phones [2]. There is perhaps one sort of Trojan that can contaminate clients' web seeking engine and change web pages or exchange online transactions. A few methodologies may be utilized to shield clients from this sort of assault, for example, online transactions approval and validity, site, of customer validation, security code advancement, and so forth. The job of security has been acknowledged to

S. Samantraj (✉) · S. Dash · P. K. Patnaik
Kalinga Institute of Industrial Technology, Bhubaneswar, Odisha, India
e-mail: 1605302@kiit.ac.in; ssambhrant3@gmail.com

S. Dash
e-mail: 1605303@kiit.ac.in

P. K. Patnaik
e-mail: patnaikprasantfcs@kiit.ac.in

© Springer Nature Singapore Pte Ltd. 2020   1
D. Swain et al. (eds.), *Machine Learning and Information Processing*,
Advances in Intelligent Systems and Computing 1101,
https://doi.org/10.1007/978-981-15-1884-3_1

wind up increasingly acute since numerous individuals, business, and government organizations store their information in the advanced digital format and share them utilizing a different kind of data exchange technology. An association is managing or putting away a client's data needs to choose fitting innovation to guarantee the security and protection of the information. Regularly, associations and especially little and medium associations purchase an item depending on the prevalence of the item. In any case, picking the wrong technology controls without breaking down the association's needs may prompt data spillage and subsequently result in customer misfortune, monetary misfortune, and harm to notoriety.

2 Need for MCDM

Multiple-criteria decision making (MCDM) is famously embraced for selecting the appropriate conceivable option from numerous choices. This is achieved by taking different aspects/criteria into account for selecting a suitable alternative. It is a method that uses the relative values for consideration rather than the absolute value so that there would be no ambiguity while relating different alternatives for criteria. This assesses numerous clashing criteria in basic decision making both in everyday life and in business. For instance, while obtaining a vehicle, we consider the distinctive angles, for example, cost, safety, comfort, and mileage. It is unoriginal that the least expensive vehicle is the most agreeable and the most secure one. There are different decision-making approaches present like ELECTRA [3], TOPSIS [4], analytical hierarchical process (AHP) [5], analytical network process (ANP) [6], GRA, and so forth which helps administrator for choosing distinctive administrations, items, assets and area, and so on.

3 Literature Review

Some reviews based on MCDM technique are discussed below:

Yigit [7] built up a product named SDUNESA for the choice of an appropriate learning object (LO) from a colossal vault dependent on specific criteria utilizing an AHP strategy, which lessens the seeking time and plays out the assignment proficiently. The product comprises of AJAX [8], XML [9], and SOA [10] web administrations for supporting certain highlights, for example, putting away, sharing, and choice of LO from the archive. Results demonstrate that the proposed framework chooses those LO, which meets the coveted and required criteria.

Roy and Patnaik [11] directed a study for discovering the distinctive ease of use measurements and assessment methods for estimating the ease of use of sites and web applications. Thinking about the significance of the web for all gatherings of clients, the creators proposed two new measurements to be specific gadget autonomy and support for the physically tested individual which means the significance of sites.

Al-Azab and Ayu [12] proposed an online multiple-criteria decision-making (MCDM) framework utilizing AHP technique to settle on the decision for picking the best option from the set of choices dependent on fundamental criteria. The proposed framework comprises of a web server, web application, a database, and web association. The judgments of the client are given to the AHP procedure to additionally handling with the end goal to choose the best option. The framework is structured utilizing PHP and the database administration framework is given by MySQL. The proposed framework gives the outcome precisely and helps in building another client decision component.

PROMETHEE I and PROMETHEE II strategies are produced by J. P. Brans and introduced without precedent for 1982 at the meeting "L'ingénièrie de la choice" sorted out at the University of Laval in Canada [13]. Around the same time, a few functional precedents of use of the strategies were displayed by Davignon and Mareschal [14], and quite a while later. J. P. Brans and B. Mareschal created PROMETHEE III and PROMETHEE IV techniques [15, 16]. Same creators additionally proposed visual, interactive modulation GAIA, which speaks to a realistic understanding of PROMETHEE technique, and in 1992 and 1995, they recommended two more adjustments—PROMETHEE V and PROMETHEE VI [3, 4]. Numerous effective executions of PROMETHEE technique into different fields are clear, and in that capacity, these strategies found their place in keeping the banking, ventures, medication, science, the travel industry, and so on [17].

3.1 MCDM Procedure

There are numerous MCDM procedures like ELECTRA [3], TOPSIS [4], AHP (analytical hierarchical process (AHP) [5], analytic network process (ANP) [6], and GRA. Our work focuses on PROMETHEE technique. PROMETHEE technique is dependent on a common examination of every elective alternative regarding every one of the chosen criteria. With the end goal to perform elective ranking by PROMETHEE technique, it has an obligation to characterize preference function $P(a, b)$ for alternatives a and b for characterizing criteria. Let $f(a)$ be the estimation of the criteria for alternative a and also $f(b)$, of similar criteria for alternative b. It is viewed as that alternative a is superior to b if $f(a) > f(b)$. The preference happens from values 0 to 1 (Table 1).

Table 1 Preference fuction table

$P(a, b) = 0$	a is indifferent to b
$P(a, b) \approx 0$	a is feebly favored over b, $f(a) > f(b)$
$P(a, b) \approx 1$	a is unequivocally favored over b, $f(a) \gg f(b)$
$P(a, b) = 1$	a is entirely favored over b, $f(a) \ggg f(b)$

Note that $P(a, b) \neq P(b, a)$

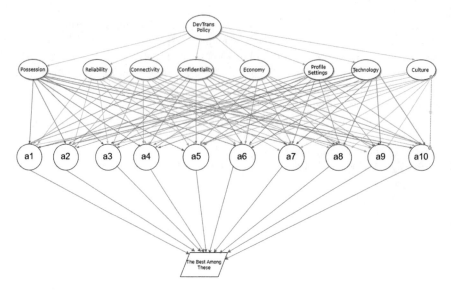

Fig. 1 Mobile transmission security policy

Higher preference is characterized by a higher value from the given interval. The administrator needs to pursue specific preference function for each criterion. The following Fig. 1 demonstrates six summed up criterion function which is gotten by selecting the extreme four inside the table of criteria from the given basis. All the six summed up capacity is mentioned by linear functions with the exception of the Gauss criterion. It is important to decide the estimation of function preference activity "a" in connection to activity "b" for every criterion and ascertains index of preference for the foundation an in connection to b. The index preference (IP) is calculated in the following way:

$$\text{IP}(a, b) = \sum_{j}^{n} W_j P_j(a, b), \quad \sum W_j = 1 \tag{1}$$

and W_j is the weight of criterion "j."

If all criteria have the same weight, i.e., $W_j = 1/n$, then index preference is:

$$\text{IP}(a, b) = \left(\frac{1}{n}\right) \sum_{j}^{n} P_j(a, b) \tag{2}$$

After deciding index preference IP (a, b), it is at long last conceivable to figure alternative imperfection index $T(a)$, the estimation of which speaks to the signification of the option.

$$T(a) = \left(\sum_{x \in A} \text{IP}(a, x) \right) \Big/ i \tag{3}$$

4 An Analogy

Considering a situation where mobile transmission security policy, rather than the four perspectives say possessions, reliability, connectivity, confidentiality, we additionally found from the feedback that profile settings, technology, culture, and economy are other huge viewpoints to be considered in defining mobile phone transmission security policy. The accompanying part depicts the eight parts of the information security arrangement connected in this investigation.

4.1 Criteria

The criteria for said analogy have been discussed in Table 2 with possession as C1, reliability as C2, connectivity as C3, confidentiality as C4, economy as C5, profile setting as C6, technology as C7, culture as C8.

Following Fig. 1 shows the model for mobile transmission security policy which consists of eight criteria and 10 alternatives with possession, reliability, connectivity, confidentiality, economy, profile settings, technology, culture as criteria and a1–a10 as different mobile brand device.

Table 2 Criteria information

C1	The phone number related to the gadget is nontransferable without the consent of the proprietor
C2	The transmission sent by one client is gotten by the predetermined beneficiary
C3	The framework is accessible to transmit and receive information
C4	The data transmission will not be intercepted
C5	Maybe the most broadly referred to mobile transmission security paper managing the financial aspects point of view is one by Anderson [16] who talks about different unreasonable motivators in the data security area. It is the expense of the gadget that is purchased
C6	It is affirmed that profile settings have turned into a required capacity by numerous cutting-edge associations that depend vigorously on the Internet to lead their tasks
C7	Innovation as far as information, equipment, and applications have been the most concerned angle since the start of the automated time
C8	Among different past angles, the cultural viewpoint is the slightest perspective talked about in scholastic papers. Then again, it is broadly demonstrated that data security ruptures are regularly caused by inner clients of an organization [15]

The feedback taken from different mobile device users is taken and the data is represented in a tabular form in Table 3. Each column represents the value of each criterion for the corresponding brand of mobile device.

The PROMETHEE technique is applied in Table 3 and the net flow is the result produced as a result in Table 4. The ranking is then depicted from the net flow.

Table 3 The table with criteria, alternatives, and weights of the respective criteria

◢	A	B	C	D	E	F	G	H	I
1	Weight	0.11	0.09	0.09	0.13	0.12	0.08	0.07	0.09
2	Alter/Crit	c1	c2	c3	c4	c5	c6	c7	c8
3	a1	1.6	7.7	7.8	5.9	6.6	7.4	6.6	7.7
4	a2	2	7.2	6.3	5.8	6.2	6.9	6.2	7.2
5	a3	2.7	6.5	5.7	4.7	5.9	6.5	5.2	7.8
6	a4	3.3	6.1	5	5	5.3	6.5	5.1	5.9
7	a5	3.2	6.6	6.1	5.4	4.6	5	5.9	6.8
8	a6	5.8	5.9	5.1	4.8	4.8	3.6	5.3	5.4
9	a7	6.6	4.6	5.3	4.6	3.8	3.4	5.3	5.2
10	a8	4.2	6.6	4.4	5.7	5.6	5.2	6	5.5
11	a9	5.6	5.8	5.3	4.6	4.2	3.7	5.9	5.6
12	a10	5.5	5.7	4.9	5	3.9	3.5	5.7	4.3

Table 4 The final table along with the criteria

◢	M	N	O	P	Q
89	Alternativ	Leaving Fl	Entry Flo\	Net Flow	Rank
90	a1	0.55	0.37	0.18	3
91	a2	0.607778	0.392222	0.2155556	1
92	a3	0.436667	0.554444	-0.117778	8
93	a4	0.403333	0.573333	-0.17	9
94	a5	0.588889	0.393333	0.1955556	2
95	a6	0.47	0.522222	-0.052222	6
96	a7	0.47	0.486667	-0.016667	5
97	a8	0.463333	0.526667	-0.063333	7
98	a9	0.49	0.466667	0.0233333	4
99	a10	0.395556	0.59	-0.194444	10

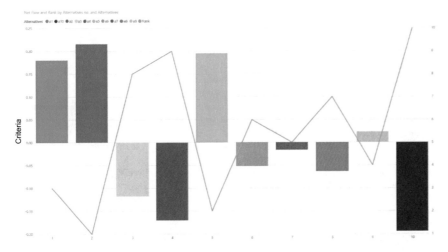

Fig. 2 Each alternative along with their net flow and rank

4.2 Result Analysis

Hence mobile brand a2 is better than others in security and economy. So, customer should prefer mobile brand a2. Figure 2 represents each alternative with values as their criteria with different legends.

In Fig. 2, the net flow value of each alternative is shown as column values (left) and the rank is shown by line values (right). The alternatives are the legends. Lower the line value more suitable is the brand off mobile device.

5 Conclusion

This paper endeavors to provide evidence that multiple-criteria decision making encourages the customer to pick the mobile phone mark for accomplishing the optimal mobile transmission security. The procedure utilized here is PROMETHEE. The point is tantamount to indicate how decision-making methods can encourage a man's security and protection to be kept up by choosing an appropriate mobile phone as decision making by his financial plan. According to our appraisal of mobile phone brands, we see that brand "a2" is most appropriate for a man contrasted with the alternate criterion. The other multiple-criteria decision-making techniques will be applied as a future scope of work.

References

1. Smartphone Security Threats and Trends. http://www.fishnetsecurity.com/News-Release/Smartphone-Attacks-and-Hacking-Security-Threats-and-Trends-2011, Description: a webpage to introduce recent security threats trends in 2011.
2. http://en.wikipedia.org/wiki/Man_in_the_Browser.
3. Hajebrahimi, Mehdi. 2014. Evaluation of financial efficiencies of cement companies accepted In Tehran Stock Exchange by multi–criterion decision methods.
4. https://en.wikipedia.org/wiki/TOPSIS.
5. Hwang, Junseok, and Irfan Syamsuddin. 2009. *Information security policy decision making.* Republic of Korea, B. C.: Seoul National University.
6. Saaty, Thomas L. 1996. *Decision making with dependence and feedback: The analytic network process.* Pittsburgh, PA: RWS Publications.
7. İnce, M., T. Yiğit, and A.H Işık. Multi criteria decision making system for learning object repository
8. http://www.w3schools.com/ajax/.
9. http://www.w3schools.com/xml/.
10. https://en.wikipedia.org/wiki/Service-oriented_architecture.
11. Roy, S., and P.K. Pattnaik. 2013. Some popular usability evaluation techniques for websites. *Advances in Intelligent Systems and Computing.*
12. Al-Azab, F.G.M., and M.A. Ayu. 2010. Web based multi-criteria decision making using AHP method. In *International Conference on Information and Communication Technology for the Muslim World.*
13. Brans, J.P. 1982. L'ingénierie de la décision; Elaboration d'instruments d'aide à la décision. La méthode PROMETHEE. In *L'aide à la décision: Nature, Instruments et Perspectives d'Avenir,* ed. R. Nadeau and M. Landry, 183–213. Québec, Canada: Presses de l'Université Laval.
14. Davignon, G., and B. Mareschal. 1989. Specialization of hospital services in Quebec—An application of the PROMETHEE and GAIA methods. *Mathematical and Computer Modelling* 12 (10–11): 1393–1400.
15. http://www.pwc.com/extweb.
16. Anderson, R. 2001. Why information security is hard: An economic perspective. In *Proceedings of 17th Annual Computer Security Applications Conference,* 10–14.
17. Kolli, S., and H.R. Parsaei. 1992. Multicriteria analysis in the evaluation of advanced manufacturing technology using PROMETHEE. *Computers & Industrial Engineering* 23 (1–4): 455–458.

Sambhrant Samantraj is a student of B.Tech, Computer Science at KIIT Deemed to be University. His area of research includes Multiple Criteria Decision-Making technique and Recommendation System.

Satwik Dash is a student of B.Tech, Computer Science at KIIT deemed to be University. His Area of research includes Multiple Criteria Decision-Making technique and Brain Computing Interface.

Dr. Prasant Kumar Patnaik Ph.D. (Computer Science), Senior Member of IEEE, Fellow IETE, is a Professor at the School of Computer Engineering, KIIT deemed to be University, Bhubaneswar. He has more than a decade and half of teaching and research experience. His area of research includes mobile computing, cloud computing, Brain computer Interface and Multiple Criteria Decision Making.

Stress Detection in Speech Signal Using Machine Learning and AI

N. P. Dhole and S. N. Kale

Abstract Individual person's speech is verbal way to have conversation with others. Speech many time probably becomes to know that individual person is in stressful condition or normal. These can lead with appropriate assessment of the speech signals into different stress types to evoke that the individual person is in a fit state of mind. In this work, stress identification and classification algorithms are developed with the aid of machine learning (ML) and artificial intelligence (AI) together with MFCC feature extraction methods. The machine learning and AI-based approaches use an intelligent combination of feature selection and neural optimization algorithms to train and to improve the classification and identification accurateness of the system. Comparison is done with approach of classical neural networks and fuzzy inference classifiers. The proposed system is suitable for real-time speech and is language and word independent. The work is implemented using MATLAB 2014 version.

Keywords MFCC · SVM · RBF · MLP · RNN · ANFIS

N. P. Dhole (✉)
Department of Electronics and Telecommunication Engineering, PRMIT&R Badnera, Badnera, Maharashtra 444701, India
e-mail: npdhole34@gmail.com

S. N. Kale
Department of Applied Electronics, Sant Gadge Baba Amravati University, Amravati, Maharashtra 444602, India
e-mail: sujatakale@sgbau.ac.in

© Springer Nature Singapore Pte Ltd. 2020
D. Swain et al. (eds.), *Machine Learning and Information Processing*,
Advances in Intelligent Systems and Computing 1101,
https://doi.org/10.1007/978-981-15-1884-3_2

Table 1 Types of stress and categorization

Stress type	Stress description	Stress category issues
Type 1	Psychological	High workload and sentiments
Type 2	Perceptual	Noise
Type 3	Physiological	Medicines and illness
Type 4	Physical	Vibration and physical workload
Normal	Normal	No stress

1 Introduction

Through speech, individuals' features like loudness, pitch, intensity, frequency, time, amplitude, etc. can be noted. These features can be calculated for both stressed and normal speeches. It becomes important to detect state of a person suffering from high workload, noises, and physical environmental factors. The various problems faced by stressed persons are psychological, perceptual, physiological, and physical [1].

Table 1 displays stress types and various issues related to it. The need has been raised to study stress in speech in the current years to divert more attention toward person's behavior and outlook. This speech identification has been strongly increased for significant contributions to national and international industries as well as societies. Detection of stress in speech has become a challenge in human perception and their intelligence. Stress and emotion go hand in hand with equal meaning. Emotions may be sentiments having fury, sorrow, cheerfulness, terror, hatred, bliss, upset, and tediousness. But the core stress issues are left uncovered by emotions. Stress is always concerned with mental and physical ascertains. Knowledge of a speaker's stressed state condition can also be beneficial in improving the performance of speech identification algorithm system which gets along with human stress mechanism. Researchers are working from decades for stress recognition from speech signal so that one can limit a person going under depression or suicide. There is a scope to work in this field for the benefits of the society. It is very essential to establish a system which will identify stress and different types of stress.

2 Databases for Stress Detection in Speech

The speech data used for the intention of this research study is selected with stressed and anxiety benchmark databases with different languages.

Table 2 Sample numbers of Berlin database

S. No.	Berlin database	No. of samples
1	Males	233
2	Females	302
	Total Samples	535

Table 3 Number of samples of TESS database

S. No.	TESS database	No. of samples
1	Females	20

2.1 Berlin Database

The Berlin disturbing and stressed speech database or corpus consists of ten sentences [2]. This database is in German language consisting of sentences audio wave files of both genders under seven stressed sentiment classes. They are all related to anxiety feelings of a terrified person. The recording is done in studio conditions using super recording equipment. Berlin emotional speech database is broadly used in emotional and stress speech recognition. This database has been chosen for the reasons; as the quality of its recording is very good, it is public and popular database of emotion recognition. The Berlin University created the emotional and disturbing speech database in Department of Communication Science consisting of 535 stressed speech wave files of 16 kHz having 3–8 s duration. The number of samples is provided in Table 2.

2.2 Toronto Emotional Speech Set Database (TESS)

This database is developed in Northwestern University has a collection of 200 words [3]. Two actresses are from the Toronto speaking English language with musical preparation. Table 3 shows the number of samples in TESS database.

2.3 Database Developed

Audacity is open-source recording software. It is comprehensible audio editor and recorder for operating systems. Audacity software is created and disseminated under the GNU General Public License (GPL) [4]. Audacity software records live audio helpful for digital recording. In this research work, the speech samples are recorded using audacity with different frequencies 8, 16, and 44.1 kHz. The recordings are for female as well as male in regional languages such as Hindi, Marathi, and English. Details of the database are shown in Table 4.

Table 4 Sample number of database developed

S. No.	Database developed	No. of samples
1	Female	59
2	Male	61
	Total samples	120

3 Machine Learning and AI-Based Stress Classifier Algorithm

The proposed classifier is based on real-time learning; it uses a combination of feature selection and neuron optimization.

3.1 Machine Learning and AI-Based Stress Classifier

The proposed classifier is based on real-time learning; it uses a combination of feature selection and neuron optimization. It has the following components,

- Feature selection unit (FSU).
- Neuron optimization unit (NOU).
- Machine learning lookup tables (MLUTs).
- AI selection unit (ASU).

The feature selection unit (FSU) pull outs the Mel-frequency cepstral components (MFCCs) through the input, i.e., speech signal. The MFCC features are extensively applied in stress speech identification. It conserves the signal characteristics without deteriorating the performance to get the best accuracy rate [5]. Using MFCC extraction, cepstral and frequency spectrum features are calculated. In a complete speech sentence, there are more than three lakh features. Using MFCC, it is able to reduce up to thousand without losing the main information from speech. MFCC includes procedures as pre-emphasis, framing, hamming window, DFT, Mel filter bank. Pre-emphasis is used for speech signal preprocessing to achieve similar amplitude for high frequency and low-frequency formants occurring in frequency domain. FIR filter is used to filter out unnecessary information or data. Usually, the pre-emphasis filter aids with first-order FIR filter having transfer function in z-domain as,

$$H(Z) = 1 - bz^{-1} \quad 0 \le b \le 1$$

where b is pre-emphasis parameter.

In windowing technique, a hamming window is the best option for stress speech detection because it maintains conciliation way out between time required and frequency resolution of the formants. The pre-emphasized signal is blocked-up on frames of samples (frame size); each of them is multiplied with a hamming window to maintain stability of the signal. Windows are chosen to knot the signal at the edges of each frame. The MATLAB syntax used for hamming window is, w = hamming (L) where w is an L point symmetric hamming window. To each frame, discrete Fourier transform is applied and is given by,

$$X[k] = \sum_{n=0}^{N-1} x[n]e^{-j(4\pi^2)/N} \quad 0 \le k \le N-1$$

The Fourier spectrum of a signal is inconsistently quantized and carries out Mel filter bank procedure. Firstly, the window frames are identically broken on the Mel-scale and further transformed back to the frequency spectrum. This spaced spectrum is then multiplied to the Fourier power band to achieve the MFCC filter bank coefficients. A Mel filter collection consists of filters that are linearly gapped at low and high frequencies capturing the phonetically vital quality of the stressed input signal while smoothening irrelevant variation in the higher frequency spectrum. The Mel-frequency coefficient is calculated using the following equation.

$$F(M) = \log\left\{ \sum_{k=0}^{N-1} [X(k)]^2 Hm(k) \right\} \quad 0 \le m \le M$$

The discrete cosine transform (DCT) equation of the filter bank providing the MFCC coefficients as,

$$d[n] = \sum_{m=0}^{M} F(M) \cos\left\{ \frac{\pi n(m-1)}{2M} \right\}, \quad 0 \le n \le M$$

The MFCCs consists of the frequency responses, the reconstructed frequency bands, and the pitch of the input speech. For a one-second speech signal sampled at 44,100 Hz, more than 22 million feature values are obtained. These feature values describe the speech independent of the spoken languages and words. This feature can be used for classification of speech in depending language and word. Due to its large feature length, these feature vectors are unusable for neural network training, and thus, there is a need of feature reduction. DB8 transform is usually used in speech detection applications. FSU uses the DB8 transform in order to produce variable length feature patterns for training the neural network classifier. The learning rule is an algorithm stating the modification of the parameters of the neural network to generate the preferred output for a given input of a network by updating the weights and thresholds of the networks.

The neuron optimization unit (NOU) handles all the operations related to neuron selection and processing. The proposed classifier uses a series of trained neural networks, each of which have intelligently selected neurons for the best-selected features and most optimum accuracy. The NOU performs neuron selection by observing the resulting accuracy and then taming the amount of neurons in each layer of the networks until a sufficient level of accuracy is achieved. The MLUTs have

- Selected features.
- Amount of neurons.
- Learning function.
- Training samples.
- Accuracy.

Neurons in the starting, i.e., input layer are MFCC selected features. These observations are then stored in the machine learning lookup tables (MLUTs).

3.2 The MLUTs Are Filled Using the Following Algorithm

1. Let the number of iterations be Ni
2. Let the number of solutions per iteration be Ns
3. Let the maximum amount of samples for training be *Nmax*, and the minimum amount of samples for training be *Nmin*
4. For each solution 's,' in Ns, perform the following steps,

   (a) Select a random number between *Nmin* and *Nmax* = Nr
   (b) Apply DB8 wavelet transform to each of the training samples, until the features of each sample are reduced to,
   Nr = dwt(Nr, 'db8')
   (c) Select random number of neurons, such that selected neurons are more than Nr
   (d) Train the neural net with Nr features, and evaluate its accuracy Ai

5. Evaluate the mean accuracy and discard all solutions where accuracy is less than the mean accuracy.
   Mean accuracy = Sum of all fitness values/total number of solutions.
6. Repeat Steps 4 and 5 for Ni iterations, and tabulate the results in the MLUT.

Once the MLUT is filled with sufficient entries, the artificial intelligence (AI) selection performs classification of the speech signal. This performs neural net evaluation for all the neural nets stored in the MLUT and obtains the classification results from each of the entries. From all these results, the most frequently occurring class is selected, and the solutions which classified the evaluation sequence to that class are marked with +1, rest others are marked with −1. After several evaluations, all the negatively marked solutions are discarded, and the MLUT filling algorithm is applied to fill in those discarded slots. During this step, the previously evaluated signals are also taken into the training set so that the system can self-learn and be adaptive to changing inputs.

The proposed algorithm is tested on three different databases, and the results and observations are described in the next section. It is also compared with existing neural network-based approaches in terms of classification accuracy and sensitivity.

The stress detection system is under five different categories namely as,

- Stress Type 1
- Stress Type 2
- Stress Type 3
- Stress Type 4
- Normal speech or no stress.

The proposed classifier approach is compared with existing neural network-based approaches and obtained an improvement in overall system performance.

3.3 Elimination of Noise from Speech

Eliminating noise from speech signal is a difficult task for stress speech recognition. An algorithm is developed using function fitting network (FitNet) for noise elimination. It is being developed and trained separately filtering out noise from stressed speech. This network has input layer in combination with hidden and output layers. The input layer has amount of neurons in speech signal where noise is added. The improved signal after removing noise is the one neuron in the output layer. The numbers of outputs are five. The FitNet separately has different architecture, effective or optimized training rules and parameters to reduce noise from speech signal. The FitNet is capable of boosting the rate and enhance the speech by obtaining elevated PSNR and less MSE [6].

Artificial neural network (ANN) approach and the FitNet algorithm for signal improvement through enhancement aid to eliminate noise in every signal of speech [7]. Optimization algorithm used in training process is Gradient descent momentum (GDM) with function fitting network. The whole process can be seen from Fig. 1

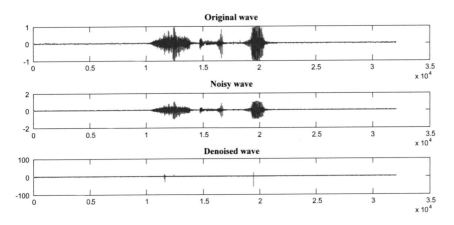

Fig. 1 Original, noisy, and denoised wave for an untrained speech signal using FitNet

that includes the original speech signal where a noise is added, and after filtering, the denoised speech signal is obtained which is untrained signal. The untrained signal is given to all seven classifiers and calculated its accuracy.

3.3.1 Signal Improvement Through Enhancement

Signal improvement through enhancement plays a significant job in stress speech identification systems applications. Enhancement of speech is a method to minimize or eradicate the noise from speech signal. This technique recovers original speech signal by enhancing the signal-to-noise ratio (SNR) preserving the original signal [8]. Here in this algorithm, observations are taken by changing SNR to 10 DB, 11 DB, and 12 DB and calculated PSNR, MSE, and time required.

3.3.2 Gradient Descent Momentum with Function Fitting Artificial Neural Network and Adaptive Filter

The training algorithm used here is Gradient descent momentum adjusted the ANN weights to decrease MSE. This is amongst the best training algorithms to reduce maximum possible errors [9]. For training the multilayer feed-forward networks, the GDM is a standard optimization algorithm to excellence the performance of neural network training. GDM training algorithm revises the weights and biases of the neural network toward the direction of degradation of the function. The FitNet mends into feed-forward networks containing hidden layer and adequate neurons, and the output is calculated directly from the input through FitNet. The process for GDM with FitNet includes,

1. Initialization of weights.
2. Compute error signal.
3. Pass error signal through the modified subband filter.
4. Update the weight based on normalized least mean square adaptive filter.
5. Perform weight stacking to get the new weights.

Adaptive filter is used in this algorithm for speech improvement through enhancement. As noise is nonstationary signal which overlies with the frequencies of speech signal which in turn reduces the identification efficiency. Adaptive filter thus becomes important which operates properly in unknown and time unstable environments. Adaptive filter improves the performance by statistical uniqueness of its operation.

Algorithm for Elimination of Noise Process

1. Insert noise to the speech signal by applying random noise.
2. Compute the outputs of hidden layer and output layer.

3. Compute the MMSE, PSNR of noisy input signal.

$$MMSE = \sqrt{\left\{ \frac{mean\left(mean(2N1 - 2N2)^2\right), 2}{20} \right\}}$$

4. Finally, calculate PSNR at the output and time required.
   PSNR = 20 log10 (1/MMSE)

4 Results

4.1 ANN Classifiers

All the ANN classifiers namely multilayer perceptron, radial basis function, support-vector machine, recurrent neural network, feed-forward back propagation neural network, adaptive neuro-fuzzy inference system, and proposed classifiers which are designed are exhaustively studied. To all these classifiers, all three databases are applied to compute the stress identification accuracy and sensitivity analysis.

4.1.1 SVM Classifier with Different Kernels

APIA percentage for SVM classifier is shown in Table 5. The APIA percentage (values of the average percentage of the identification accuracy APIA values are calculated using different kernels such as linear, quadratic, polynomial, multilayer perceptron, and Gaussian for SVM. Their accuracy in identifying stress types is obtained.

Table 5 Stress APIA percentage of SVM classifier using various SVM kernels

Databases	APIA percentage				
	Linear kernel	Quadratic kernel	Polynomial kernel	Multilayer perceptron kernel	Gaussian kernel
Berlin database	58.6	55.3	54.3	54.6	57.5
Database developed	57.9	54.2	53.2	57.6	56.5
Toronto emotional speech database	58.9	53.2	56.7	58.9	54.9

20

4.1.2 Feed-Forward Back Propagation Neural Network (FFBPNN) Classifier

APIA percentage for FFBP NN for three different databases namely Berlin, Toronto Emotional, Database developed are displayed in Tables 6, 7, and 8. Here, the observations are considered for 12 training algorithms as Levenberg–Marquardt (L M),

Table 6 APIA (percentage) of Berlin database

Training algorithm	Nos. of neurons for hidden layer				
	1	5	10	15	18
LM	63.53	62.50	62.10	61.20	60.80
BR	67.20	66.20	65.48	64.23	63.40
BFGS QN	61.20	60.29	57.76	58.50	57.60
RB	60.01	60.40	57.80	61.40	61.23
SCG	64.20	65.29	66.30	63.50	62.48
CG	67.70	66.70	65.71	64.40	64.39
PRCG	64.50	63.21	62.90	61.17	60.25
OSS	70.17	69.80	68.70	67.15	66.40
VLRGD	61.60	60.54	57.31	57.65	62.40
GD with M	60.23	61.40	60.05	57.20	57.60
GD	62.20	61.05	60.20	58.01	57.80
FPCG	57.20	56.40	55.40	57.20	55.40

Table 7 APIA (percentage) of Toronto emotional speech set

Training algorithm	Nos. of neurons for hidden layer				
	1	5	10	15	18
LM	64.50	63.89	62.40	62.21	61.71
BR	65.32	66.51	64.92	63.22	64.32
BFGS-QN	63.19	63.81	63.02	61.47	60.11
RB	61.23	60.31	63.50	57.80	57.69
SCG	67.20	66.32	65.40	64.30	65.48
CG	65.40	64.93	63.20	62.10	61.54
PRCG	66.90	65.82	64.05	64.88	63.50
OSS	66.35	65.71	64.59	62.50	61.47
VLRGD	64.89	63.17	62.39	61.80	60.40
GD with M	67.80	66.40	65.40	66.89	63.20
GD	66.50	65.41	64.19	63.90	57.80
FPCG	60.50	60.80	57.70	57.80	58.60

Table 8 APIA (percentage) of database developed

Training algorithm	Nos. of neurons for hidden layer				
	1	5	10	15	18
LM	61.10	60.86	57.70	58.13	56.10
BR	63.20	62.17	61.18	60.99	60.49
BFGS-QN	57.30	58.70	57.39	54.55	52.15
RB	60.30	57.80	58.70	57.30	55.20
SCG	62.30	62.87	63.40	60.16	59.69
CG	63.40	62.10	63.19	61.09	60.08
PRCG	66.40	65.11	64.45	63.31	61.22
OSS	68.80	66.42	65.70	64.30	63.20
VLRGD	65.40	64.23	63.91	62.20	61.10
GD with M	64.01	63.00	62.89	61.46	60.40
GD	63.08	62.33	61.88	60.54	57.42
FPCG	62.40	62.10	62.20	60.11	57.80

Bayesian Regularization Artificial Neural Network (BRANN), Broyden–Fletcher–Goldfarb–Shanno Quasi-Newton (BFGS QN), Resilient Back propagation (RB), Scaled Conjugate Gradient (SCG), Conjugate Gradient with Powell/Beale Restarts (CG), Fletcher–Powell Conjugate Gradient (FPCG), Variable Learning Rate Gradient Descent (VLRGD), Polak–Ribiere Conjugate Gradient (PRCG), Davidon–Fletcher–Powell (DFP), Gradient Descent with Momentum (GDM), Gradient Descent (GD), One-Step Secant (OSS).

4.1.3 Adaptive Neuro-Fuzzy Inference (ANFIS) Classifier

APIA percentage for ANFIS for three databases is shown in Tables 9 and 10. Accuracy is calculated for the stress by entering number of membership functions and by varying number of training epochs to ANFIS classifier (Table 11).

4.2 Stress Detection

The MFCC features are extracted for the classifiers SVM, RBF, RNN, MLP, ANFIS, FFBPNN, and proposed classifier. Table 12 shows percentage accuracies of all neural network classifiers for three databases (Fig. 2).

Table 9 APIA (percentage) of Berlin database

Database	Entering membership functions	Number of training epochs	Accuracy percentage
Berlin database	3	100	57.80
		250	60.18
		500	63.12
	4	100	57.10
		250	57.91
		500	61.15
	5	100	60.32
		250	61.43
		500	62.20

Table 10 APIA (percentage) of Toronto emotional speech set

Database	Entering membership functions	Number of training epochs	Accuracy percentage
Toronto Emotional Speech Set (TESS)	3	100	63.50
		250	65.41
		500	67.36
	4	100	64.48
		250	66.23
		500	67.20
	5	100	66.10
		250	67.80
		500	71.00

Table 11 APIA (percentage) of database developed

Database	Entering membership functions	Number of training epochs	Accuracy percentage
Database developed	3	100	72.42
		250	73.49
		500	75.55
	4	100	71.45
		250	74.58
		500	75.56
	5	100	73.18
		250	75.03
		500	77.16

Table 12 Accuracy (percentage) of classifiers

Databases	SVM	RBF	RNN	MLP	ANFIS	FFBP NN	Proposed classifier
TESS database	61.23	70.37	80.01	84.56	62.76	77.22	97.41
Berlin database	60.3	68	77.83	86.15	65.67	76.66	96.55
Database developed	61.56	67.88	77.99	85.66	64.01	77.45	97.52

Fig. 2 Graphical representation of accuracy of different classifiers

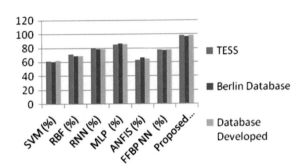

4.3 Detection of Stress Types

The proposed classifier is used to classify the stress type defined as Type 1, Type 2, and Type 3, Type 4, and normal. For this classifier, stress types and normal speech accuracies are computed for three databases namely Berlin, Toronto emotional speech, and database developed. Table 13 depicts the proposed classifier stress types of average APIA (percentage).

Table 13 Stress type APIA (percentage) for proposed classifier

Stress type	Databases		
	Berlin database	Toronto Emotional Speech	Database developed
Type 1	91.50	91.34	91.19
Type 2	92.23	92.78	97.17
Type 3	92.34	97.65	92.17
Type 4	95.12	95.12	95.16
Normal	97.34	95.00	94.15

Table 14 Male and female stress accuracy for proposed classifier

Database	Stress accuracy in percentage (male)	Stress accuracy in percentage (female)
Database developed	97.22	95.33
Berlin	93.22	94.20

Table 15 Performance parameters of GDM algorithm

Languages	MMSE	PSNR (DB)	Time required (s)
Berlin	0.023	27.98	61.56
TES database	0.016	34.76	64.80
Databases developed	0.020	37.24	60.12

4.4 Stress Detection Based on Gender

In this research works, the proposed classifier builds a strong evidence of the stress levels that depend on gender. The gender-specific stress accuracy is shown in Table 14 for databases Berlin and database developed. These databases comprise male and female speech signal.

4.5 Network Results of GDM with FitNet and Adaptive Filtering

The stress classification results are represented by removing noise. The training of the FitNet using GDM training algorithm is shown in Table 15. It consists of MMSE, PSNR in DB, and the time required in seconds for training GDM algorithm. These three different parameters are calculated for all three databases. The PSNR is computed at the output, where the signal is completely free from noise.

4.6 APIA Percentage Sensitivity of Classifiers for Stress Classification

Table 16 experiments are carried out based on testing the GDM training algorithm. Sensitivity of all seven classifiers against noise is calculated.

Table 16 APIA percentage sensitivity of classifiers

Languages	SVM	RBF	RNN	MLP	ANFIS	FFBP NN	Proposed classifier
TES database	38.77	27.63	17.99	15.44	37.24	20.78	2.59
Berlin database	37.7	32	22.17	13.85	34.33	23.34	3.45
Database developed	38.44	30.12	20.01	14.34	35.99	22.55	2.48

5 Conclusions

This paper presented a proposed classifier for detection of stress from speech signal. Various classifiers such as SVM, RBF, RNN, MLP, ANFIS, FFBP NN, and proposed classifier are implemented to identify the stress into speech signal and compared. For the SVM classifier, linear kernel is possessing highest accuracy. The FFBPNN classifier is having the highest APIA percentage obtained by OSS learning rule amongst all twelve learning rules. For ANFIS classifier, the results obtained from the APIA percentage are highest when the numbers of training epochs are increased. Performance parameters MMSE, PSNR, and time of GDM algorithm are determined for every database wherein ANN model FitNet is constructed to enhance speech signal and to eliminate noise from any clattered signal of stressful speech. Finally, these works conclude that amongst all the classifiers, the proposed classifiers possess the highest accuracy of 97.52% including the gender stress identification. Again the sensitivity of the proposed classifier is least 2.48% than all classifiers.

References

1. Steeneken, H.J.M., and J.H.L. Hansen. 1999. Speech under stress conditions: overview of the effect on speech production and on the system performance. In *IEEE International Conference on Acoustics, speech, and Signal Processing (ICASSP), Atlanta, Georgia*, vol. 4, 2079–2082.
2. Burkhardt, F., A. Paeschke, M. Rolfes, W.F. Sendlmeier, and B. Weiss. 2005. A database of German emotional speech. In *International Conference on Speech Communication (INTER-SPEECH)*, vol. 5, 1517–1520.
3. Toronto Emotional Speech Set. Database available on http://tspace.library.utoronto.com.
4. Database Developed: By using audacity software download from http://audacity.sourceforge.net/download.
5. Bors, A.G., and G. Gabbouj. 1994. Minimal topology for a radial basis function neural network for pattern classification. *Digital Signal Processing: A Review Journal* 4 (3): 173–188.
6. Omaima, N.A., and A.L. Allaf. 2015. Removing noise from speech signals using different approaches of artificial neural network. *International Journal of Information Technology and Computer Science* 7: 8–18.
7. Karam, M., H.F. Khazaal, H. Aglan, and C. Cole. 2014. Noise removal in speech processing using spectral subtraction. *Journal of Signal and Information Processing* 1026–1033.
8. Widrow, B., et al. 1975. Adaptive noise cancelling: Principles and applications. *Proceedings of the IEEE* 63: 1692–1716.

9. Chatterjee, Kalyan, et al. 2013. Adaptive filtering and compression of bio-medical signals using neural networks. *International Journal of Engineering and Advanced Technology (IJEAT)* 2 (3): 323–327.

10. MathWorks, Neural Network Toolbox 7.0, MathWorks Announces Release 2014a of the MATLAB and Simulink Product Families, MathWorks Inc. 2014.

e-Classroom Using IOT to Revolutionize the Classical Way of Knowledge Sharing

Nikhil Vatwani, Shivani Valecha, Priyanka Wadhwani, Vijay Kataria and Sharmila Sengupta

Abstract Advancement in technology has not only led to exchange of information between machines and objects but also led to reduced human intervention. With this, everything is being made smart. A classroom with the functionality to track attendance using an application, change the slides of a presentation, emailing important notes with voice access commands and managing power of the lecture hall automatically can be termed as e-Classroom. Conventional methods of knowledge sharing (or session delivery) and the use of technology are not mutually exclusive, but they complement each other. The classroom will incorporate new innovative aids for teaching which are possible only in an electronic environment. The e-Classroom project aims at flexible, interactive conduction of oral sessions, e-Records of list of sessions conducted by instructors and attendees, maintenance of e-Notes, resources management such as power management and many such modules.

Keywords Speech recognition · Artificial intelligence · Face recognition · e-Deliverables

N. Vatwani (✉) · S. Valecha · P. Wadhwani · V. Kataria · S. Sengupta
Department of Computer Engineering, Vivekanand Education Society's Institute of Technology, Mumbai, India
e-mail: nikhil.vatwani@ves.ac.in

S. Valecha
e-mail: shivani.valecha@ves.ac.in

P. Wadhwani
e-mail: priyanka.wadhwani@ves.ac.in

V. Kataria
e-mail: vijay.kataria@ves.ac.in

S. Sengupta
e-mail: sharmila.sengupta@ves.ac.in

© Springer Nature Singapore Pte Ltd. 2020
D. Swain et al. (eds.), *Machine Learning and Information Processing*,
Advances in Intelligent Systems and Computing 1101,
https://doi.org/10.1007/978-981-15-1884-3_3

1 Introduction

The impact of technology on our day-to-day lives is exponentially increasing. Even the *"Digital India"* campaign aims at enhancing the lives of people using electronic technology. Due to an increasing enhancement in technology, new and more efficient ways of doing things are being discovered. The e-Classroom project is a step toward digitization era. It makes use of various technologies such as speech recognition, etc., to provide presenters and attendees with an effective knowledge sharing environment.

Every educator thinks how they can reduce their efforts and time in performing these miscellaneous tasks and dedicate more time of the lecture for teaching. We tried to modify some of the traditional methods like taking attendance using pen and paper and to take attendance on a Google sheet, but it did not make much difference. So in order to overcome this, we are developing a system, wherein all the tasks of educators are automated, and educator can focus and utilize full time for session delivery.

PowerPoint presentation is commonly used for knowledge delivery sessions. During presentation delivery, it is troublesome to control the navigation of slides and simultaneously explain the topic, solve doubts. For smooth conduction of presentation, various methods using gesture recognition technology are developed, but there are many constraints such as presenter has to be in fixed area, proper lighting for gesture detection, etc. To overcome these constraints, speech recognition is used. The PowerPoint presentation is smoothly controlled by voice. It appears natural and hassle-free with smooth interaction within the classroom.

For certain seminars and especially for lectures, it is usually required to maintain records like list of participant's duration and nature of the course, etc., which can be later utilized for generation of annual reports, newsletter, etc. Various ways for attendance have been developed, but they have not proved efficient enough to be used in real-time scenarios. Our system (application) can take the attendance of an entire class precisely which will also be helpful for power management.

Learners/students usually find it difficult to cope up with the pace of certain session, and therefore, it is also proposed that the written matter either on blackboard or on slides would be shared as e-Deliverables/e-Notes for record keeping or for further studies. Capturing of notes is voice controlled.

2 Literature Survey

Many attendance systems are developed using RFID. Recently, a lot of work has been done to improve it. In [1], authors have proposed an RFID-based attendance system. Students have to carry RFID tags, and readers are installed in every class. A server application is installed on laptop which process information from all readers. Readers use Wi-fi module to transmit information. The proposed system may require additional hardware cost to install the RFID system. Another drawback is that tag collision will occur if more than one person enters the room at a time. Moreover, there are chances of fraudulent access.

Biometric-based attendance system makes individual identification easy without using any cards or tags, and they also provide enhanced accuracy and security. This paper [2] describes portable attendance management using fingerprint biometric and Arduino. This approach has a drawback of additional hardware and maintenance costs. Another limitation of this approach is that passing fingerprint scanner during lecture can lead to distraction.

Hand gesture recognition system has received great attention in the recent few years because of its manifold applications and its ability to make human–computer interaction easier and natural. Gestures can be static which are less complex or dynamic with high complexity. Different methods have been proposed for acquiring information necessary for recognition of gestures. Some methods used hardware devices such as data glove devices for extraction of gesture features. Other methods are based on the appearance of the hand using the skin color to segment the hand and extract process image for identifying the gesture [3], and based on it, respective action is taken. Proper training is required for identification of gestures and storage of images to recognize hand gestures.

In this paper [4], interactive projector screen is controlled using hand gestures. In the proposed system, camera detects hand gesture and applies image processing techniques to identify the respective command stored in the database which acts as a virtual mouse and controls the OS of the projector screen. The downside of this methodology is that the presenter has to be in a particular area, i.e., in the range covered by camera for proper gesture recognition. Further, it requires complex image processing techniques for diagnosing the gestures and processing the action of the virtual mouse which is slower than actual performance of touch screen. Separation of gesture from background is troublesome. The proposed method requires high definition camera to capture good quality of image which increases the cost, and sometimes extreme environment lighting will create disturbance in the system, and orientation of gestures also increases the complexity in the system.

3 Proposed System

The e-Classroom comprises of four modules.

3.1 Attendance Record

The main aim of this module is to automate the traditional method of keeping records of attendees. Two android applications are implemented, one for teacher and other for students using IEEE 802.11a technology.

Fig. 1 Flowchart for
educator app

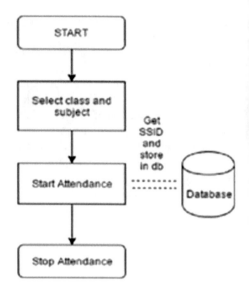

The flowcharts, as shown below, explains the working of Android app for the speaker and audience.

Steps for educator:

A. Login into the system.
B. Fill the session details.
C. Start attendance.

Steps for attendees:

A. Login into the system.
B. Scan for available Wi-fi networks.
C. Select the SSID of educator (Fig. 1).
D. Verification of SSID.
E. Authentication of user using face recognition.

In order to overcome the problem of false attendance, it has to be ensured that the attendance is marked from lecture room, and authenticated attendee is giving attendance.

The first step, i.e., scanning of available APs and selecting educator's SSID, verifies that attendance is recorded from room. User authentication is accomplished using face recognition API. The image is compared with the image copy stored in user's application at the time of registration. If and only if authentication is successful, the credentials are sent to server for marking attendance (Fig. 2).

Fig. 2 Flowchart for
attendees app

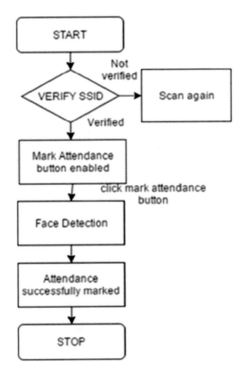

3.2 Speech Controlled Slide Navigation

Speech recognition has gained lots of importance in recent years. It is used for various purposes such as dictation tool and offline speech recognition tool.

The purpose of this module is to make the session delivery appear natural and hands free. This module is explained as follows:

The database of various instructions such as Next Slide, Go ahead 3 slides, etc., is maintained. The command given by the speaker is amplified using wireless microphone; thus, the speaker need not to be loud and can access the software from anywhere in the room (Fig. 3).

For further processing of this command, Google speech recognition API is used. The output Python string from speech recognition tool (Google API) is compared with data in the database. When a match is found, appropriate command is executed using keypress event. The keypress event is simulated using ctypes library of Python. Pseudo Code:

A. Set the block size to be processed at once.
B. Set the sampling frequency.
C. Set the buffer size.
D. Record audio using speech recognition module of Python.
E. Send the recorded speech to Google speech recognition API.

Fig. 3 Navigation of slides

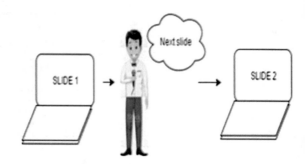

F. Fetch the output sent by Google as Python.
G. Compare the output string with dataset.
H. If a match occurs, call the execute command associated.

3.3 Resource Management

The power management system is implemented using Raspberry Pi and count of attendees from attendance module. Raspberry Pi fetches the count of attendees from the server. Pi publishes this count to the ESP8266 using MQTT protocol. ESP8266 toggles electric appliances subject to count. The system is configured to fetch the count at regular intervals using cron job. Cron job is nothing but a job scheduler. It is used to automate tasks like managing user subscription.

To fetch the data from server and execute appropriate scripts, we have used Apache server and DNS port forwarding in Raspberry Pi.

Steps for Raspberry Pi configuration:

A. Set up Apache server on Pi.
B. Install phpmyadmin and Sql packages.
C. Enable DNS and port forwarding to access it from anywhere.

Pseudo code:

A. Connect to server.
B. Fetch the latest value of count.
C. Check count value.
D. If greater than threshold value, then respective electric appliances are turned on.
E. Else turn/keep it off.
F. Repeat steps A to E after specified interval.

3.4   Record Keeping of Lecture Notes (e-Notes)

The design goal of this module is to digitize the sharing of notes. For implementing this, speech recognition module on Raspberry Pi is used. The working of this module is as described as

A. Content of blackboard is captured on the command given by the speaker.
B. The captured content is shared as deliverables via cloud or is directly sent to students as an attachment.
C. The PowerPoint presentations are also shared as an email attachment.

4   Experimental Setup and Results

Figure 4 shows the experimental setup of the proposed methodology.

The following screenshots illustrate the flow of activities in attendance module. Figure 5 demonstrates the GUI for entering session details. Start button is enabled after entering lecture details. When start attendance button is clicked, the AP mode is turned on (see Fig. 6).

Session participant has to scan for available Wi-fi networks (Fig. 8) and select educator's SSID. Mark attendance button is enabled after selection of correct SSID (Fig. 7).

After verification, succeeding step, i.e., authenticating user is shown in Fig. 9. For authenticating, face recognition is used. If the authentication is successful, then credentials are sent to server, and attendance is marked (see Fig. 10).

By default the count is taken as zero, so initially all appliances will be turned off. When the count is 10, i.e., $0 <$ Count ≤ 10, the center light and fan will be turned on. As count becomes equal to 20, center right fan and light is turned on (Fig. 11).

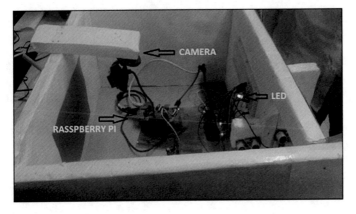

Fig. 4 Experimental setup

Fig. 5 Enter lecture details

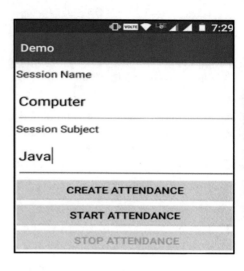

Fig. 6 Hotspot started
(attendance created)

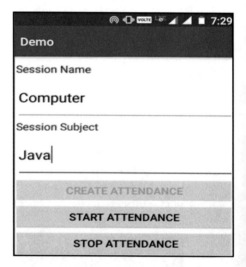

NEXT SLIDE command transitions current slide to the next slide on the voice command given by the user. Similarly, the system is tested for various commands such as PREVIOUS SLIDE, GO BACK 3 SLIDES, GO FORWARD 3 SLIDES, etc. The system responds correctly according to commands.

For capturing notes, camera is placed over the projector, and the snaps of blackboard are captured on command given by the speaker. The images captured through the camera are shared with the attendees via cloud or email.

Fig. 7 Scan for available SSID

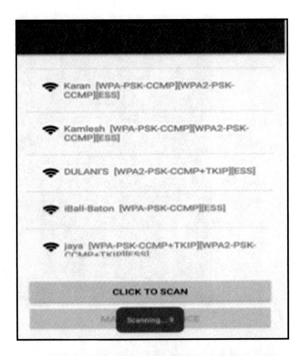

Fig. 8 Mark attendance button enabled (only if correct SSID is selected)

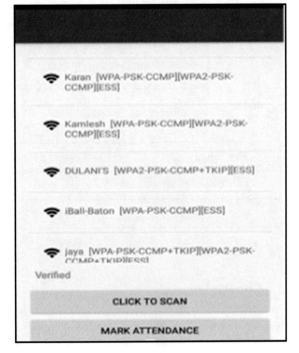

Fig. 9 Verifying attendees' identity

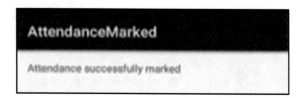

Fig. 10 Attendance marked

50 < Count < 65	10 < Count < 20	Count≥65
35 > Count ≤40	0 < Count ≤ 10	30 < Count ≤ 35
40 < Count ≤ 45	20 < Count ≤ 30	45 < Count ≤ 50

Fig. 11 Matrix arrangement of light and fan

5 Conclusion

In this paper, an advanced approach has been proposed to revolutionize the classical methods of conducting a knowledge sharing seminar (or lecture). The proposed technique has overcome the drawbacks of tracking attendance using RFID. By replacing it with face recognition, a drastic save in power consumption, easy availability of e-Notes and a lot more save in time and efforts of lecturer (or speaker) has been observed. Face recognition can be made to detect faces even in low light conditions with the use of advanced algorithms. A compression algorithm can be used to efficiently manage storage of e-Notes.

References

1. Nguyen, H.K., and M.T. Chew. 2017. RFID-based attendance management system. In *2017 Workshop on Recent Trends in Telecommunications Research (RTTR)*. Palmerston North, New Zealand: IEEE Publications.
2. Zainal, Nur Izzati, Khairul Azami Sidek, Teddy Surya Gunawan, Hasmah Manser, and Mira Kartiwi. 2014. Design and development of portable classroom attendance system based on Arduino and fingerprint biometric. In *The 5th International Conference on Information and Communication Technology for The Muslim World (ICT4M)*. Kuching, Malaysia: IEEE Publications.
3. Jadhav, Dnyanada, and L.M.R.J. Lobo. 2014. Hand gesture recognition system to control slide show navigation. *International Journal of Application or Innovation in Engineering & Management (IJAIEM)* 3 (1).
4. Sharma, Rishabh, Raj Shikher, Nutan V. Bansode, and Prachi R. Rajarapollu. 2016. Interactive projector screen with hand detection using gestures. In *International Conference on Automatic Control and Dynamic Optimization Techniques (ICACDOT)*. Pune, India: IEEE Publications.

An Efficient Heart Disease Prediction System Using Machine Learning

Debabrata Swain, Preeti Ballal, Vishal Dolase, Banchhanidhi Dash
and Jayasri Santhappan

Abstract With the shifting gears of lifestyle, large populations of the world are getting prone to the heart diseases. It is becoming one of the topmost reasons for loss of life. As the death rate is increasing due to coronary diseases, the people of healthcare department depend largely on the patient's data to predict if the patient may have a risk of heart disease. Not every time can the doctors go through every minute detail of the data and predict accurately. It is time consuming and risky. The aim of the paper is to find best predicting algorithm which can help the non-specialized doctors or medical technicians in predicting the risk of disease. The prediction system uses different machine learning algorithms like logistic regression, support vector machine, k-nearest neighbor, Gaussian naïve Bayes, decision tree classifier and random forest classifier. The prediction accuracy for logistic regression is found to be the highest among all with 88.29% accuracy.

Keywords Logistic regression · Support vector machine · K-nearest neighbor · Gaussian naïve Bayes · Decision tree

1 Introduction

According to the World Health Organization(WHO), cardiovascular heart-related diseases are the number one cause of death worldwide. In the year 2016, 31% of the global deaths, which estimated to 17.9 million people, were due to heart attack

D. Swain (✉)
Department of IT-MCA, Vishwakarma Institute of Technology, Pune, India
e-mail: debabrata.swain7@yahoo.com

P. Ballal · V. Dolase
Department of Masters in Computer Application, Vishwakarma Institute of Technology, Pune, India

B. Dash
School of Computer Engineering, K.I.I.T. University, Bhubaneswar, Odisha, India

J. Santhappan
NorthBay Solutions, Houston, TX, USA

© Springer Nature Singapore Pte Ltd. 2020
D. Swain et al. (eds.), *Machine Learning and Information Processing*,
Advances in Intelligent Systems and Computing 1101,
https://doi.org/10.1007/978-981-15-1884-3_4

39

and stroke. By 2030, the number of deaths caused by heart diseases is expected to reach 23.3 million [1]. Coronary heart disease is an abnormality in the blood vessels that connect to the heart muscles. The most common reason behind an individual suffering from a heart disease is unhealthy diet, physical inactivity, tobacco, alcohol and cigarettes. The effect of these reflects in a very harmful way as increased blood pressure, increased glucose, increased blood lipids, obesity and increased body mass index (BMI). These dangerous effects on heart are measured in any primary medical care facility or hospitals. They indicate the chances of developing risk of heart attack, heart stroke, heart failure and many other such heart complications.

Unhealthy lifestyle leads to obesity which is one of the major reasons behind heart stroke or attack. Because of such fatty deposit on the inner lining of blood vessels, the blood vessels become weak and fragile. The walls of the heart thicken with an increased pressure of blood in the arteries which causes blockage. If the walls become thick, it slows down the flow of blood to the heart eventually leading to heart disease which can be life threatening [2]. Heart disease risk can be avoided or prevented by ceasing the use of cigarettes, alcohol and tobacco. Reducing the use of salt, reducing the consumption of fried products, consuming fruits and vegetables, daily exercise or small physical activity can also reduce the risk behind heart disease. Medical treatment of diabetes, blood pressure and glucose can also be done to be safe. These medical treatments can be given by medical authorities after detecting the presence or risk of heart disease. The detection of coronary disease depends upon the clinical information of the patient. If the clinical information is present accurately, only then can the risk of heart disease be predicted accurately. The health care and medical industry is regularly collecting clinical information of patients on a large scale which can be processed. The data processing technique leads to discovery of unknown or concealed information. This clinical information will assist the doctors or medical authorities in taking more accurate and effective decisions which can be lifesaving [9].

Heart disease can be prevented or addressed in an appropriate way by early detection of the risk of disease. People who have been detected early in life about the risk of heart disease can be counseled about many ways in which they can prevent it or be prepared for the same. Doctors or medical technicians can analyze the patient and his/her lifestyle characteristics and predict if the patient can be prone to heart disease or no. But, it is a time consuming and risky process to follow [10].

To predict if a person may have heart disease, immense understanding of data and its attributes relationship with the target variable has to be studied. Machine learning gives a wide range of algorithms which help in co-relating this relationship and predicting the value of the target variable. The dataset that is used in this research work is *Framingham Dataset* which includes 16 attributes. The factors which affect the risk prediction of heart disease are age, number of cigarettes smoked per day, the presence of blood pressure, diabetes, cholesterol level, body mass index (BMI), heart rate and glucose level.

2 Literature

Aakash Chauhan, Aditya Jain et al. have studied the rules for prediction of the risk of coronary disease among patients using evolutionary learning. A deep analysis of data mining had been carried out to achieve higher accuracy. Computational intelligence approach is used to find the relation among disease and patients. The dataset used for this research is *Cleveland Dataset* with 13 attributes included [2]. Data preprocessing is done to get the data clean and in quality. Keel is a Java tool to simulate evolutionary learning, and hence, it is used [2]. An accuracy of 0.53 has been achieved on test dataset.

Frequent pattern growth association mining is applied on dataset to obtain better association rules. The study has evaluated to the conclusion that more the rules, better is the prediction of coronary disease.

A. H. Chen, S. Y. Huang et al. have developed a system using artificial neural network that evaluates the patient's clinical data to help healthcare professionals in predicting the heart-related risk factors. The dataset used by authors is from UCI machine learning repository. The approach used included mainly three steps. First, 13 out of the 14 features were selected which are sex, age, trestbps, cholesterol, chest pain type, blood sugar, number of vessels colored, thal, resting ECG, heart rate, exercise-induced angina, slope and old peak [3]. Second, the author proposed a classification-based algorithm which uses artificial neural network based on the 13 features. An accuracy of nearly 80% is achieved. In the third and last step, they developed a graphical user interface to get user-friendly access.

Ankita Dewan and Meghna Sharma designed an algorithm that is efficient and hybrid with backpropagation approach for prediction. The author proposed use of neural networks from set of different classification techniques for nonlinear data. In this proposal, the focus is on backpropagation (BP) algorithm which is considered as the best classifier of artificial neural network using updating technique of weights by propagating the errors backward [4]. A drawback is also identified; that is, the risk of getting stuck in local minima solution while solving an efficient optimizing technique for further improvement of accuracy is used. The author has implemented various kinds of techniques which predict heart disease; also, a methodology is proposed for hybrid technique to implement in the future with accuracy of almost 100% or with least error [4].

Aditi Gavhane, Gouthami Kokkula et al. have proposed a heart disease prediction system using multilayer perceptron (MLP) in machine learning that gives a result if a user may get CAD. MLP gives a better accuracy and efficiency than other algorithms [5]. If the usability of the system increases, it will spread an awareness to reduce the heart disease death rate. The system uses Cleveland dataset which is available in UCI data repository. It is developed in PyCharm IDE using Python coding. Finally, a graphical representation of the result is shown to conclude that multilayer perceptron gives better accuracy [5].

Shashikant U. Ghumbre and Ashok A. Ghatol have suggested the implementation of SVM for diagnosis of heart disease. A detailed description of heart disease and a support system for decision making which diagnosis the heart disease using support vector machine algorithm and radial basis functional network structure is proposed by them. The data is collected from various patients. A comparison of algorithms like support vector machine (SVM), multilayer perceptron (MLP), BayesNet, rule, radial basis function (RBF) and J46 has been shown in the paper in tabular format [6]. The diagnosis is performed on this data, and results are obtained; thus, concluding that support vector machine (SVM) using sequential minimizing optimization is as good as the artificial neural network (ANN) and some other models of machine learning. The overall accuracy obtained is of 85.51% with fivefold cross-validation and 85.05% with tenfold cross-validation [6]. The performance of SVM classification is high. Hence, it can be considered for the heart disease diagnosis.

R. Kavitha and E. Kannan have shown the analysis of feature selection and feature extraction in this paper. A framework is created for the proposed system. The framework creation includes various steps which are feature extraction using principal component analysis and outlier detection. In order to obtain better performance, wrapper filter along with feature subset selection is implemented. The scoring functions like Pearson's correlation coefficient and Euclidean distance show an average or below average performance in comparison with the proposed system [7]. The proposed framework not only predicts the presence of risk of heart disease but also shows the reduction in the attributes that are required to diagnose a heart disease. The author presented a scope in the future work. It has addressed the presence of outlier in the class boundary, which has shown a noticeable improvement in the performance of the system. The system can also be used for the prediction of those class labels which are unknown [7].

C. Sowmiya and Dr. P. Sumitra have conducted an assessment of five different classification algorithms for heart disease prediction. The five algorithms are decision tree classifier, naïve Bayesian neural network, artificial neural network, KNN, support vector machine. The authors have proposed an algorithm of support vector machine (SVM) and Apriori algorithm [8]. The authors have also performed data mining technique. Data mining techniques mainly involve four sub-techniques mainly classification, cluster, feature selection and association rule. They have concluded that the techniques used in classification-based algorithm are more effective and show better performance compared to the previous implemented algorithms. The paper also includes brief description of the different types of heart diseases which are angina, arrhythmia, fibrillation, congenital heart disease, coronary artery disease, myocardial infarction and heart failure [8].

3   Logistic Regression

Classification is done on a daily basis in many ways by people. For example, checking the classification type of tumor and cancer, checking if email received is promotional, social media or spam, etc. Logistic regression is a type of classifier algorithm in machine learning. Logistic regression is quite identical to linear regression. It is a very simple and elementary level of algorithm which solves many real-world classification problems easily. It can be used for binomial classification as well as multi-valued classification.

Logistic regression's main purpose is to relate the data and interpret the relation that is shared among the dependent variable and single or multiple ordinary variable which are independent. It is an appropriate classifier to use when the target or dependent variable is divided into two values, that is, binary. There are three types of logistic regression, namely binary, multi-valued and ordinary classifier.

4 Dataset Description

The dataset used in this proposed system is *Framingham Dataset*. It is an open-source dataset that is available on Kaggle dataset repository. It consists of a total of 16 columns with 15 independent variables and one dependent target variable. It has 4239 rows. The variable's description is given in Table 1.

5 Proposed System

The objective of the system proposed by the paper is to detect risk of heart disease using patient's data. The dataset contains one dependent variable and 15 independent variables, and the logistic regression has proven the best performance among other algorithms. The logistic regression is originated from statistics. Binary classification is done using logistic regression, i.e., for problems with two class values. Logistic function, that is, sigmoid function is the core of logistic regression. This was originally developed for the purpose of evaluating growth of population and calculation of rising quality of population.

5.1 Data Cleaning

The dataset used for training and testing purpose may contain errors and duplicate data. In order to create a reliable dataset and to maximize the quality of data, the data cleaning procedure is important. The following steps for identification and classification of errors have been used.

1. The first step used to clean the data is finding incomplete data or the null values and dealing with them to improve the performance of model. The missing values found are 489 out of 4239 rows which are approximately 12% of actual data.
2. In this case, appropriate solution is to remove such rows that contain null values.

Table 1 Description of dataset columns

S. No.	Variable name	Variable type	Description
1	Male	Discrete	This gives a value 1 if patient is male and 2 if female
2	Age	Discrete	The age of the patient
3	Education	Discrete	It takes values as: 1: high school 2: high school or GED 3: college or vocational school 4: college
4	currentSmoker	Discrete	It gives binary value 0 for nonsmoker and 1 for smoker
5	cigsPerDay	Continuous	It gives an average of the number of cigarettes smoked per day
6	BPMeds	Discrete	It tells if patient is on BP medicines 0: not on medications 1: is on medications
7	prevalentStroke	Discrete	1: patient has suffered from stroke before 0: patient has not suffered from stroke
8	prevalentHyp	Discrete	1: hypertension 0: no hypertension
9	Diabetes	Discrete	If patient has diabetes then value is 1 else 0
10	totChol: mg/dL	Continuous	Cholesterol value present in a patient
11	sysBP: mmHg	Continuous	Value of systolic blood pressure
12	diaBP: mmHg	Continuous	Value of diastolic blood pressure
13	Body mass index	Continuous	Body mass index is calculated by weight (in kg)/height (meter)
14	Heartrate	Continuous	It gives ventricular heart rate which is calculated by beats/min
15	glucose: mg/dL	Continuous	Level of glucose in blood
16	TenYearCHD	Discrete	It is the target variable with 0 as no risk of disease and 1 means risk of disease is present

5.2 Data Dimensionality Reduction

The large amount of data sometimes produces worst performance in data analytics applications. As the algorithm is based on column-wise implementation, this makes algorithm slower in performance as the number of data columns grows. Here, the first step in data reduction is to minimize the number of columns in dataset and lose minimum amount of information possible at the same time.

For this, backward feature elimination is used, in which the process is started with all features, and the least significant features are removed with each iteration to improve the performance of model. This process is repeated until no improvement is detected on removal of features. The eliminated data column is BPMeds as it did not perform well. The remaining 14 features which gave better performance and high results are selected.

5.3   Correlation Identification

Dataset may contain complex and unknown relations between variables. The discovery and classification of feature dependency are important for any model. Pearson coefficient is applied to identify the correlation between two features.

$$\text{Pearson's coefficient} = \text{cov}(X, Y)/(\text{stdv}(X) * \text{stdv}(Y))$$

The calculated value of Pearson's correlation coefficient is found to be between 0.7 and 0.8 for the features 'age', 'Sex_male', 'currentSmoker', 'cigsPerDay', 'prevalentStroke', 'prevalentHyp', 'diabetes', 'totChol', 'sysBP', 'diaBP', 'BMI', 'heartRate' and 'glucose'. The Pearson's correlation coefficient for 'Education' is 0.37 which shows neutral Correlation. Hence, 'Education' column is eliminated.

5.4   Flowchart

See Fig. 1.

5.5   Algorithm

1. Data cleaning and preparation.
2. Data splitting.
3. Model training and testing.
4. Accuracy calculation.

Data Cleaning and Preparation. The process of systematically and accurately cleaning the data to make it ready for analysis is known as data cleaning. Most of times, there will be discrepancies in the gathered data like wrong data formats, missing data values, errors while collecting the data. The dataset has been cleaned using null value removal method, backward feature elimination and Pearson's correlation coefficient.

Data Splitting. For training and testing of algorithm, the data need to be split in two parts. The training set contains well known classified output and the model trains on this data to be generalized to other data later. Later, the testing data is used to test the model's performance on this testing data. While performing data splitting two things

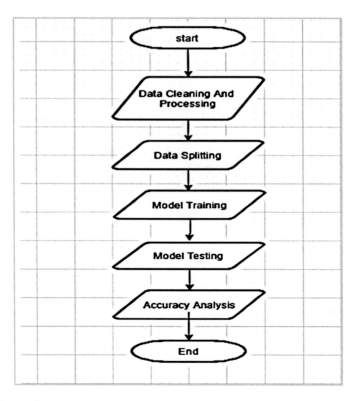

Fig. 1 System flowchart

may happen: The model can overfit or underfit which is harmful for the results of system. In order to prevent that, the proportion of splitting is taken as 90–10%.

Model Training and Testing. In model training, relationship between a label and its features is identified. This is done by showing the model a set of examples from dataset. Each example helps to define how each feature affects the label. This is known as model training.

Logistic regression has performed highest among other algorithms. Logistic regression is considered to be a supervised classifier which trains the model with labeled data. In this, the output is the target variable. It can take only binary discrete values for a set of features which are the input. The target variable is binomial, based on number of class values. Logistic regression has the following types:

1. Multinomial classifier: The dependent target variable can have more than two classes which are unordered like 'Class 1' and 'Class 2' versus 'Class 3.'
2. Binomial classifier: The dependent target variable can have exactly two classes: 0 and 1 representing true and false, respectively.
3. Ordinal classifier: It deals with dependent target variable with ordered class. A test outcome can be sectioned as: very poor, poor, good and very good. Here, each section is given an outcome values like 0, 1, 2 and 3 (Fig. 2).

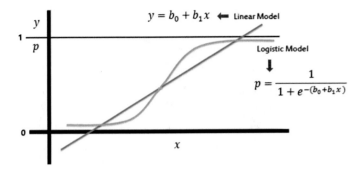

Fig. 2 Diagrammatic representation of logistic regression

In order to define curve steepness, the b_0, which is constant, moves the curve in both left and right direction. Hence, this slope b_1 states the steepness of curve. With few modifications, the following equation can be derived in terms of odds ratio.

$$\frac{p}{1-p} = \exp(b_0 + b_1 x)$$

After taking log of both sides, equation in log-odds is a sigmoid function of the prediction. The coefficient b_1 is the amount the logit changes with a one-unit difference in x.

$$\ln\left(\frac{p}{1-p}\right) = b_0 + b_1 x$$

Capacity of logistic regression is not limited. It can perform on any number of categorical variables.

$$p = \frac{1}{1 + e^{-(b_0 + b_1 x_1 + b_2 x_2 + \cdots + b_p x_p)}}$$

Accuracy Calculation. Predictive modeling is used to create models which have good performance in predicting over unseen data. Therefore, using a robust technique for training and evaluation of model on available data is crucial. As more reliable estimate of performance is found, further improvement scope can be identified.

The confusion matrix is used to evaluate accuracy. The confusion matrix gives an output matrix and provides the description of performance of system. The samples are from two classes: 'True' and 'False,' and the implementation of confusion matrix can be done.

There are 4 terms:

- **TP**: The cases predicted TRUE and actual output TRUE.
- **TN**: The cases predicted FALSE and actual output FALSE.
- **FP**: The cases predicted TRUE and actual output FALSE.

Out[39]: Text(0.5,257.44,'Predicted label')

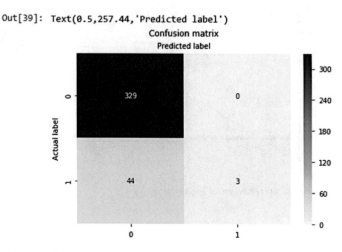

Fig. 3 Confusion matrix

Fig. 4 Accuracy score of
logistic regression

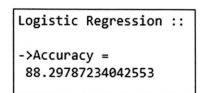

- **FN**: The cases predicted FALSE and actual output TRUE.

    Accuracy can be calculated by:

$$\mathbf{Accuracy} = \frac{TP + TN}{TP + TN + FP + FN}$$

The testing data contains 376 rows whose classification values are shown in Figs. 3 and 4.

6   Performance Analysis

The suggested flowchart was followed to calculate and analyze the accuracy of different classification models. Table 2 shows the accuracy percentage of each classifier.

Table 2 Accuracy score of different algorithms

Algorithm	Accuracy (%)
Logistic regression	88.29
Random forest	87.23
Gaussian naïve Bayes	85.37
KNN—K-nearest neighbor	87.51
SVM—Support vector machine	87.50
Decision tree	78.72

7 Conclusion and Future Scope

In the paper, several binary classification algorithms that are very useful in detecting heart disease are analyzed. Logistic regression has proven to be the best classification algorithm to classify the risk of heart disease with 88.29% accuracy. Logistic regression is a classification-based technique that contributes high effectiveness and obtains high accuracy compared to other algorithms. The future work can be analyzing the performance for parameter tuning in order to increase accuracy. The future work can be set as- (1) Analyzing the performance (accuracy) of the system using parameter tunning. (2) Implementation of ensemble based classifier to improve the performance of the system.

References

1. Ali, S., S.M. Adnan, T. Nawaz, M. Obaidullah, and Sumair Aziz. 2017. Human heart sounds classification using ensemble methods.
2. Chauhan, A., A. Jain, P. Sharma, and V. Deep. 2018. Heart disease prediction using evolutionary rule learning. In *2018 4th International Conference on Computational Intelligence & Communication Technology (CICT)*, Ghaziabad, 1–4.
3. Chen, H., S.Y. Huang, P.S. Hong, C.H. Cheng, and E.J. Lin. 2011. HDPS-Heart disease prediction system. In *2011 Computing in Cardiology*, Hangzhou, 557–560.
4. Dewan, A., and M. Sharma. 2015. Prediction of heart disease using a hybrid technique in data mining classification. In *2015 2nd International Conference on Computing for Sustainable Global Development (INDIACom)*, New Delhi, 704–706.
5. Gavhane, Aditi, Gouthami Kokkula, Isha Pandya, and Kailas Devadkar. Prediction of heart disease using machine learning, 1275–1278. https://doi.org/10.1109/iceca.2018.8474922.
6. Ghumbre, S.U. and A.A. Ghatol. 2012. Heart disease diagnosis using machine learning algorithm. In *Proceedings of the International Conference on Information Systems Design and Intelligent Applications 2012 (INDIA 2012) held in Visakhapatnam, India, January 2012*, vol 132, ed. S.C. Satapathy, P.S. Avadhani, and A. Abraham. Advances in Intelligent and Soft Computing. Berlin: Springer.
7. Kavitha, R., and E. Kannan. 2016. An efficient framework for heart disease classification using feature extraction and feature selection technique in data mining. In *2016 International Conference on Emerging Trends in Engineering, Technology and Science (ICETETS)*, Pudukkottai, 1–5.

8. Sowmiya, C., and P. Sumitra. 2017. Analytical study of heart disease diagnosis using classification techniques. In *2017 IEEE International Conference on Intelligent Techniques in Control, Optimization and Signal Processing (INCOS)*, Srivilliputhur, 1–5.
9. Swain, Debabrata, Santosh Pani, and Debabala Swain. 2019, April. An efficient system for the prediction of Coronary artery disease using dense neural network with hyper parameter tuning. *International Journal of Innovative Technology and Exploring Engineering (IJITEE)* 8 (6S). ISSN: 2278–3075.
10. Swain, Debabrata, Santosh Pani, and Debabala Swain. 2019, July. Diagnosis of Coronary artery disease using 1-D convolutional neural network. *International Journal of Recent Technology and Engineering (IJRTE)* 8 (2). ISSN: 2277–3878.

Automatic Helmet Detection in Real-Time and Surveillance Video

Shubham Kumar, Nisha Neware, Atul Jain, Debabrata Swain and Puran Singh

Abstract In current world, number of vehicles growing day by day results in higher accidents. So, "helmet" is the one key element for ensuring safety of bike riders. Knowing the fact, people tend to avoid wearing helmet. Government had imposed various rules making it compulsory to wear helmet and fine being levied on offenders. But it is not possible to track each rider in current manual tracking and video surveillance system. A model for detection and classification of bike riders who are wearing the helmet as well as those who are not wearing it is proposed in this paper. The proposed model trained on COCO dataset uses only one neural network per image and that is quicker than R-CNN and Fast R-CNN as they use multiple neural networks.

Keywords YOLO · Darknet · COCO data model

1 Introduction

Nowadays, two-wheelers are popular because of ease of handling and affordability but results in higher road accidents. Stats by Transport Ministry and Times of India survey show 28 two-wheeler riders die daily on Indian roads in 2016 because of

S. Kumar (✉) · N. Neware · A. Jain · D. Swain · P. Singh
Vishwakarma Institute of Technology, Pune, India
e-mail: shubham.kumar17@vit.edu

N. Neware
e-mail: nisha.neware17@vit.edu

A. Jain
e-mail: atul.jain17@vit.edu

D. Swain
e-mail: debabrata.swain7@yahoo.com

P. Singh
e-mail: puranjsingh@gmail.com

© Springer Nature Singapore Pte Ltd. 2020
D. Swain et al. (eds.), *Machine Learning and Information Processing*,
Advances in Intelligent Systems and Computing 1101,
https://doi.org/10.1007/978-981-15-1884-3_5

not wearing helmets. Additionally, about 98 two-wheeler users without helmets lost their lives in road accidents every day.

As per the survey of 2018, over 57% of the Indian motorcyclists do not wear helmets. For enforcement, government has introduced many systems which used real-time CCTV to capture riders without helmets in order to punish them and issue an e-Challan. But this system still needs human intervention to manually identify riders without helmets from the control room hence reduces the efficiency and effectiveness of the model as it will not be available 24 * 7.

Problem of identifying whether the person is wearing helmet or not can be categorized as the problem of detecting an object. Object detection can be treated as a classification or a regression problem. All the techniques such as Faster R-CNN, Fast R-CNN, R-CNN treat problem as classification one, but the main issue with these methods is that process is divided into layers of convolutional (on an average divided into thousands of layers), which takes longer time to process the data and not well suited for real-time processing.

Whereas, you only look once (YOLO), single-shot detector (SSD) treat the problem as regression problem. As their name suggests, they process over the image only once, thus not need to process over again and again as it was done in R-CNN and other related methods, making YOLO and SSD methods faster and best suited for real-time processing as used in real-time traffic surveillance system.

The aim of this paper is to find a method with less human intervention and achieve accuracy in recognizing riders without helmet. The proposed approach is divided into five phases: **background subtraction, object detection, classification of vehicles, segment ROI (region-of-interest)** and **detection of helmet**.

2 Related Work

Noticing to the seriousness of this problem, researchers have carried out a lot of work in this domain. Prasad and Sinha [1] proposed a method which was implemented to detect the objects in complex, light variant, changing background. Used to solve the problem of detecting object of same colour as that of its background. The pre-processed frames achieved using *contrast stretching* and *histogram equalization* were provided as an input to the object detection module. Here, they went through various operations like background subtraction, colour segmentation, morphological operations (dilation and erosion). The recognition was done by analysing the dimensions of the shapes, filtration of objects by area in binary image and analysing the estimated perimeter.

Chiverton [2] proposed a method of helmet detection from a video feed. The classifier (SVM) was equipped with training on the histograms. These histograms were derived from the head regions of image data. Then, the motorcycle rider was automatically segmented from the video by extracting the background. The results for the motorcycle acquired were in the form of sequence of regions called tracks. These tracks were further classified, and an average result was calculated. But drawback is that every circular object around the bike was recognized as "helmet" resulting in very low accuracy. Additionally, the vehicles in speed were also identified which need high computation hence leading it to a very expensive model.

Gujarati et al.'s [4] model finds number of traffic offenders. Classification of commodity objects was done using COCO model. Developing training dataset followed by classification of person riding the bike and the helmet was done using TensorFlow. Meanwhile, number plate recognition by OCR optical character recognition (OCR) was performed using tesseract.

Aires et al. [3] have implemented the model of helmet detection using hybrid descriptor based on local binary pattern, which is a visual descriptor, histograms of oriented gradients, i.e. a feature descriptor, the Hough transform which is a feature extraction technique and support vector machine (classifier). Since it is a classification algorithm, it uses large number of layers of convolution neural network, hence making it more time consuming.

Desai et al. [5] aimed to avoid the number of road accidents and developed a system to detect helmet. They have carried out this work by dividing it into two tasks: (a) *Fall detection* (b) *Helmet detection*.

Fall detection was done using background subtraction and OCR, whereas helmet detection was achieved using background subtraction followed by HOG transform descriptor, popularly used for detecting curves. This system is linked with GPS, i.e. if any fall is detected, the system will extract the license number plate using OCR, find owner's information and will report to the nearby hospital and owner's family.

In [6], Duan et al. suggested a robust approach to track the vehicles in real time from single camera. In order to obtain faster results, integrated memory array processor (IMAP) was used. However, it proved to be an expensive model due to its dedicated hardware requirements and thus inefficient.

In [7], K Dahiya et al. divided their work into two phases: (a) Detection of bike riders (b) Detection of bike riders without helmet. For detecting bike riders, the features were extracted using HOG, scale-invariant feature transform (SIFT) and LBP. SVM was used to classify the objects into two categories, i.e. "bike riders" and "others". Once the bike riders were detected, again the features were extracted for the upper one-fourth part of the image using HOG, SIFT and LBP. Then, using SVM, the classification whether "helmet" or "without helmet" was carried out.

3 Proposed Methodology

The flow of the proposed system is depicted in Fig. 1.

1. **Input Frames**: This is the initial step where the traffic video is taken from camera, and the frames from the video are passed to this system.
2. **Image Processing**: After getting frames, image pre-processing takes place. Following are the steps:

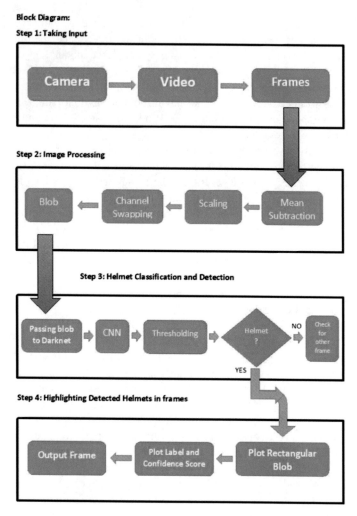

Fig. 1 Proposed methodology

a. **Mean Subtraction**: This is for combat illumination change in input. Taking out the mean for each BGR colour and subtract it from each pixel intensity.

$R = R - \mu_R$

$G = G - \mu_G$

$B = B - \mu_B$

b. **Scaling**: Scaling is used for normalization. In this paper, value of scaling $\sigma = 1/255$.

$$R = (R - \mu_R)/\sigma, \quad G = (G - \mu_G)/\sigma, \quad B = (B - \mu_B)/\sigma$$

Fig. 2 A visual representation of mean subtraction of RGB value

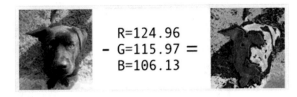

$$R=124.96$$
$$- \ G=115.97 \ =$$
$$B=106.13$$

c. **Channel Swapping**: Since OpenCV assumes that image is in BGR format, but mean value assumes that image is in RGB format (As shown in Fig. 2). As we need to use BGR format in OpenCV there is a need of swapping of channel R and B.

d. **Blob**: Above procedure is applied for getting a blob.

3. **Helmet Classification and detection**: The blob is passed to Darknet, which implements a convolutional neural network (CNN). The Darknet is already fed with pre-trained weight values for helmet and a configuration file containing different convolutional layers (whose hyperparameters like strides, number of filters, padding, etc., are already set). The Darknet then classifies whether the given image contains helmet or not along with its confidence score. Only those objects with confidence scores of more than the threshold value of 30% are considered as helmets. It also returns the coordinates of the region in which helmet was detected.

Convolutional Neural Network (CNN):

CNN is just like another neural network. It consists of different neurons with learned weights (**W**) and biases (**B**). Every neuron gets input, takes weighted sum of them and passes them to an activation function and provides output. As shown in Fig. 3, convolutional neural network has neurons organized in three dimensions [W (width), H (height) and D (depth, i.e. BGR colour)].

CNN operates over volume. Each one layer revamps the 3D input volume into 3D output volume neurons activation.

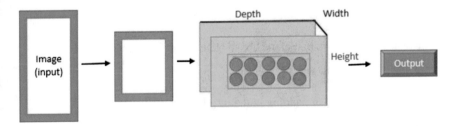

Fig. 3 Convolutional neural network

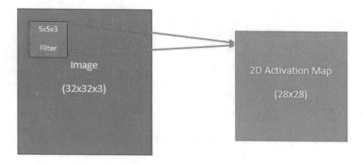

Fig. 4 Filter (5 × 5 × 3) over image (32 × 32 × 3) generates 2D activation map (28 × 28)

Layers used to build CNN:

Since CNN is a collection of layers and every layer of CNN converts one volume of activation into other with the help of differentiable function. CNN architecture consists of (1) Convolutional Layer (2) Fully Connected Layer.

Convolutional Layer

Figure 4 shows the main building block which does lots of computation is explained here. It consists of a filter. A filter is a small spatial (height, width), although expand to the extent of volume. For **example**, **(as shown in** Fig. 4), take a filter of 5 × 5 × 3 (5 is pixel width, 5 is pixel height, and 3 is colour channel BGR). During forward pass, the filter slides from the top left corner of the input image and calculates dot product of filter matrix and portion of image matrix over which filter is applied. As filter is sliding over the input image, it creates 2D activation map. From this, here the network interprets the type of visual features like edges, corners, etc., on the first layer or entirely honeycomb structure, wheel pattern or any other pattern. As a result, we got all set of filters in each layer of CNN, and each layer produces different 2D activation map. The output volume is produced by arranging all the activation map along with the depth of input image. 75 CNN layers are configured in this project with different filter sizes and number of filters.

Summary,

1. Accept W1xH1xD1
2. Require four hyperparameters: "N"—No. of filters

   - "SE"—Spatial Extent
   - "S"—Strides
   - "PD"—Amount of padding in the resultant matrix

3. Produce volume of size W2xH2xD2, where W2 = (W1 − SE + 2PD)/S + 1

   - H2 = (H1 − SE + 2PD)/S + 1
   - D2 = N

4. SE.SE.D weights/filter, for total of (SE.SE.D1). N weights and N biases.

5. The Dth depth slice of size W2xH2 is the result of valid convolutional in the output.

No Pooling Layer is used

Fully Connected Layers:

Neurons having fully connected layers, i.e. the neurons are associated with all other neurons belonging to preceding layer.

4. **Highlighting Detected Helmets in frame:** Darknet gives the coordinates of the helmet containing regions. With the help of those coordinates, rectangular blob is plotted on the frame with the label "helmet" and confidence score.

4 Activation Function

The task of object detection consists of identifying the image site where an object is available, in addition also classifying those objects. Preceding technique for this, alike the R-CNN and its various forms, this way makes the use of pipeline which allowed task completion in multiple strides. The indicated can be sluggish in operation and too tough to hone (optimize) because every single component needs to be trained independently. We try to accomplish it all by using a single neural network. A neural network is fed with lots of information, and here, activation function comes to help as it suppresses the irrelevant data points. Results obtained after applying activation function contribute towards making the model more optimized and less prone to produce error or output far away from the expected one.

In the proposed model, following mentioned is the activation function. But there is lot to cover in the activation function, so explained in parts.

A. *First part*:

$$\lambda_{cod} \sum_{a=0}^{L_1} \sum_{b=0}^{L_2} I_{ij}^{obj} \left(x_a - \hat{x}_a\right)^2 + \left(y_a - \hat{y}_a\right)^2 \tag{1}$$
(Loss function)

Equation (1) helps to compute the loss which is related to previously predicted position bounding box (x, y). The function figures out a sum corresponding to each one bounding box predictor $(j = 0, ..., L1)$ of the grid cell $(i = 0, ..., L2)$. $I obj$ here is defined as below mentioned:

1, for object being found in grid cell a and the bth related to bounding box predictor is "answerable" for the same prediction 0, else.

B. *Second part*:

$$\lambda_d \sum_{a=0}^{L_1} \sum_{b=0}^{L_2} I_{ij}^{obj} \left(\sqrt{w_a} - \sqrt{\hat{w}_a} \right)^2 + \left(\sqrt{v_a} - \sqrt{\hat{v}_a} \right)^2 \tag{2}$$

(Predicted box dimension)

Using Eq. (2), the loss is linked to the predicted box dimensions, i.e. width or height.

C. *Third part*:

$$\sum_{a=0}^{L_1} \sum_{b=0}^{L_2} I_{ij}^{obj} \left(C_a - \hat{C}_a \right)^2 + \lambda_{nd} \sum_{a=0}^{L1} \sum_{b=0}^{L2} I_{ab}^{nd} \left(C_a - \hat{C}_a \right)^2 \tag{3}$$

(Confidence level)

Using Eq. 3, loss related to confidence level in accordance with each bounding box is being calculated. *C* here depicts the confidence level, and the $\hat{C}$ presented here as **intersection over union (IOU)** of the predicted bounding box. *1*obj is equivalent to the "one" when an object being found inside the cell and else 0. *1*no obj is the adverse.

The λ parameter is utilized to individually weight components of the function of loss. Above process is paramount to enhance the stability of model. The maximum cost related to coordinate predictions (λ d $= 5$), and the minimum cost related for confidence predictions in case where no object is found (λ nd $= 0.5$).

D. *The last section is related to classification loss*:

$$\sum_{a=0}^{L_2} I_{ij}^{obj} \sum_{c \in cls} \left(p_i(c) - \hat{p}_i(c) \right)^2 \tag{4}$$

(Classification loss)

5   Performance Analysis and Result

The videos those are used for testing were captured using a CCTV camera installed on public roads and which records the video in day as well as night. The videos were captured at **30** frames per second (FPS) and contained resolution of image is 1280×720 pixels. The videos we considered were on an average of 3 min length. Python and the OpenCV library included as the key tools used to implement all the algorithms.

YOLOv3 is a good performer when it is compared with another model, and it almost outperforms 80% of the existing model by using Formula (5). The neural network framework we used is Darknet, written in C language, i.e. a framework allowed to work on high level of neural network and focusing on functionality, rather than on complex structure of neural network.

With respect to COCOs metric AP, it is on level of average with the SSD variants but is 3 × faster.

However, when compared for detection metric of mAP at IOU = 0.5 (or AP50 in the chart), YOLOv3 is very well built. It is almost on par with RetinaNet and beyond the SSD variants. Thus, YOLOv3 is a powerful detector, which outdo at producing befitting boxes for objects (Fig. 5).

Following are the things that do not coordinate without our expectation, i.e. **errors**

A. **Anchor box** x, y **offset predictions**. Tested using normal anchor box prediction method, where x, y predicted, offset as a multiple of the box width or height using a linear activation. But this decreased model stability.

B. **Linear** x, y **predictions instead of logistic**. With linear activation to precisely predict the x, y offset instead of the logistic activation. But seen drop in mAP.

C. **Focal loss**. Using focal loss, resulted in mAP drop about two points, YOLOv3 has separate objectless predictions and conditional class predictions, thus making it robust for focal loss. But in results, under some cases, there is no loss from the class predictions? Or something? Cannot be completely sure on this.

Fig. 5 Detection of helmet and its confidence score after implementation of algorithm in the video frames

Calculations:

$$\textbf{Accuracy} = \frac{TP}{TP+FP} X100 \tag{5}$$

where TP (True Positive), FP (False Positive)

Video sequence	TP	FP	Accuracy (%)
1	16	1	94
2	19	4	82
3	18	3	85

Resultant: 87% accuracy.

6 Conclusion

Considering the importance of wearing helmet while riding two-wheeler, proposed model identifies and counts number of riders without helmet. Using Darknet CNN framework and YOLOv3, i.e. a regression model, here, we try to minimize the disruption of classification which may arise due to bad climate condition, improper lighting condition and quality of video captured. With integrated use of open source and free technologies like anaconda navigator, OpenCV and Darknet CNN framework, resulted in software relatively less expensive. This model can be extended further for detecting other rule violations like report number plates of violators. Thus, it is making the future scope very prosperous and promising.

References

1. Prasad, Shashank, and Shubhra Sinha. 2011. Real-time object detection and tracking mentioned at: proposed solution, proposed object detection and tracking technique for unknown environment.
2. Chiverton, J. 2012. Helmet presence classification with motorcycle detection and tracking, 8–14.
3. Silva, Romuere, Kelson Aires, Thiago Santos, Kalyf Abdala, and Rodrigo Veras. 2013. Automatic detection of motorcyclists without helmet, 5–11.
4. Devadiga, Kavyashree, Pratik Khanapurkar, Shubhankar Deshpande, and Yash Gujarathi. 2018. Real time automatic helmet detection of bike riders, 2–4.
5. Desai, Maharsh, Shubham Khandelwal, Lokneesh Singh, and Shilpa Gite. 2016. Automatic helmet detection on public roads, 1–2.
6. Duan, B., W. Liu, P. Fu, C. Yang, X. Wen, and H. Yuan. 2009, February. Real-time on road vehicle and motorcycle detection using a single camera, 1–6.
7. Dahiya, Kunal, Dinesh Singh, C. Krishna Mohan. 2016. Automatic detection of bike-riders without helmet using surveillance videos in real-time, 2–5.

Camera Model Identification Using Transfer Learning

Mohit Kulkarni, Shivam Kakad, Rahul Mehra and Bhavya Mehta

Abstract Wide series of forensic problems can be solved by detecting the camera model from copyright infringement to ownership attribution. There are many proposed methods for detection of the camera model. A method to identify the camera model of any image is proposed in this paper. It involved feature extraction and classification. CNN-based architectures are best suited for the task of image classification.

Keywords Camera model identification · Convolutional neural networks · ResNet · Transfer learning

1 Introduction

An investigator wants the footage of crime which was captured on tape. If somehow the investigator manages to get the footage, the question remains if the footage is real. The basic aim of this project is to classify images in which the camera model was taken. This can thus help the investigation team and the general crowd to know the un-tampered truth. Further, this can stop the media from producing fake news for the sake of publicity and business. It can also impede political parties from using false methods in the campaigning to cheat the common man and come to power. This problem can be solved with machine learning using convolutional neural networks [1].

M. Kulkarni (✉) · S. Kakad · R. Mehra · B. Mehta
Electronics and Telecommunication Department,
Vishwakarma Institute of Technology, Pune, India
e-mail: mohitvkulkarni@gmail.com

S. Kakad
e-mail: shivamkakad05@gmail.com

R. Mehra
e-mail: mehrar12@gmail.com

B. Mehta
e-mail: bhavya.y.mehta@gmail.com

© Springer Nature Singapore Pte Ltd. 2020
D. Swain et al. (eds.), *Machine Learning and Information Processing*,
Advances in Intelligent Systems and Computing 1101,
https://doi.org/10.1007/978-981-15-1884-3_6

This paper proposes a unique solution to this problem by identifying the camera model of the input image by using metadata and deep learning. Metadata can be a source to identify the camera model from which the image/video has been captured as it contains detailed information such as DPI, ISO number, shutter speed, and date and time when the image was clicked.

The issue with this method, i.e., using metadata, which contains information about make and model of camera and timestamp when the image was captured, in the image is that it can be easily tampered and manipulated by using simple computer software. This makes it an unreliable source and cannot be trusted.

The second method uses deep learning to classify the input image according to the camera model based on the factors such as compression ratio, pixel density, color contrast, and other image properties which are unique to the lens of the camera. Deep learning uses convolutional neural networks to train the model and classify images. To increase the performance of the trained model, we used transfer learning. The dataset consists of images from three different camera models [2]. The data had to be pre-processed and cleaned to boost the computation and to increase the accuracy.

2 Literature Review

2.1 SmallNet Based Neural Network

A SmallNet consists of three convolutional layers [3]. The image is taken as a matrix of dimensions 256×256 which is further pre-processed by passing through a kernel of size 5×5. The filtered image is of size 252×252. The output is then given as an input to three convolutional layers with kernel stride of 64, 64, and 32 feature maps with sizes of 126×126, 63×63, and 63×63, respectively. The activation function is ReLU. The output is max pooled with 3×3-sized kernels. Now the output is classified using three full connection layers, with 256, 4096, and 3 nodes, respectively, with softmax activation function.

The illustration is shown in the image. This model gave an accuracy of about 49.04% for training data after 10 epochs. Three types of camera models were used in the training data (Fig. 1).

2.2 Feed-Forward Neural Network

A cycle is not formed between the layers of this type of neural network. Information flows in a single direction, from the input layers through the hidden layers followed by output layer. There is no back-propagation—information is not fed back to the input. Each layer consists of nodes, and the path between nodes has certain weights. Product of weight and the input are calculated at each node, and the resultant value

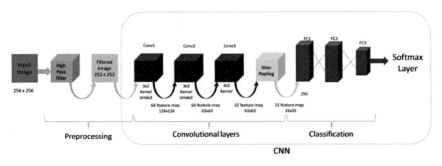

Fig. 1 Conv. SmallNet architecture

is compared to a certain threshold. The neuron gets fired if the value is greater than the threshold. There are three layers, viz. input layer, hidden layer, and output layer. A neural network must have at least one hidden layer but can have as many of them as necessary.

Following are the details of architecture referred:

Input layer—8192 nodes
Hidden layer—64 nodes
Output layer—3 nodes.

3 CNN Architecture

For the image recognition and classification areas, nowadays CNN has become the effective method of machine learning. The basic skeleton of a CNN is: Input → Convolution layer → Pooling layer → Convolution layer → Pooling layer → Fully Connected Layer → Output [4].

Extracting features from an input image is the primary goal of convolution in case of a CNN. Every image is considered as a matrix of pixel value. A small matrix of size $n \times n$ with pixel values 0 and 1 is considered, and another smaller matrix of size $m \times m$ (where $m < n$) is considered. The convolution of the first matrix and the second matrix also called as feature detector matrix is computed. The resultant matrix is called as 'Feature Map' [5].

The following three characteristics control the dimensions of resultant Feature Map:

(a) Depth: The number of filters used for convolution is the Depth [4].
(b) Stride: When scanning the feature horizontally and vertically, the number of pixels which we skip is the Stride [4].
(c) Zero-padding: Zeros are added to the matrix of Feature Map to tweak its dimensions if required [4].

Pooling layer helps to optimize the dimension of each feature map and keeps only the most important information which makes computations faster. Max pooling is performed only after ReLU is applied individually on every feature map. All these layers fetch important features from input images, introduce nonlinearity, and make the features equivariant to scale and translation. The overall output of convolution layer and pooling layer is the extracted features from input images [6].

Fully connected layer gets its input from the output of latter pooling layer. It uses these features to classify the input images into various classes, and it uses the softmax activation function in the output layer. All neurons in the previous layer are connected to all neurons on the next layer.

The whole process of classification using convolutional neural networks can be recapitulated as shown below:

1. Initialize all the filters with some arbitrary values.
2. Input to the network is an image which is then passed to the above-discussed steps—convolution, ReLU, and the pooling layers followed by the fully connected layer. Based on the calculations, the output probability is calculated for every class.
3. Error is calculated at the output layer, which is nothing but the square of summation of differences of expected probability and output probability for each class. The goal is to minimize this error so that the output accuracy is maximum. This is done using back-propagation technique in convolutional neural network [7].
4. Using back-propagation, the gradient of error w.r.t all the weights is calculated. Now by using gradient descent algorithm, the initially assigned arbitrary values are updated according to the value of error.
5. Now Step 2 to Step 4 is repeated for every image in the training dataset.

Thus, this requires heavy computation, and usually, a graphics processing unit (GPU) is preferred to boost the computation power and reduce the required overall training time.

3.1 Transfer Learning

Transfer learning [3] is a machine learning method in which a model created for some task is again used to develop a model for another task. The weights and biases of a pre-trained model epitomize the features for that respective dataset. Very few neural network models are built from scratch (random initialization of weights as discussed in Sect. 3) because it is very difficult to get huge data for a particular use case. Hence, using the already available relevant pre-trained network's weights, starting point weights can be very useful and beneficial to build a strong and accurate neural network for classification [8].

For example, if a model is trained for a huge dataset with car images, then that model will contain features to classify edges or horizontal/vertical lines that can be used for some other problem statement [3].

There are numerous benefits of a pre-trained model. Huge amount of training time can be saved with transfer learning. Our model will benefit from someone else's model; there is no point in spending time and resources if that same model can be reused, unless there is a completely new problem statement or there are some other constraints considering data [3].

In this case, we are using Microsoft's ResNet-50 model which was trained on popular ImageNet dataset. We then tweaked the weights according to our requirement by adding more dense layers and finally softmax with three outputs.

3.2 Experimental Details

The data, i.e., the images, were pre-processed by cropping each image in a matrix of 512×512 pixels. The training data used for the ResNet-50 [9] CNN model contain 825 files of three different camera models of which 80% have been used for training and 20% have been kept for validation. For testing, different images captured by camera were used. We used keras framework in python with TensorFlow [10] backend to create convolutional neural network for classification.

Residual networks provide much better results than many other image classification models. It prevents the saturation of accuracy as we go deeper into the network by making jump connections between layers, which in turn reduces the time complexity.

The version used in this project is ResNet-50, where '50' stands for 50 layers. Each jump connection is after skipping two layers of 3×3 convolution with last layers being unfrozen fully connected layers with output layers of three nodes representing the three camera models, which are the three classes.

We achieved a significant improvement of 20% in our accuracy by using ResNet as the transfer learning model with initial layers frozen. ResNet gives much faster results with reduced error, thus solving the problem of saturation [11].

If we use many layers in a residual network, then there is a chance that gradient might get vanished. This can have negative effect on the expected accuracy. To increase the accuracy, skip connections have been used in ResNet-50. We fine-tuned the parameters of fully connected layer to increase the overall accuracy of our model.

The ResNet-50 architecture is shown in Fig. 2.

4 Conclusion

The trained model has been tested, and the following results were obtained. The accuracy obtained after 25 epochs was 80.61%, and the loss has decreased to 0.6345 after using ResNet-50 model for transfer learning.

Following results were obtained for three classes (Table 1).

From the results, it can be seen that ResNet-50 gave most desirable results when compared to other training models for camera model identification. It can also be seen

Fig. 2 VGG-19 model and ResNet with 34 parameter layers' architecture [12]

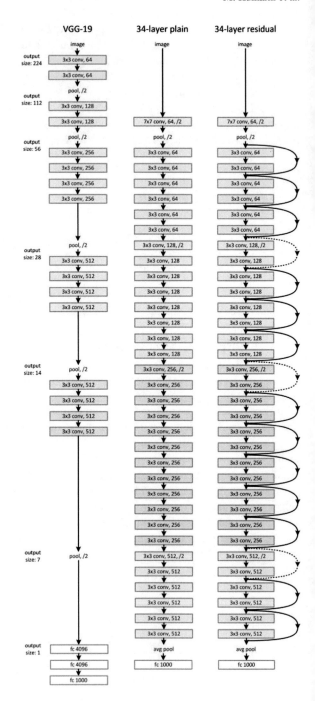

Table 1 Comparison of results from different models

S. No.	Training model	Image size	Weight size (MB)	Accuracy (%)
1	ResNet-50	224 × 224	99	80.61
2	MobileNet	224 × 224	17	65.03
3	VGG-19	224 × 224	528	71.46
4	SmallNet	252 × 252	11	49.04
5	Inception V3	299 × 299	92	76.09

that the size of ResNet-50 is quite less as compared to other models with accuracy closer to that of ResNet-50.

The performance of the model rises quickly during the training period when we use transfer learning as compared to model being trained without transfer learning. With the use of skip connection, there is very less possibility that deeper layers will worsen the result provided by the previous layers.

The power of ResNet can only be utilized if we use proper and clean input data. The accuracy can be improved significantly by using more images in the training dataset, i.e., by increasing the data and by increasing the number of epochs.

Acknowledgements The authors feel grateful to and they wish their profound indebtedness to their guide **Prof. Milind Kamble**, Department of Electronics and Telecommunication, Vishwakarma Institute of Technology, Pune. The authors also express their gratitude to **Prof. Dr. R. M. Jalnekar**, Director, and **Prof. Dr. Shripad Bhatlawande**, Head, Department of Electronics and Telecommunication, for their help in completion of the project. The authors also thank all the anonymous reviewers of this paper whose comments helped to improve the paper.

References

1. Tuama, Amel, Frédéric Comby, Marc Chaumont. 2016. Camera model identification with the use of deep convolutional neural networks. Available via http://www.lirmm.fr/~chaumont/publications/WIFS-2016_TUAMA_COMBY_CHAUMONT_Camera_Model_Identification_With_CNN_slides.pdf. Accessed January 13, 2018.
2. Antorsae. 2018. *IEEE's Signal Processing Society—Camera Model Identification Kaggle Competition*. Available via https://github.com/antorsae/sp-society-camera-model-identification. Accessed March 25, 2018.
3. Shah, Anuj. 2018. Transfer learning using Keras. Available via https://github.com/anujshah1003/Transfer-Learning-in-keras—custom-data. Accessed February 3, 2018.
4. Anonymous. 2018. Convolutional neural networks for visual recognition. Available via http://cs231n.github.io/convolutional-networks/. Accessed February 21, 2018.
5. Bondi, Luca, Luca Baroffio, David Guera, Paolo Bestagini, Edward Delp, and Stefano Tubaro. 2015 First steps toward camera model identification with convolutional neural networks. *Journal of Latex Class Files* 14: 1–4. arXiv:1603.01068.
6. Ujjwalkarn. 2016. An intuitive explanation of convolutional neural networks. Available via https://ujjwalkarn.me/2016/08/11/intuitive-explanation-convnets/. Accessed March 17, 2018.
7. Kuzin, Artur, Artur Fattakhov, Ilya Kibardin, Vladimir Iglovikov, and Ruslan Dautov. 2018. Camera model identification using convolutional neural networks, 1–4. arXiv:1810.02981.

8. Jay, Prakash. 2017. Transfer learning using Keras. Available via https://medium.com/@14prakash/transfer-learning-using-keras-d804b2e04ef8. Accessed February 21, 2018.
9. Kaggle. 2018. ResNet-50 Pre-trained Model for Keras. Available via https://www.kaggle.com/keras/resnet50. Accessed February 26, 2018.
10. Google. 2015. Tensorflow. Available via https://www.tensorflow.org/. Accessed January, February 9, 2018.
11. Koustubh. 2018. ResNet, AlexNet, VGGNet, inception: Understanding various architectures of convolutional networks. Available via http://cv-tricks.com/cnn/understand-resnet-alexnet-vgg-inception/. Accessed March 23 2018.
12. He, K, X. Zhang, S. Ren, and J. Sun. 2015. Deep residual learning for image recognition, 1–8. arXiv:1512.03385.

Sentiment Classification of Online Mobile Reviews Using Combination of Word2vec and Bag-of-Centroids

Poonam Choudhari and S. Veenadhari

Abstract Sentiment classification is a technique to understand the feeling/attitude/sentiment toward a written piece of text by analyzing and then classifying the text as positive, negative, or neutral. One of the important aspects of classification is data that should be handled and represented carefully in the classification process, which affects the performance of the classifier. In the sentiment classification process, feature vector is used as the representation of data to work on. In the paper, we have experimented with the combination of Word2vec and Bag-of-Centroids' feature vector in the sentiment classification process of online consumer reviews about different mobile brands. The feature vector is tested on different well-known machine learning classifiers used for sentiment analysis and compared with Word2vec feature vector. We also investigated the performance of a feature vector as the size of the dataset is increased. We found that the proposed feature vector performed well in comparison with Word2vec feature vector.

Keywords Sentiment classification · Word2vec · Bag-of-Centroids · K-means clustering

1 Introduction

Sentiment classification is an automated way of classifying opinions expressed about some topic in a given piece of text by analyzing it and then finding whether the overall text data reflects positive, negative, or neutral polarity [11, 15]. In today's digital world where people often take the help of Internet to make their life easier in various aspects, for example, when making an opinion about whether to buy a particular product or not, people surf the online sites such as Amazon, Flipkart to view the online

P. Choudhari (✉) · S. Veenadhari
Department of Computer Science and Engineering, Rabindranath Tagore University, Bhopal, Madhya Pradesh, India
e-mail: choudhari.poonam@gmail.com

S. Veenadhari
e-mail: veenadhari1@gmail.com

© Springer Nature Singapore Pte Ltd. 2020
D. Swain et al. (eds.), *Machine Learning and Information Processing*,
Advances in Intelligent Systems and Computing 1101,
https://doi.org/10.1007/978-981-15-1884-3_7

customer's reviews or ratings given by them for that product [3, 7]. The problem here is that it becomes difficult to manually analyze thousand of reviews posted daily on various digital platforms and then to decide about shopping the product. The solution to this problem is sentiment classification. Sentiment classification mines those large numbers of reviews and classifies them into different classes. Feature vector extraction process is an important phase of sentiment classification. The choice of the features used for classification can improve classification performance. A context-based feature vector, Word2vec is used in our approach which is more efficient from traditional feature vectors like Bag-of-Words, term-based feature vector [5].

The rest of the paper is divided as follows. Section 2 gives a brief review of the related work in sentiment analysis. Section 3 explains the proposed work. Section 4 discusses and analyzes the experimental results. It also gives the conclusion of the paper and the future aspects of the proposed work.

2 Related Work

A lot of work has been done in the field of sentiment classification during the last couple of years. Pang et al. [13] used movie review data for sentiment classification and used unigrams as the feature vector. The author has also experimented with another feature vector like bigrams, POS-based features their combination as features. The author found that unigram features performed best among all the features with SVM. Pang and Lee [12] introduced a novel machine learning method for sentiment classification. They extracted only the subjective sentences as feature vector from the whole text data by applying minimum cut in graph technique. The author found significant improvement in sentiment classification accuracy. Narayanan et al. [10] applied negation handling, for feature extraction phase used bag of n-grams, and for feature selection, mutual information was used. They found a significant improvement in accuracy with naive Bayes classifier and achieved an accuracy of 88.80% on IMDB movie reviews dataset. Bansal et al. [2] used Word2vec as feature vector with different dimensions of window size and found improvement when the dimensions were increased. They found the combination of CBOW with random forest has performed the best among all the classifiers with 90.6622% accuracy and AUC score of 0.94. Alshari et al. [1] experimented with Word2vec feature vector by lowering their dimensions. This was done by clustering of word vectors obtained based on the opinion words of sentiment dictionary. The experimental results proved that the proposed feature vector performed well in comparison with simple Word2vec, Doc2vec, and Bag-of-Words' feature vectors. Zhang et al. [17] proposed a method in which a candidate feature vector is obtained by clustering of the Word2vec feature vector. A group of synonyms that are near to product feature was obtained by clustering. Feature selection method, lexicon-based, and POS-based were applied on the clustered Word2vec feature vector obtained. They compared their results with TF-IDF feature vector and obtained improved results.

3 Proposed Method

In this section, the workflow of the proposed method is given, and then the steps of the method are explained in detail (Fig. 1).

3.1 Data Description

The corpus is taken from the Amazon unlocked mobile phone reviews publicly available on Kaggle [6] which consists of online customer reviews of mobile phones sold on Amazon. For the analysis purpose, we took only reviews and corresponding rating field from the dataset. As we are doing machine learning-based classification, the dataset is divided into training and testing set in the ratio of 70:30.

3.2 Steps of the Proposed Method

3.2.1 Data Preprocessing

The first step consists of getting the cleaned data from the raw data. Unwanted digits, symbols, HTML tags, stop words that do not contribute to the classification process are removed from the data reviews. It also involves stemming and conversion of all the words from upper case to lower case.

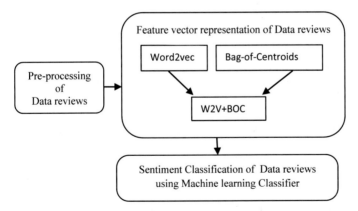

Fig. 1 Workflow of the proposed method

3.2.2 Feature Vector Representation of Preprocessed Data

The second step involves transforming the preprocessed data into the feature vector. Here the preprocessed data is tokenized into words and converted to numerical representation so that it can be given as input to the machine learning classifier.

In our proposed method, feature vectors used are

- Combination of Bag-of-Centroids and Word2vec (Unigrams)
- Combination of Bag-of-Centroids and Word2vec—with phrases (Unigrams + Bigrams).

Unigrams refer to single word, whereas bigrams refer to a sequence of two words. Word2vec—with phrases—means all the unigrams and bigrams formed are not considered since many of them are not really useful. So those phrases (Unigrams + Bigrams) are taken into consideration which satisfies a threshold value.

The calculation of the feature vector involves the following steps. First, the Word2vec feature vector is calculated from preprocessed data. Then Bag-of-Centroids' feature vector is calculated. Then both Word2vec and Bag-of-Centroids are combined, which will be given as input to the sentiment classification phase.

Word2vec Feature Vector Transformation

In this sub-step, the preprocessed data is converted into Word2vec feature vector. First, a brief description of Word2vec with its mathematical representation is explained and then the implementation details of Word2vec are given.

Word2vec

Word2vec is a context-related feature vector representation developed by Mikolov [8] at Google in 2013. They also enhanced their work by introducing N-gram phrases with Word2vec [9]. The word vectors produced are located in the vector space such that the words with similar context are close to each other in the vector space There are two models of Word2Vec–CBOW (Continuous Bag-of-Words) in which the target word is predicted using the surrounding context words, Skip-gram uses the opposite technique as compared to CBOW, and the surrounding context words are predicted from the target word.

Mathematical Representation of Word2vec

Word2vec is a shallow two-layer neural network where there is an input layer, hidden layer, and output layer [14]. The network is trained based on the maximum likelihood principle; i.e., in case of CBOW model, it finds the probability of the target word given the neighboring words and then tries to maximize the probability.

The input layer consists of neurons of V dimension equal to the number of words of text corpus or vocabulary. The hidden layer consists of neurons of D dimension which is equal to the dimension of the resulting word vector. The input to hidden layer connection can be represented by W_{VD} matrix ($V * D$ size) that maps the words of the input layer to the hidden layer. We can calculate the output of the hidden layer as:

$$H = X * W_{VD}. \tag{1}$$

The output layer consists of neurons of V dimension which is equal to the number of words in the vocabulary containing Softmax values. The hidden-to-output layer connection can be represented as W_{DV} matrix ($D * V$ size) that maps the result of hidden layer to the output layer. The output layer produces the output as:

$$O = H * W_{DV}. \tag{2}$$

Since we need the probabilistic values for the words in the vocabulary, the values of the output layer are converted to corresponding probabilistic values by using Softmax function. The Softmax function to calculate probability can be represented as:

$$P\left(\text{Word}_{\text{target}} \middle| \text{Word}_{\text{context}}\right) = \frac{\exp\left(\text{score}\left(\text{Word}_{\text{target}}, \text{Word}_{\text{context}}\right)\right)}{\sum_{w \in \text{vocab}} \exp(\text{score}(\text{Word}_w, \text{Word}_{\text{context}}))} \tag{3}$$

The numerator calculates the similarity between the target word and the context word. The maximum is the similarity, and the maximum is the probability. The denominator normalizes the probability of the other words in the vocabulary with the context word so that the sum of all is equal to one. As discussed above, the model is trained by maximizing its log-likelihood on the training set. The training can be further extended for other target–context pairs from the vocabulary. In this way, the Word2vec model learns the relationship between the words and finds the vector representation of the words in the vocabulary.

The implementation is done as follows. First, we parse each of the reviews of the training dataset into sentences and obtain the corresponding Word2vec feature vector vocabulary for each word present in training dataset. Four forms of Word2vec are considered, i.e., CBOW, CBOW with phrases, Skip-gram, and Skip-gram with phrases. In the second step, the transformation of the training and testing data reviews to Word2vec feature vectors is done; i.e., for each training/testing review, we get its corresponding Word2vec feature vector. In this process, a review is transformed into a feature vector by averaging feature vectors of words appeared in that review and in the vocabulary list created. This process is done for each of the four forms of Word2vec that were considered in the previous step.

Bag-of-Centroids' (BOC) Transformation

In this step, the training and testing sets are transformed into Bag-of-Centroids' (BOC) feature vector. It means that the training and testing data reviews are transformed into clusters of Word2vec feature vector obtained in the previous step, and those clusters are used as feature vectors. As Word2vec places semantically related word vectors near each other, so we have used BOC feature vector to utilize this property of Word2vec. The clustering algorithm used is K-means clustering algorithm. Initially, the number of clusters is obtained by partitioning five-word vectors per cluster from Word2vec vocabulary. K-means is then applied on the CBOW Word2vec vocabulary to get the clusters. The training and testing sets of reviews are then transformed into BOC. This transformation is done as follows: First, an empty array is created whose size is equal to the number of clusters, and then while looping over the words of the review, if the words in the review are found in the vocabulary, then we need to find in which cluster the word belong to and increment that cluster by one.

Mathematical Representation of Bag-of-Centroids

Bag-of-Centroids' transformation involves making clusters of Word2vec feature vector. K-means clustering algorithm [16] is used to form clusters. The input to the algorithm is $O_1, O_2, ..., O_V$ Word2vec feature vector as data points and K, i.e., the number of clusters needed. The steps involved are as follows:

- Choose K data points as initial centroids from $O_1, O_2, ..., O_V$ Word2vec feature vector.
- Calculate Euclidean distance of each data point from $O_1, O_2, ..., O_V$ with the k-cluster centroids. The Euclidean distance is calculated as:
  If $v = (v1, v2)$ and $w = (w1, w2)$, then distance is given by:

$$d(v, w) = \sqrt{(w1 - v1)^2 + (w2 - v2)^2} \tag{4}$$

- Find the minimum Euclidean distance and assign each of the data point from $O_1, O_2, ..., O_V$ to the nearest cluster centroid.
  If the cluster centroid is denoted as cc_n, then each of the data point O is assigned to a cluster by calculating

$$\arg\min_{cc_n \in CC} d(cc_n, O)^2 \text{ where } d() \text{ is the Euclidean distance} \tag{5}$$

- Take the average of the data points in each cluster and find the new cluster centroid. The new cluster centroid is calculated as:

$$cc_n = \frac{1}{|P_i|} \sum_{O_i \in P_i} O_i \tag{6}$$

where p_i is the group of data points assigned to the ith cluster.
- The above steps are repeated for a number of times, till we get the unchangeable cluster centroid points.

Combining Bag-of-Centroids' Feature Vector and Word2vec Feature Vector

In this step, the Bag-of-Centroids' feature vector and Word2vec feature vector are combined to get the final feature vector which is then fed to the machine learning classifiers. The combination of feature vector can boost the performance of classification in terms of accuracy and F1 score as compared to Word2vec which is used individually. The combination is done by adding the clusters obtained in the previous step to each of the four forms of Word2vec that were considered previously, i.e.,

- CBOW Word2vec + Clusters of CBOW.
- CBOW with phrases (CBOW-PH) Word2vec + Clusters of CBOW.
- Skip-gram (SG) Word2vec + Clusters of CBOW.
- Skip-gram with phrases Word2vec (SG-PH) + Clusters of CBOW.

3.2.3 Sentiment Classification

Finally, machine learning-based sentiment classification is done. The machine learning-based classifiers are trained on the training dataset. After that, the testing dataset is used for evaluating the performance of trained classifiers. The classifiers used are logistic regression CV (LRCV), multilayer perceptron (MLP), random forest (RF), decision tree (DT), and Gaussian naïve Bayes (GNB).

4 Experimental Results and Discussion

The performance of the feature vector on sentiment classification is evaluated by comparing the proposed feature vector with our previous work's experimental results [4], where we had used Word2vec individually (Fig. 2).

Actual	Negative	TN =3092	FP = 688
	Positive	FN =930	TP =7359

Negative   Positive

Predicted

Fig. 2 Confusion matrix of MLP classifier (W2V + BOC) with 40,277 instances and skip-gram (unigrams + bigrams) phrases

Tables 1, 2, 3 and 4 show that the results obtained with the proposed feature vector are better than our previous work results, i.e., with Word2vec. LRCV and MLP classifiers both performed well in case of accuracy and F1 score for all the dataset sizes. The best accuracy and F1 score are obtained with Skip-gram phrases (Unigrams + Bigrams) model. As the number of reviews increases in the dataset, the feature vector performed well.

There are many aspects of proposed work which can be investigated in the future. The feature vector can be experimented with other product domain. In the proposed method, we have used K-means clustering, so another clustering algorithm like BIRCH clustering in which incremental and dynamic clustering is possible and can be implemented in the future.

Table 1 Comparison of accuracy on different classifiers with Word2vec on different database sizes

No. of reviews	Classifier used	CBOW (unigrams)	CBOW-PH (uni + bigrams)	SG (unigrams)	SG-PH (uni + bigrams)
6655	MLP	0.808713	0.792188	0.810215	0.803205
	LRCV	0.819229	0.812719	0.824737	0.826740
	RF	0.794692	0.805709	0.753130	0.704056
	DT	0.754632	0.799199	0.708563	0.661492
	GNB	0.621432	0.665999	0.571357	0.553330
16,678	MLP	0.784572	0.788569	0.765987	0.712030
	LRCV	0.844924	0.837730	0.842326	0.840727
	RF	0.784772	0.799161	0.757994	0.738209
	DT	0.767986	0.740807	0.743805	0.709233
	GNB	0.603717	0.611511	0.585132	0.562150
26,764	MLP	0.837983	0.801494	0.815567	0.781320
	LRCV	0.850560	0.838356	0.850311	0.841843
	RF	0.785181	0.784433	0.780448	0.731258
	DT	0.739975	0.768742	0.748319	0.720672
	GNB	0.610336	0.601494	0.600623	0.573350
40,227	MLP	0.848123	0.817135	0.813903	0.819869
	LRCV	0.852846	0.848206	0.851852	0.851106
	RF	0.787223	0.787555	0.768001	0.740244
	DT	0.773469	0.750352	0.751346	0.701135
	GNB	0.603778	0.591515	0.596404	0.576187

Table 2 Comparison of accuracy on different classifiers with Word2vec + Bag-of-Centroids on different database sizes

No. of reviews	Classifier used	CBOW (unigrams) + BOC	CBOW-PH (uni + bigrams) + BOC	SG (unigrams) + BOC	SG-PH (uni + bigrams) + BOC
6655	MLP	0.809715	0.835754	0.842263	0.841763
	LRCV	0.834752	0.832749	0.832248	0.834752
	RF	0.787181	0.790686	0.736605	0.711067
	DT	0.801202	0.791688	0.729594	0.707061
	GNB	0.686530	0.723585	0.635954	0.611417
16,678	MLP	0.828737	0.828737	0.825739	0.832734
	LRCV	0.850320	0.845124	0.850719	0.851519
	RF	0.788569	0.802758	0.752198	0.735412
	DT	0.774580	0.764588	0.768585	0.742806
	GNB	0.691247	0.696843	0.671063	0.643285
26,764	MLP	0.833499	0.860274	0.861395	0.866501
	LRCV	0.863512	0.858281	0.864010	0.858780
	RF	0.788045	0.802491	0.773973	0.752802
	DT	0.741594	0.766376	0.718431	0.704981
	GNB	0.721295	0.713450	0.716065	0.686177
40,227	MLP	0.840500	0.841412	0.863369	0.865938
	LRCV	0.864198	0.860966	0.864860	0.861463
	RF	0.796669	0.776286	0.773884	0.723175
	DT	0.787223	0.752258	0.755324	0.771646
	GNB	0.729058	0.717707	0.728975	0.705858

Table 3 Comparison of F1 score on different classifiers with Word2vec on different database sizes

No. of reviews	Classifier used	CBOW (unigrams)	CBOW-PH (uni + bigrams)	SG (unigrams)	SG-PH (uni + bigrams)
6655	MLP	0.849488	0.834727	0.849663	0.846544
	LRCV	0.861314	0.855263	0.866003	0.866615
	RF	0.838455	0.851227	0.802088	0.751367
	DT	0.803055	0.849756	0.756485	0.698752
	GNB	0.655738	0.707584	0.597744	0.576046
16,678	MLP	0.825510	0.828747	0.805190	0.744549
	LRCV	0.881091	0.874730	0.878485	0.876911
	RF	0.831112	0.843580	0.806085	0.789321
	DT	0.822123	0.789482	0.798110	0.764677
	GNB	0.638469	0.646031	0.618523	0.590850
26,764	MLP	0.875108	0.840088	0.853468	0.819638
	LRCV	0.885584	0.875456	0.885349	0.878213
	RF	0.830600	0.830576	0.827411	0.779481
	DT	0.788364	0.821253	0.798283	0.779167
	GNB	0.642849	0.633280	0.633108	0.600420
40,227	MLP	0.884768	0.856437	0.851749	0.858739
	LRCV	0.888105	0.883948	0.887064	0.886775
	RF	0.833398	0.834538	0.814594	0.790118
	DT	0.823271	0.802335	0.799358	0.752962
	GNB	0.638221	0.623836	0.632294	0.607775

Table 4 Comparison of F1 score on different classifiers with Word2vec + Bag-of-Centroids on different database sizes

No. of reviews	Classifier used	CBOW (unigrams) + BOC	CBOW-PH (uni + bigrams) + BOC	SG (unigrams) + BOC	SG-PH (uni + bigrams) + BOC
6655	MLP	0.853055	0.875758	0.882331	0.881381
	LRCV	0.876312	0.873676	0.873728	0.875940
	RF	0.833006	0.836973	0.787045	0.763621
	DT	0.853775	0.842065	0.784517	0.757563
	GNB	0.736532	0.774141	0.681838	0.655417
16,678	MLP	0.866573	0.867399	0.863835	0.870973
	LRCV	0.886601	0.882129	0.887109	0.887781
	RF	0.833176	0.845176	0.798177	0.783519
	DT	0.819751	0.810733	0.818779	0.789465
	GNB	0.746347	0.751271	0.728024	0.698632
26,764	MLP	0.871058	0.894151	0.896455	0.900997
	LRCV	0.897282	0.892944	0.897944	0.893681
	RF	0.833333	0.846734	0.820564	0.801718
	DT	0.791016	0.816259	0.761069	0.749762
	GNB	0.775796	0.769138	0.772637	0.743224
40,227	MLP	0.878111	0.879029	0.898204	0.900955
	LRCV	0.898583	0.895867	0.898765	0.896214
	RF	0.841698	0.824447	0.820212	0.773353
	DT	0.839760	0.805288	0.803957	0.836303
	GNB	0.786581	0.776663	0.788216	0.766540

References

1. Alshari, E.M., A. Azman, S. Doraisamy, N. Mustapha, and M. Alkeshr. 2017. Improvement of sentiment analysis based on clustering of word2vec features. In *28th International Workshop on Database and Expert Systems Applications*.
2. Bansal, Barkha, and Sangeet Shrivastava. 2018. Sentiment classification of online consumer reviews using word vector representations. *Procedia Computer Science* 132: 1147–1153. In *ICCIDS 2018. International Conference on Computational Intelligence and Data Science*, Apr 2018, ed. V. Singh and V.K. Asari. Elsevier.
3. Cambria, E., B. Schuller, Y. Xia, and C. Havasi. 2013. New avenues in opinion mining and sentiment analysis. *IEEE Intelligent Systems* 28 (2): 15–21.
4. Poonam, Choudhari, and S. Veenadhari. 2019. Sentiment analysis of online product reviews with word2vec n-grams. *Journal of Emerging Technologies and Innovative Research* 6 (4): 555–559.
5. Fang, Xing, and Justin Zhan. 2015. Sentiment analysis using product review data. *Journal of Big Data*. https://doi.org/10.1186/s40537-015-0015-2.

6. https://www.kaggle.com/PromptCloudHQ/amazon-reviews-unlocked-mobile-phones/. Accessed 12 Mar 2019.
7. Liu, Bing. 2012. Sentiment analysis and opinion mining. Synthesis Lectures on Human Language Technologies.
8. Mikolov, T., K. Chen, G. Corrado, and J. Dean. 2013. Efficient Estimation of Word Representations in Vector Space. arXiv: 1301.3781v3 [cs.CL].
9. Mikolov, T., I. Sutskever, K. Chen, G. Corrado, J. Dean. 2013. Distributed Representations of Words and Phrases and their Compositionality. arXiv:1310.4546v1 [cs.CL].
10. Narayanan, Vivek, Ishan Arora, and Arjun Bhatia. 2013. Fast and accurate sentiment classification using an enhanced naive bayes model. In *Intelligent Data Engineering and Automated Learning, IDEAL 2013*. Lecture Notes in Computer Science 8206, 194–201.
11. Pang, B., and L. Lee. 2008. Opinion mining and sentiment analysis. *Foundations and Trends in Information Retrieval* 2: 1–135.
12. Pang, B, and L. Lee. 2004. A sentimental education: Sentiment analysis using subjectivity summarization based on minimum cuts. Association for Computational Linguistics.
13. Pang, B., L. Lee, and S. Vaithyanathan. 2002. Thumbs Up? sentiment classification using machine learning techniques. In *Annual Conference on Empirical Methods in Natural Language Processing, Association for Computational Linguistics*.
14. Rong, Xin. 2016. word2vec parameter learning explained. arXiv: 1411.2738v4 [cs.CL].
15. Van Looy, A. 2016. Sentiment analysis and opinion mining (business intelligence 1). In *Social media management, Springer texts in business and economics*. Cham: Springer International Publishing Switzerland. https://doi.org/10.1007/978-3-319-21990-5_7.
16. Zalik, K.R. 2008. An efficient k-means clustering algorithm. *Pattern Recognition Letters* 29: 1385–1391.
17. Zhang, Dongwen, Hua Xu, Zengcai Su, and Yunfeng Xu. 2015. Chinese comments sentiment classification based on word2vec and SVMperf. *Expert Systems with Applications* 42: 1857–1863.

Emotion Analysis to Provide Counseling to Students Fighting from Depression and Anxiety by Using CCTV Surveillance

Sheona Sinha, Shubham Kr. Mishra and Saurabh Bilgaiyan

Abstract Around 18.5% of the Indian population of students suffer from depression and around 24.4% of students suffer from anxiety disorder. Depression and anxiety are treatable through counseling and certain medicines, and thus, to avail to this huge percentage of students, the help that they require is provided in many reputed colleges and universities. These colleges and universities hire professional counselors to cater to the needs of these students. But, as all problems are not easy to overcome, in this situation also, there is a huge problem. That problem is of students not venting out their need for counseling due to various reasons, and hence, they do not go to counselors to get themselves back in happy life. To conquer such problems, a solution is proposed in this paper, that is, the use of CCTV surveillance recording that is now readily available in various colleges and universities, along with sentiment analysis of each and every student. Their emotional well-being will be monitored through their facial landmark recognition, and if certain students are showing signs of depression through his or her activities, then their names are given to their respective counselors, so as to provide them with care and support and right guidance to start their life afresh. This paper makes use of computer vision, image processing, and convolutional neural network to complete the above-mentioned task.

Keywords Sentiment analysis · Facial landmark recognition · Computer vision · Image processing · Convolutional neural network

S. Sinha · S. Kr. Mishra · S. Bilgaiyan (✉)
School of Computer Engineering, KIIT Deemed to be University, Bhubaneswar, Odisha, India
e-mail: saurabh.bilgaiyanfcs@kiit.ac.in

S. Sinha
e-mail: 1606508@kiit.ac.in

S. Kr. Mishra
e-mail: 1606511@kiit.ac.in

© Springer Nature Singapore Pte Ltd. 2020
D. Swain et al. (eds.), *Machine Learning and Information Processing*,
Advances in Intelligent Systems and Computing 1101,
https://doi.org/10.1007/978-981-15-1884-3_8

1 Introduction

Every phase of a person's life is affected by grim conditions like depression and anxiety which are together termed as mental pain and has impacted one's career, social life, sense of self-worth and interpersonal relationships [1]. Physical pain is more dramatic than mental pain but it is harder to acknowledge mental pain than physical pain. Mental pain is hard to cope with and is frequently attempted to be kept under wraps which only enhances the impact of mental pain. There is no definite definition for depression as all its victims have unique stories to tell about how depression and anxiety have engulfed them and left them live in dark [1].

To fight against depression and anxiety, one needs to acknowledge and seek help and advice from counselors who are expert in the fields of psychological problems. Students in schools and colleges are provided with this facility of seeking advice from counselors as and when they would require. But because of concealing habits, these facilities are often left unused and many students find themselves dragged in the deepest oceans of depression. To overcome such problem, this paper is initiated with the main motive of making the counselors reach out to the students who are in need of their advice and care. This motive has further raised an issue on the determination of those students who need help which our proposed system tries to overcome by studying the emotions of every student in classrooms by making use of CCTV cameras which are installed in almost all schools and colleges. Every student's emotions will be monitored in real time which will be used for creating a feedback on what was the emotion that was predominant on monthly basis. These feedback records will be sent to the counselor appointed by particular school or college and a special mail will be sent to the counselor containing the names of students who showed predominantly negative emotions so that the counselor can keep track of those students and help them with right guidance.

The proposed system is divided into four segments: starting with face recognition to determine the names of students who are under surveillance by making use of open CV [2]. Facial recognition is automation for identifying people and used for verifying a person with help of existing databases. Facial recognition is used in this system to identify individual student and update database consisting of student's name and roll number with respect to the identification done in the first segment. Second segment is facial expression recognition in which convolutional neural network and machine learning are intensively used [3]. In facial expression recognition, emotion analysis is done on all students who are under surveillance by making use of facial landmarks in multiple frames. For emotion analysis, emotions are categorized in seven parts, namely happy, sad, neutral, depressed, surprised, angry and disgust. The assessment done on the basis of emotions is updated in the database in real time for all the students and then comes the third segment which is all about feedback generation in which Python libraries are used. Feedback is generated from the updated databases on daily basis which will be sent to the counselors for their further study and names of the students who are showcasing negative emotions like anger, disgust, sadness or depressed emotions are sent via mail to counselors so as to keep an eye on their activities and reach out to them to help them out of their miseries. These segments will be discussed in detail in later sections.

This paper consists of five sections. Second section is about works already done in this field. Third is about few basic concepts that should be known before proceeding further. Fourth section is about the proposed system architecture in which four segments mentioned above will be discussed in detail along with the results obtained after completion of all four segments. Finally, fifth section will be about future endeavors and conclusion along with shortcomings of the system proposed.

2 Related Works

These tough tasks have been made easy by immense advancement in the field of machine learning. In some difficult visual tasks, the emergence of deep convolutional neural network has made it possible for the machines to match and sometimes, even exceed human performance. This section displays few research papers that demonstrate similar kind of machines.

For making use of LBPH-based improved face recognition at low resolution, a reference was taken from the work done by Aftab Ahmed et al. Their paper employs the local binary pattern histogram (LBPH) algorithm architecture to address the human face recognition in real time at the low level of resolution [4].

Byeon and Kwak proposed a video-based emotion recognition that was developed by making use of three-dimensional CNN. Their machine achieved an accuracy of 95% with a data set consisting of 10 people [5].

Monika Dubey and Prof. Lokesh Singh gave a review in Automatic Emotion Recognition Using Facial Expression to introduce needs and applications of facial expression recognition. They worked on six basic facial expressions [6].

3 Basic Concepts

This paper makes use of words like positive images for images containing face and negative images for images not containing face [7]. Some of the concepts that should be known before moving forward in this paper to proposed system architecture are as follows.

3.1 Haar Cascade Algorithm

It is part of machine learning which acts as a classifier for determination of positive images and negative images. It works on grayscale images as color information does not provide any details about face being in a picture or not. This algorithm was given

Fig. 1 Selection of Haar
features

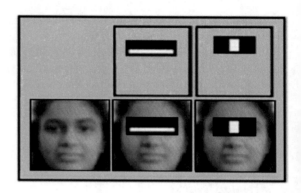

by Viola and Jones [8] and is trained on millions of negative and positive images.
This trained algorithm is now used to detect faces in other images. The algorithm
has four stages:

Selection of Haar feature. Haar features share its similarity with convolutional
kernel [8] but differ slightly in terms of the images. To determine a single feature,
each window is placed over the picture. The obtained feature is usually single value
which is computed by subtraction of the summation of pixels under the window's
white from the summation of pixels under the window's black part. It will not be
easy to perform this step on large scale and hence comes in the application of integral
images [9]. Figure 1 shows selection of Haar features.

Creation of integral images. Those images in which the current pixel value is the
sum of pixel values before that pixel are called integral images [10]. An example of
integral image along with normal image is shown in Fig. 2.

AdaBoost training. AdaBoost is a lavish approach of keeping only relevant features
and discarding irrelevant features from the image [11]. It helps in improving the
accuracy of classification of positive/negative images of classifier as it is a training
task for detection of face, which chooses only known features. AdaBoost refers to a
particular method of training a boosted classifier. A boost classifier is a classifier in
the AdaBoost refers to a particular method of training a boosted classifier. A boost
classifier is a classifier in the form

$$G_T(a) = \sum_{t=1}^{T} g_t(a) \tag{1}$$

2	4	8	1	6
9	7	4	8	5
5	9	0	3	7
7	6	1	4	3
3	1	7	9	2

2	6	14	15	21
11	22	34	40	54
16	36	48	57	78
23	49	62	75	99
26	53	73	95	121

Fig. 2 On left is normal image and on right is an integral image

where each g_t is a weak learner that takes an object a as input and returns a value indicating the class of the object.

Cascading classifiers. In the concept of cascading classifiers, features are assembled in diversified platforms of classifiers as a replacement of using all the features in a window at once [12]. These assembled features are applied in one by one fashion after determining the success of various platforms. If a window is unsuccessful in first platform, then it is discarded and no further features are applied, and if it is successful, second platform of features is applied and hence the procedure continues. When a window passes all the platforms, then window is termed as a face region.

3.2 Local Binary Patterns (LBP)

Faces are composed of microvisual patterns and LBP makes use of these patterns to extract feature vector that will classify a positive image from a negative image. Thus, LBP is also known as visual texture descriptor [4]. In training period, images are divided into blocks as shown in Fig. 3.

LBP makes use of nine pixels at a time, i.e., 3 × 3 window with a prime focus on pixel that is located at the window's center for every block. Then comparison is done between center pixel value and every neighboring pixel's value existing in that window. If the value of neighboring pixels is equal to or greater than the value of center pixel, then its value is set to 1, and for rest, set it to 0. A binary number is formed by reading pixel values in a clockwise direction which is then converted into decimal number and that decimal number serves as the new value of the center pixel. This task is done for every pixel contained in a block. Figure 4 shows an example of how LBP works.

Fig. 3 LBP windows

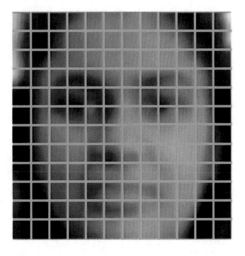

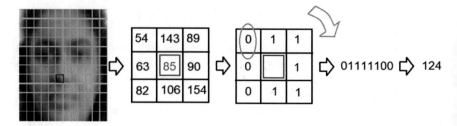

Fig. 4 LBP conversion to binary

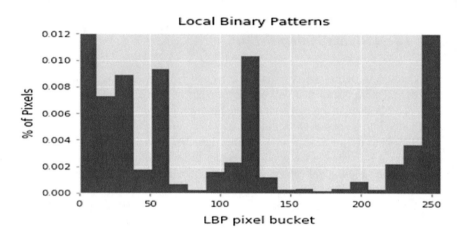

Fig. 5 LBP histogram

Histogram is made by making use of all the block values. Thus, for each block of an image, a histogram is obtained. Figure 5 depicts one such histogram.

Eventually, one feature vector for single image is obtained after concatenating all the block histograms. One feature vector consists of all the features relevant for an image.

4   Proposed System Architecture

The most effective form of visual communication is facial expression and it provides proclamation about emotional state, mindset [13]. Reaction to emotional states, intentions or social communications of a person can be determined by changes in facial appearance is known as Facial expressions To communicate the emotions and express the intentions, the facial expression is the most powerful, natural, nonverbal and instant way for humans. Figure 6 shows the block diagram of the proposed system architecture. Each step will be explained in detail.

Fig. 6 System architecture
block diagram

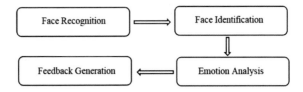

4.1 Face Recognition

The first stage of any vision system is acquisition of image or video. Acquisition of video is required for the proposed system. Open CV lightens this task of acquiring video by providing a built-in library function:

$$cap = cv2.videoCapture(0)$$

Once a video is captured, a frame from a video which will be referred to as image is taken into account to classify whether the image under observation is a positive image or a negative image. Definition of positive and negative image is mentioned in basic concepts. For the purpose of classifying the image into positive or negative image, the proposed system works on Haar cascade algorithm. Open CV again facilitates the functioning of the proposed system by providing a built-in library function:

```
Face_cascade =
cv2.CascadeClassifier("Haarcascades/haarcascade_frontal
face_default.xml")
faces =face_cascade.detectMultiScale(image_array,
scaleFactor = 1.5, minNeighbors = 5)
```

The above code snippets help in detection of multiple faces in an image and this whole process is repeated for a large set of image constituting a video.

Results obtained after recognizing faces in an image using Haar cascades are given in Figs. 7 and 8.

4.2 Face Identification

Images of various people are captured and stored in database with their unique college/university ID number. After the creation of data set consisting of all the images of students and faculty members, LBP features, as explained earlier in basic concepts, are extracted from each test image. The information obtained from the study of LP features is used to classify and recognize various faces. For the proposed system, a total of 2500 images were taken, 100 images per person, which means that faces of 25 people were taken in for creating data set. The images, after recognition of whether they are positive images or negative images, are compared with the database

Fig. 7 Face recognition for
person 1

Fig. 8 Face recognition for
person 2

images. For comparison, LBP features of the testing image are compared with the
LBP features of the images available in the database. If sufficient amount of features
matches, then the face is said to be identified or else termed as 'unknown face.' The
classifier helped in drawing similarity between stored images and the testing images
[3]. If the testing image matches any one of the stored images, then ID assigned
against that image is displayed on the frame. Local binary pattern histogram (LBPH)
face recognizer is used in the proposed system.

```
model = cv2.face.createLBPHFaceRecognizer()
```

Results obtained after face identification done by comparing the LBP features of
the testing image with the images in the data set are shown in Figs. 9 and 10.

Fig. 9 Face identification for person 1

Fig. 10 Face identification for person 2

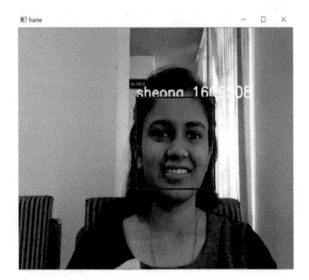

4.3 Emotion Analysis

The natural, most powerful, non-verbal and instant way for humans to communicate is by facial expressions which give an insight into one's emotional state and intentions and also of the social interactions one is involved in. Facial expressions are the result of changing facial appearance caused by the movement of facial features like eyebrows, mouth, nose and eyes. Of all the ways of communication, facial expression is the fastest mode of communication [14]. Conversations are more effective because of facial expressions than those conversations that involve just spoken words. Emotions play a vital role as it conveys one's feelings about the subject and is held as the essential thing for effective interactions among the society. Humans are filled with a lot of emotions, but six of them are termed as basic facial expressions by modern psychology. These six universal emotions are: happiness, sadness, surprise, disgust, fear and anger [15]. But in this research paper, another expression is taken into account which is the most common expression portrayed a person when he/she is not communicating with other people, and that emotion is termed as neutral in the proposed system. The facial muscle movement helps in determining which type of emotion is being depicted by a person.

In this paper, we make use of emotions_classifier.predict, a given method in Open CV.

```
preds = emotion_ classifier.predict(roi)[0]
```

Table 1 gives out proper definitions of the emotions and also gives a brief description about the movement of the facial features that result in a particular emotion [6].

Figures 11 and 12 have shown results acquired during run time emotion analysis.

4.4 Feedback Generation

After getting the percentage of all the six emotions, we inserted them in a data set along with the id of each student as primary key for identification and then the data about their emotions is taken for the generation of line graphs for easy understanding of the emotions shown by the respective individual. The graphs further help in feedback generation which is done by creating list which consists of average of all the emotions shown by the individual whole day and then the max operation is used to determine the maximum value and the name of the corresponding emotion is given as the overall emotion shown by the individual. Graphical representation of emotion analysis is displayed in Figs. 13 and 14.

Final feedback obtained after study of the visualization of emotions done by the proposed system by making use of few calculations is displayed in Figs. 15 and 16.

5 Conclusion and Future Work

The paper proposes an emotion analyzing system that will help in analyzing person's emotions to help them in case of negative emotions portrayed by them for substantially long time. The system is created by the means of Haar cascade, linear binary pattern histogram, 3D convolutional neural networks and machine learning. The proposed system works for a single face detection and identification in a frame

Table 1 Emotion identification

Emotion	Description	Motion of facial features
Anger	This emotion is the one which humans refrain from having and is termed as the most dangerous emotion Hate, dislike, frustration and annoying are secondary emotions of anger	Teeth shut and lips tightened, open eye, upper and lower lids pulled up and eyebrows pulled down
Fear	Danger of physical or psychological pain is termed as fear. Horror, panic, nervousness, dread and worry are secondary emotions of fear	Jaw dropped, outer eyebrow down, inner eyebrow up and mouth open
Happiness	Most desired emotions by humans are happiness. Cheerfulness, relief, pride, pleasure, hope and thrill are the secondary emotions of happiness	Lip corner pulled up, open eyes, wrinkles around eyes, mouth edge up, open mouth, and cheeks raised
Sadness	Opposite of happiness is sadness. Suffering, despair, hurt, hopelessness and pity are secondary emotions of sadness	Lip corner pulled down, closed eye, mouth edge down, outer eyebrow down and inner corner of eyebrows raised
Surprise	Emotions displayed by humans when something unexpected happens are termed as surprise. Astonishment and amazement are secondary emotions of surprise	Jaw dropped, open eye, mouth open and eyebrows up
Disgust	When humans do not like something, they tend to portray an emotion of disgust. They usually feel disgusted from the surrounding sound, smell and taste	Lip corner depressor, nose wrinkle, lower lip depressor and eyebrows pulled down
Neutral	This emotion is a one a person normally keeps when not in communication with other people	Eyebrows, eyes and mouth are relaxed in this emotion

as of now but in the near future it will be made to recognize and identify multiple faces in a single frame. The system is working with the accuracy of 83%.

The proposed system works on unique data set created for a college and will recognize their emotional well-being and will provide a single word remark for study of a person's overall emotional state which can then be sent as a mail to the counselor for their further study and to let them help students who need but never say. The system is suitable for an organization that has its entire campus under CCTV surveillance. This paper tries to evaluate the emotion showcased by students and faculty of the universities and colleges so as to determine if they would need any type of counseling by the counselor assigned. The feedback generated by the system is given to the counselor for their study of the behavior of students and providing faculties so that they can reach out their helping hand to the required students when required.

Fig. 11 Emotion analysis for person 1

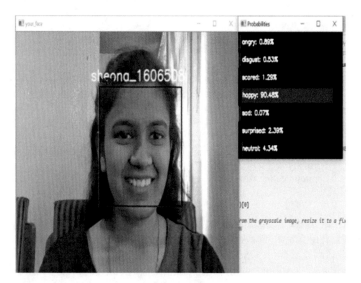

Fig. 12 Emotion analysis for person 2

Fig. 13 Graph of emotions for person 1

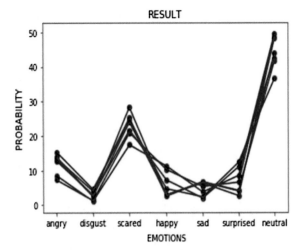

Fig. 14 Graph of emotions for person 2

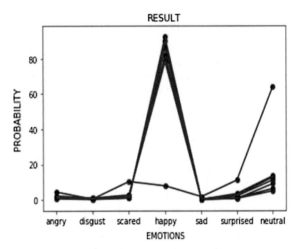

Fig. 15 Feedback generated for person 1

```
Average of anger= 15.338368713855743
Average of disgust= 2.6977885913636004
Average of scare= 23.064019211700984
Average of happiness= 5.974590325994152
Average of sadness= 4.725709876843862
Average of surprise= 7.04411513039044
Average of neutral= 44.37106336866106
```

```
Value is 44.37106336866106
Emotion displayed throughout the day:   neutral
```

Fig. 16 Feedback generated
for person 2

```
Average of anger= 4.490938410162926
Average of disgust= 1.0063385582486117
Average of scare= 5.744043000403939
Average of happiness= 58.55271696177385
Average of sadness= 0.7516283320001094
Average of surprise= 5.82388543231278
Average of neutral= 24.201864823025016
```

```
Value is 58.55271696177385
Emotion displayed throughout the day:  happiness
```

References

1. Balasubramanian, S. 2017. #LetsTalk, Health and Life, Mental Health, Taboos: 'I didn't want to Say I'm Depressed. Help Me.' A Student Opens Up. https://www.youthkiawaaz.com/2017/04/depression-college-students-india.
2. Xu, Z., Y. Jiang, Y. Wang, et al. 2019. Local polynomial contrast binary patterns for face recognition. *Neurocomputing* 355: 1–12.
3. Revina, I.M., and W.R.S. Emmanuel. 2018. A survey on human face expression recognition techniques. *Journal of King Saud University-Computer and Information Sciences* 1–8.
4. Ahmed, A., J. Guo, F. Ali, et al. 2018. LBPH based improved face recognition at low resolution. In *2018 International Conference on Artificial Intelligence and Big Data (ICAIBD)*, 144–147.
5. Byeon, Y.H., and K.C. Kwak. 2014. Facial expression recognition using 3D convolutional neural network. *International Journal of Advanced Computer Science and Applications (IJACSA)* 5 (12): 107–112.
6. Dubey, M., and P.L. Singh. 2016. Automatic emotion recognition using facial expression: a review. *International Research Journal of Engineering and Technology* 03 (02): 488–492.
7. Dang, L.M., S.I. Hassan, et al. 2019. Face image manipulation detection based on a convolutional neural network. *Expert Systems with Applications* 129: 156–168.
8. Viola, P., and M. Jones. 2001. Rapid object detection using a boosted cascade of simple features. *Accepted Conference on Computer Vision and Pattern Recognition.*
9. Park, K.Y., and S.N. Hwang. 2014. An improved Haar-like feature for efficient object detection. *Pattern Recognition Letters* 42: 148–153.
10. Badgerati. 2010. Computer vision—the integral image. https://computersciencesource.wordpress.com/2010/09/03/computer-vision-the-integral-image/.
11. Chai, S., B. Kisačanin, and N. Bellas. 2010. Special issue on embedded vision. *Computer Vision and Image Understanding* 114 (11): 1115–1316.
12. Kun, B., L.L. Zhao, L. Fang, and S. Lian. 2012. 3D face recognition method based on cascade classifier. *Procedia Engineering* 29: 705–709.
13. Khan, R.A., A. Meyer, H. Konik, et al. 2013. Framework for reliable, realtime facial expression recognition for low resolution images. *Pattern Recognition Letters* 34 (10): 1159–1168.
14. Lv, C., Z. Wu, X. Wang, and M. Zhou. 2019. 3D facial expression modeling based on facial landmarks in single image. *Neurocomputing* 355: 155–167.
15. Babu, D.R., R.S. Shankar, G. Mahesh, et al. 2017. Facial expression recognition using Bezier curves with hausdorff distance. In *2017 International Conference on IoT and Application (ICIOT)*, 1–8.

Reconstructing Obfuscated Human Faces with Conditional Adversarial Network

Sumit Rajgure, Maheep Mahat, Yash Mekhe and Sangita Lade

Abstract In today's era of advanced forensic and security technologies, the problem of identifying a human face from a low-quality image obtained from low-quality hardware or other reasons is a major challenge. Trying to extract meaningful information from these images is very difficult. These low-quality images have mainly two kinds of distortion in it, i.e. blurring and pixelation. Prior attempts have been done using different machine learning and deep learning techniques, but the desired high-quality images are not obtained. In this paper, we have used the conditional adversarial network to reconstruct highly obfuscated human faces. Various previous works on the conditional adversarial network have suggested it as a general-purpose solution for image-to-image translation problems. The conditional adversarial network is able to learn mapping from the provided input image to resulting output image. It is also able to learn the loss function to train the mapping. We have examined the result of this model using pixel loss function which gives the exact mapping of obtained high-quality human face with ground truth; furthermore, we have examined the capabilities of this model with very high-level obfuscated images.

Keywords Generator · Discriminator · Generative adversarial network · ReLu · U-Net · PatchGAN · CNN

1 Introduction

We often encounter images where either a part of an image is hidden or it is obfuscated in some way. This obfuscation process is done by removing details from the photograph in order to reduce the features that would be used to identify the image. Even though these images are obfuscated, they may still contain enough details such that we might be able to get back the original features.

Two common methods of obfuscation are pixilation and blurring. These methods are often provided in image editing software and are used to obfuscate faces. Also

S. Rajgure · M. Mahat (✉) · Y. Mekhe · S. Lade
Department of Computer Engineering, Vishwakarma Institute of Technology, Pune, India
e-mail: maheepmahat1@gmail.com

© Springer Nature Singapore Pte Ltd. 2020   95
D. Swain et al. (eds.), *Machine Learning and Information Processing*,
Advances in Intelligent Systems and Computing 1101,
https://doi.org/10.1007/978-981-15-1884-3_9

to note, it is often just the face that is obfuscated because it has the most identifying features, with the rest of the image might be not obfuscated and in better quality.

Also, when we capture images using low-quality cameras, we often do not get enough details on the image when we zoom in. For example, in case of a closed circuit, cameras were used for security. In this case also, we can take the low-quality image and using conditional generative adversarial network, and we can reconstruct the image with better quality.

In this project, we examine the use of conditional adversarial nets to reconstruct obfuscated human face having two types of obfuscation, i.e. pixilation and blurring. This technique is widely accepted as the general-purpose technique for image-to-image translation. These types of works are usually done using convolutional neural networks (CNN), but we found that using generative adversarial networks, the results obtained are much better.

Furthermore, we also explore the model's capabilities by using images with high levels of obfuscation and evaluating its performance on reconstructing the original images.

2   Related Work

Similar work has been done in the past focusing on obfuscated images and super-resolution. We can see this demonstration where subjects can still recognize the faces of famous people in static images despite being blurred [1]. Higher accuracy can be achieved by showing the obfuscated faces in motion. Lander et al. also show that obscured faces using methods such as pixilation can still be recognized [2].

Gross et al. also show that privacy is not protected when using low pixilation levels as obfuscated faces can still be recognized and their image quality enhanced [3]. McPherson et al. examine the topic of classifying images that underwent various forms of obfuscation highlighting issues with obfuscation [4].

In this paper, they show how neural networks can classify images that underwent obfuscation processes such as pixilation and blurring by examining different datasets such as MNIST, CIFAR-10, AT&T, and FaceScrub. They demonstrate that neural networks can be used to accurately classify images despite obfuscation processes since these images, while unrecognizable to humans, still contain enough information to enable appropriate classification.

There are four main approaches to the super-resolution task as shown by Yang et al.: prediction models, edge-based models, image statistical models, and patch-based models [5]. Interpolation-based methods are used by prediction models to generate smoother regions in high-resolution images. Edge-based models try to learn priors to generate sharp images, but they usually lack texture.

Image statistical models use image properties to predict the details in higher resolution images. Lastly, patch-based models learn mapping functions from smaller lower resolution patches to their corresponding high-resolution version.

3 Method

Generative adversarial networks (GANs) are a type of generative networks that are able to learn an association, given a noise vector z to resulting image y, $G: z$ implies y. Conversely, these conditional GANs are able to understand a mapping from given images x and a noise vector z. This mapping is mapped to y $G:\{x, y\}$ implies y. We train the generator to obtain outputs which cannot be differentiated from "real" images by a discriminator D that is trained in an adversarial manner. We train the discriminator for the purpose of it detecting the "fakes" produced by the generator.

3.1 Objective

The main purpose of our conditional GAN is given as

$$\mathcal{L}_{cGAN}(G, D) = \mathbb{E}_{x,y}[\log D(x, y)]$$
$$+ \mathbb{E}_{x,z}[\log(1 - D(x, G(x, z)))], \tag{1}$$

Here, generator G is trying to reduce the given objective in opposition to an adversarial D who is trying to increase it, i.e.

$$G = \arg \text{minimize}_G \text{ maximize}_D \; L_{cGAN} \; (G, D).$$

For the purpose of testing the significance of tuning our discriminator, a comparison is made to a variant that is without condition, where input x is not taken into account by the discriminator:

$$\mathcal{L}_{GAN}(G, D) = \mathbb{E}_y[\log D(y)]$$
$$+ \mathbb{E}_{x,z}[\log(1 - D(G(x, z)))] \tag{2}$$

$$\mathcal{L}_{L1}(G) = \mathbb{E}_{x,y,z}\left[\|y - G(x, z)\|_1\right] \tag{3}$$

Our final objective is

$$G^* = \arg \min_G \max_D \; \mathcal{L}_{cGAN}(G, D) + \lambda \mathcal{L}_{L1}(G) \tag{4}$$

Excluding z, the network is still able to understand the association from x to y. If that is done, it will make the network produce deterministic y. Then, it will not match any other distribution except a delta function. Conditional GANs have previously taken this into account. They have taken noise z as an input and along with x fed them to the generator.

3.2 Network Architecture

For the construction of our generator and discriminator, we have taken a reference from ones present in [6]. Convolution BatchNorm ReLu [7] is taken as the preferred module by generator and the discriminator (Fig. 1).

3.2.1 Generator Network Using Skips

One of the most prominent attributes of problems in translating one image to another image is that a high-resolution input image is being translated to a high-resolution output image. Also, for the problem we have considered, the input image and output image are different in exterior viewing. But both of them are results of the same fundamental form. So, the form in the given input is meticulously in line with form in the output. We try to model the generator design considering these things.

A lot of former findings [8–12] for tasks in this field have made use of an encoder–decoder network [13]. In models like that, we feed the input through different layers that increasingly down sample. This happens until we reach a bottleneck layer. After that, the process moves backward. In such implementation, the model needs for all of the information to go through all the layers. This includes the bottleneck. In a lot of translation problems involving images, there is a substantial amount of basic data passed between the input and output layers, and it would be useful to pass this data

Fig. 1 Network architecture

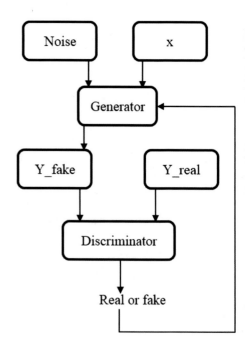

straight without any intermediary. Taking an example, considering the colorization of image, the location of prominent edges is shared between input and output. We add skip connections so that the generator can circumvent the bottleneck for such information, in accordance with the usual shape of a "U-Net" [14]. Respectively, skip connections are added between each layer x and layer $n - x$. Here, n is the aggregate number of layers. Every skip connection just combines every channels present at layer x with the channels at layer $n - x$.

3.2.2   Discriminator Network Using PatchGAN

For the purpose of modeling high frequencies, focusing on the structure in local image patches is enough. Therefore, we use a discriminator architecture called PatchGAN that only scans the image in scale of $N * N$ patches. The discriminator attempts to determine if everyone of the individual $N * N$ patch of given image is genuine or not. We operate the discriminator in a convolutional manner throughout the image, and we take the mean of all responses and get the output of D.

N can be of any size but smaller than the image. Given so, it can produce high excellent outputs. The reason it is beneficial is because, a small PatchGAN contains less number of parameters, performs quicker, and can also be used in case of bigger inputs. That type of discriminator successfully determines the image as a Markov random field. Assumption is made that the independence in the middle of pixels is divided by not less than a patch diameter. This relation is also a usual supposition in architectures of texture and style. Therefore, the PatchGAN could be considered as a type of texture/style loss.

4   Dataset

For our experiments, we use the CelebA human face dataset. The features of the CelebA dataset are as follows:

- 202,599 face images of various celebrities.
- 10,177 unique identities, but names of identities are not given.
- 40 binary attributes annotations per image.
- 5 landmark locations.

We perform preprocessing on this dataset. For each image, we first crop the image in 128 * 128 maintaining the face at the center of the image. The purpose of this is so we can focus on the construction of the face and ignore the majority of the background in the images.

Then for each image, we obfuscate it. For each obfuscation method, we copy the original image and then run the given obfuscation method (pixelation and Gaussian blur) (Fig. 2).

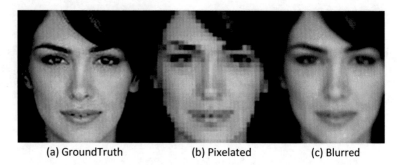

(a) GroundTruth   (b) Pixelated   (c) Blurred

Fig. 2 Variety of images present in the dataset

5   Experiment

For our experiments, we trained the dataset of 200 k human faces for 10 epochs with batch size of 100 images. As discussed earlier, we used generator as unet_256 and basic discriminator. The images given as input and the output images are both 1–3 channel.

We have manually blurred the dataset using Gaussian blur and similarly pixelated the images for our training purpose. The generator generates a random image at first. The first image generated is just a noise. Then that image is sent to discriminator along with its original image, and the discriminator tells the generator whether the image generated is real or fake.

In first attempt, the discriminator is going to say its fake and the generator is going to try to improve the image it generates. This happens multiple times until the discriminator cannot tell the difference between real and fake images and tells us that the fake image is real.

Once our model's generator generates an image that the discriminator thinks it is real, we can pass a sample image to it to see the reconstructed face of the obfuscated face.

5.1   Analysis of Generator Architecture

Making use of U-Net model, basic data is able to get across the network very fast. We see if this produces improved outputs. The encoder–decoder is made by cutting off the connections that are skipped in U-Net. The encoder–decoder model is not able to understand to produce pragmatic outputs in our tests.

The edge of the U-Net did not turn out to be fixed to conditional GANs; when $L1$ is used to train both U-Net and the encoder–decoder network, it is observed that U-Net produces superior outputs.

We have two layers present in the generator architecture, and we reduce our filter from initial numbers to 1 by dividing it by 2, repetitively.

5.2 Analysis of Discriminator Architecture

In discriminator, we have three convolution layers and each layer has 64 filters. Here, a PatchGAN is used. What this does is, it takes a patch of size $N * N$ in the generated image and penalizes structure at the scale of patches. It tells if the $N * N$ patch is real or not. We operate this discriminator in convolutional manner across the image and taking the mean of all responses to get the final result of discriminator.

6 Evaluation

We evaluate the accuracy of our generated image by comparing each pixel value of the generated image with the original image and calculating the accuracy. We do this for all the pixels in our image and take the average. The accuracy we are getting is around 95%.

For blurred images, using Gaussian blur, keeping the value of sigma at 4 and varying the kernel size, we achieved different degrees of blurred images.

The result obtained is the average of 10% of total samples that we have taken from our dataset as part of testing dataset.

Here are the results obtained.

6.1 For 7 × 7 Kernel Size

See Fig. 3.

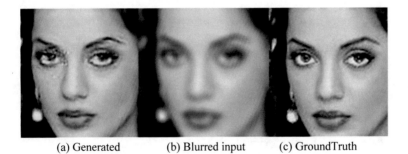

(a) Generated　　　(b) Blurred input　　　(c) GroundTruth

Fig. 3 Kernel size 7 × 7

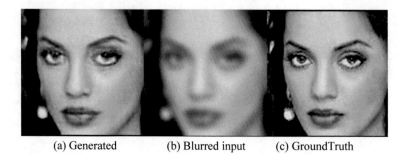

(a) Generated   (b) Blurred input   (c) GroundTruth

Fig. 4 Kernel size 9 × 9

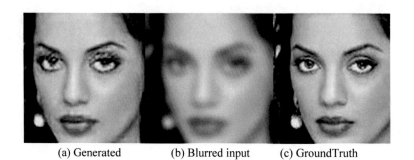

(a) Generated   (b) Blurred input   (c) GroundTruth

Fig. 5 Kernel size 11 × 11

6.2 For 9 × 9 Kernel Size

See Fig. 4.

6.3 For 11 × 11 Kernel Size

See Fig. 5.

6.4 For 7 × 7 Kernel Size

See Fig. 6.

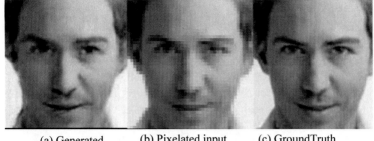

     (a) Generated   (b) Pixelated input   (c) GroundTruth

Fig. 6 Kernel size 7 × 7

6.5 For 9 × 9 Kernel Size

See Fig. 7.
Here is the result for blurred images.

	Kernel size		
	7 × 7	9 × 9	11 × 11
Accuracy (%)	97.2	96.4	95.1

Here is the result for pixelated images.

	Kernel size	
	7 × 7	9 × 9
Accuracy (%)	96	95.6

     (a) Generated   (b) Pixelated input   (c) GroundTruth

Fig. 7 Kernel size 9 × 9

7 Conclusion

Using conditional adversarial network, we were successfully able to achieve the desired results for both blurred and pixelated images.

Observing the above results, we can imply that using conditional adversarial networks is a good approach for performing tasks that require an image to be translated from one form to another, mainly for the ones that involve outputs that are highly structured and graphical. Using conditional networks, we can do a wide variety of tasks. The conditional property of this network helps to make our results more accurate compared to a non-conditional network.

References

1. Lander, K., and L. Chuang. 2005. Why are moving faces easier to recognize? *Visual Cognition* 12 (3): 429–442.
2. Lander, K., V. Bruce, and H. Hill. 2001. Evaluating the effectiveness of pixelation and blurring on masking the identity of familiar faces. *Applied Cognitive Psychology* 15 (1): 101–116.
3. Gross, R., L. Sweeney, F. de la Torre, and S. Baker. 2006. Model-based face de-identification. In *2006 Conference on Computer Vision and Pattern Recognition Workshop (CVPRW'06)*, 161–161, June 2006.
4. McPherson, R., R. Shokri, and V. Shmatikov. 2016. Defeating image obfuscation with deep learning. CoRR, abs/1609.00408.
5. Yang, C.-Y., C. Ma, and M.-H. Yang. 2014. Single-image superresolution: A benchmark. In *Proceedings of European Conference on Computer Vision*.
6. Radford, A., L. Metz, and S. Chintala. 2016. Unsupervised representation learning with deep convolutional generative adversarial networks. *ICLR* 2 (3): 16.
7. Ioffe, S., and C. Szegedy. 2015. Batch normalization: Accelerating deep network training by reducing internal covariate shift. *ICML* 3: 4.
8. Pathak, D., P. Krahenbuhl, J. Donahue, T. Darrell, and A.A. Efros. Context encoders: Feature learning by inpainting. In *CVPR*, 2, 3, 13, 17.
9. Wang, X., and A. Gupta. 2016. Generative image modeling using style and structure adversarial networks. *ECCV* 2 (3): 5.
10. Johnson, J., A. Alahi, and L. Fei-Fei. 2016. Perceptual losses for real-time style transfer and super-resolution. *ECCV* 2: 3.
11. Zhou, Y., and T.L. Berg. 2016. Learning temporal transformations from time-lapse videos. *ECCV* 2 (3): 8.
12. Yoo, D., N. Kim, S. Park, A.S. Paek, and I.S. Kweon. 2016. Pixellevel domain transfer. *ECCV* 2: 3.
13. Hinton, G.E., and R.R. Salakhutdinov. 2006. Reducing the dimensionality of data with neural networks. *Science* 313 (5786): 504–507.
14. Ronneberger, O., P. Fischer, and T. Brox. 2015. U-net: Convolutional networks for biomedical image segmentation. *MICCAI* 2: 3.

A Proposed Wireless Technique in Vehicle-to-Vehicle Communication to Reduce a Chain of Accidents Over Road

Shekharesh Barik, Surajit Mohanty, Rajeev Agarwal, Jitendra Pramanik and Abhaya Kumar Samal

Abstract Vehicle-to-vehicle (V2V) communication is used to transmit data between vehicles to prevent chain of accidents. Chain of accident refers to serial collision of vehicles one behind another, which happens in a busy traffic. This paper proposes an effective technique for V2V communication to deal with chain of accidents. A vehicle can communicate with its rear vehicle by a wireless transmitter and receiver system. The same procedure can be used in each vehicle. So, we can create a chain of vehicles for intercommunication among them. Parameters related to the vehicle such as speed, steering, brake, indicator are fed to the microcontroller of this system. A critical condition arising due to the front vehicle is reported or transmitted to rear vehicle through this system. Critical condition can be: collision of a front vehicle due to fog, direction change by a front vehicle without indicator. This message transmission along with automatic braking system (or speed control system) can avoid the danger of chain of accidents. Automatic braking system (ABS) is used to control the speed of vehicle. The current system is mostly sensor-based, and corrective actions are taken place on the basis of data from sensors. The problem arises when sensor does not work in critical conditions like dust or mud deposited over its surface, high-temperature condition, etc. Another disadvantage of sensor-based system is they cover a very short range of distance, so an affected vehicle may not get sufficient response time to take corrective actions. But, our technique is mainly communication-based, which provides a larger range of distance compared to sensor-based system. Hence, the system provides more response time to avoid accidents. Also, the rear vehicle gets message for the future estimated accident from front vehicle. Another aspect is when a vehicle needs to move toward left or right direction, it should forward this respective data to the rear vehicle. This technique will

S. Barik · S. Mohanty (✉) · R. Agarwal
Computer Science & Engineering, DRIEMS, Cuttack, Odisha, India
e-mail: mohanty.surajit@gmail.com

J. Pramanik
Centurion University of Technology and Management, Bhubaneswar, Odisha, India

A. K. Samal
Computer Science & Engineering, Trident Academy of Technology, Bhubaneswar, Odisha, India

© Springer Nature Singapore Pte Ltd. 2020   105
D. Swain et al. (eds.), *Machine Learning and Information Processing*,
Advances in Intelligent Systems and Computing 1101,
https://doi.org/10.1007/978-981-15-1884-3_10

be helpful to avoid serial collision of vehicles, because knowing the forward vehicle movement the rear vehicle can slow down their speed in order to avoid accident.

Keywords Vehicle-to-vehicle · Auto-braking system · Speed control system

1 Introduction

Due to the rapid increase in population, the use of vehicle is rapidly increasing. It leads to an increase in accident. This technique is helpful in critical condition where the sensors failed to measure the data. Critical condition can be dust or mud over sensor, physical damage of sensor, very high temperature etc. V2V communication was started in early 1980. V2V communication should be intelligent and effective to solve the traffic problem, and it should have features to avoid the accident too. The communication should involve messages regarding lane change, speed slow down. This information can be shared with a set of vehicle so that they can take corrective measures to avoid any type of hazard or accident. Safety driving system is collaborative in nature that means safety driving should consider a set of vehicles. Collaborative driving will be hazard-free when the communication between them is effective, because accident in a busy traffic not only hamper the culprit vehicle but also all the vehicle nearby to it. Collaborative driving focuses on five strategies: (a) cruise, (b) tracking, (c) combination, (d) splitting, and (e) lane changing. These strategies control the vehicle movement. The parameters regarding vehicle movement is stored in the vehicle storage system. This information along with the above five strategies can be used to develop an effective control mechanism for a vehicle communication system. This technique will be helpful to avoid collision in un-signaled zone. This technique can predict the collision time between intermediate vehicles.

2 Background Study

Wireless communication

The communication established without wires is termed as wireless communication. That means one device can transfer data to another without using wires or cables. Radio waves are used in wireless communication. The range of radio waves varies for different technologies to be used. In Bluetooth, radio waves can range up to 10 m. For long-range communication, radio waves can range up to 100 km [1].

Vehicle-to-Vehicle (V2V) communication

Vehicle-to-vehicle communication interconnects the vehicles using a wireless medium. The purpose of this communication is to send important warning messages from one vehicle to the nearby rear vehicle. This is received by the rear vehicle driver before an accident happens. So, corrective measures can be taken to avoid accident such as automatic braking to slow down the speed [2].

WAVE

WAVE stands for wireless access in vehicular environment. It supports the applications based on the intelligent transportation system. IEEE 802.11p is an amendment to 802.11 standard for the inclusion of WAVE. It defines the enhancement to 802.11 standard such as data transmission between two high-speed vehicles or data transmission between vehicle and infrastructure [3].

Zigbee Technology

Zigbee communication network is based on IEEE 802.15.4 standard. This standard defines the MAC layer and physical layer for the communication that to be established. Zigbee can be used for the applications whose coverage distance is within 10–100 m. It is a wireless personal area network. Zigbee has features like low-power consumption, low cost. It uses different operation modes for efficient power utilization. It is suitable for different embedded system applications. Zigbee can work under types of configurations such as master-to-master or master-to-slave. Depending on the requirement, Zigbee can be extended to a wider area network [4].

3 Related Work

Zunyi Zhang et al. have given a model to avoid the accident near the traffic junction. He has given a thoughtful study on the accelerating and de-accelerating behavior at these points. A warning message in voice or image is to be provided at the downhill road section of the intersection. Dynamic information is found to be more appropriate than that of static information. It is found that voice only warning is more effective than that of voice and image warning technique.

Saleh Yousefi et al. had analyzed the effect of inter-vehicle communication over the chain accident. They had used Markov chain for modeling the problem. The model can evaluate the influence of different road traffic parameters (i.e., vehicle density, velocities, and decelerations) and network parameters (i.e., delay, penetration rate) on the probability of chain collisions. They have given a study on the effect of driver behavior as well as vehicle characteristics on the probability of accidents. The effect of technology penetration rate is also considered in the proposed model. Simulation result reveals that the proposed model has an acceptable accuracy.

Mahmoud Taghizadeh et al. had given report regarding chain of accident for a group of vehicles traveling in a convoy. Collaborative communication is used to avoid collision. Various parameters such as reaction time, time headway, and recommendation speed are used to calculate the number of collisions under different situations. An analytical model combined is designed to explore the effect of number of collisions over low market penetration rate.

Gaur Saurabh Kumar et al. had given a vehicle application model to be deployed over a networking framework by considering some important vehicle parameters. This requires a productive research work. Vehicular ad hoc network (VANET) can be used for an alert system. That means it can give alert signal for traffic jam. This will be helpful to reduce the traveling time by road. When an accident occurs, the rear vehicle's drivers can be given emergency signals. Traffic police and ambulance can be called by sending emergency messages.

Jaiswal Siddhant and Dr. D. S. Adane had proposed an algorithm which is based on routing mechanism to be used in a hybrid situation. This technique uses clustering methods for transfer of packets. It is designed to run in adverse situations. It supports static as well as dynamic infrastructure.

Tejpreet Singh et al. had provided thoughtful study over VANET. It focuses mainly on the dynamic nature arising due to the change in network topology. This can lead to the security and communication issues. He suggested that wormhole attack is one of the very crucial attacks to be taken care of because this can create problems in the communication system. These attacks can result in denial-of-service (DoS) attack, masquerading, data tampering, etc. Different routing protocols (AODV, OLSR, and ZRP) are analyzed on the basis of performance metrics like thoughtful, delay, jitter. Performance is done in two cases: one is with wormhole attack and the other is without wormhole attack.

Lee et al. proposed a protocol called wireless token ring MAC protocol (WTRP) in order to communicate a set of vehicles. The vehicles drive cooperatively. R-ALOHA, slot reservation MAC protocols are discussed for inter-vehicle communication.

Bin Hu et al. have proposed vehicle-to-vehicle/vehicle-to-roadside communication protocol to avoid collision in vehicular ad hoc networks (VANETs). Emergency warning messages are transmitted via vehicle-to-vehicle (V2V) and vehicle-to-roadside (V2R) communications to achieve multipath diversity routing. A multichannel (MC) technique is proposed to improve communication reliability and low latency. Simulation results show that the proposed technique is capable of improving the message delivery ratio and obtaining low latency.

E. Abinaya and R. Sekar had given a model to optimize the traffic to be used at traffic intersection points. It needs vehicle data such as speed, position in real time. The delay can be minimized by using the oldest job first (OJF) algorithm. A comparison study is given with other techniques such as Webster method, Pre-timed signal control method, and Vehicle-actuated method.

Xue Yang et al. have proposed vehicle-to-vehicle (V2V) and vehicle-to-roadside (V2R) communication technology for cooperative collision warning. It can avoid collision by giving warning messages. Achievement of low latency is a challenge here. The simulation shows that low latency condition is met. Also, the bandwidth is used effectively in the different road environments.

4 Proposed Architecture

Microcontroller receives the following set of data

1. Speed
2. Steering
3. Brake
4. Indicator (Fig. 1).

1. *Speed*: Speed of a vehicle represents current speed of vehicle. Speed of a vehicle can change during the course of time, but when the speed decrease suddenly, then it indicates that critical condition has been reached. So, this information must be transmitted to the rear vehicle for corrective action to be taken by rear vehicle [5].
2. *Steering*: While moving to left or right direction, the driver turns the steering toward left or right accordingly and also respective indicators are 'ON.' But, when the driver moves to left or right without showing any indication, then it can cause an accident. This condition must be informed to the rear vehicle so that it can take necessary safety action to avoid accident [6].
3. *Brake*: The driver applies brake in order to slow down the speed. When the driver applies brake, we represent it as 'ON' condition. When the driver does not apply brake, we represent it as 'OFF' condition.
4. *Indicator*: Indicator is used when the driver moves the vehicle toward left or right direction. This information about the movement is fed to microcontroller (Fig. 2).

ATmega 328 microcontroller

A microcontroller can be considered as a PC. It integrates all the units required for the processor in a single microchip. It includes RAM, clock circuits, ROM, Serial port,

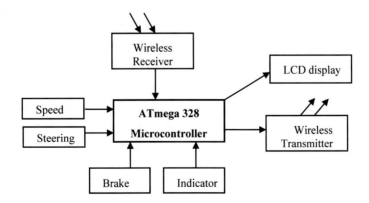

Fig. 1 Proposed architecture for V2V communication

Fig. 2 ATmega 328 microchip

I/O ports. ATmega 328 provides high processing speed with low power consumption. It can achieve throughput of about 1 MIPS per Mhz. The chip operates within 1.8–5.5 V. It provides serial communication techniques and supports SPI protocol. The memory unit includes 2 KB SRAM, 1 KB EEPROM, 32 KB ISP flash memory, and 32 general-purpose registers. The other units are SPI serial port, timer with internal oscillator, serial programmable USART, internal and external interrupts, etc.

ATmega-328 receives the data from front vehicle through a wireless receiver. It can also transmit data to the rear vehicle using the wireless transmitter. The input data such as speed, steering, brake, and indicator are taken from the vehicle system. The microcontroller ATMEG-328 recognizes the critical condition by analyzing these parameters.

When the microcontroller recognizes a critical condition, it informs it to the rear vehicle which is displayed on a LCD display system. LCD display system can be combined with buzzer sound as a warning signal.

The critical condition can be

(i) Speed of a vehicle suddenly reaches to zero.
(ii) Left and right movements of vehicle without indicator.

We use IR sensor to detect the rear vehicle. After detection, the front vehicle sends a password in order to communicate with the vehicle using Bluetooth communication technique. This technique pairs two communication and proceeds for communication of data. After every two minutes, IR sensor resumes its operation. If a rear vehicle has changed the lane and a new vehicle has been replaced in its position, then IR sensor detects it in its nest iterative phase [4].

LCD

Liquid-crystal display (LCD) is used as an output unit in the proposed system. LCD is connected to the microcontroller. LCD displays the required message that the microcontroller wants to give. The microcontroller decides the type of message to be generated, which is forwarded to transmitter for transmission.

Ultrasonic Sensor

Ultrasonic sensor uses ultrasonic sound waves to compute the distance to an object. The best thing about ultrasonic sensor is that they can sense all types of materials. It can work within 4–10 m range. Ultrasonic uses a transducer to send and to receive the sound pulse. The time gap between sending and receiving the pulse is used to calculate the distance to an object. Obviously, object in our case is the vehicle. When an object is detected, the ultrasonic sensor sends signal to the microcontroller. Now, the microcontroller through the LCD found that distance is within a critical range. In the proposed architecture, there will be two ultrasonic sensors: one at the front side and another is at the rear side of vehicle.

For the system design to be carried out, we have to consider different issues that will be considered as its challenges and the factors which will be considered as its challenges.

Advantages: The system creates a group of vehicles connected with each other that means it creates a combined traffic control system where the critical or urgent information related to the accident is shared in real time.

It increases the number of vehicles to be controlled in traffic.

It helps the drivers to organize their time in other productive areas since the vehicle is self-organized and well-coordinated to avoid collision.

Challenges: Natural calamities and extreme weather conditions such as heavy rainfall or heavy snowfall can disrupt the communication system.

In near future, self-communicating vehicles along with artificial intelligence are going to lead the transportation sector. This will result in stealing the job of many drivers.

The additional unit of communication has to be added in the design unit of vehicle. So, the extra cost is to be added in the manufacturing process. Hence, the vehicle becomes expensive.

5   Conclusion

We found from a survey that nearly 1.25 million people die in road accidents each year. As the population is increasing day by day, the use of vehicle is increasing which is putting a great impact on road accident. The proposed approach is useful to control the chain of accident. We will use IR sensor and ATMEGA-328 microcontroller in our device along with Bluetooth mechanism to accomplish wireless transmission and receive data. The main objective of our system is:

(a) To provide an effective communication between two vehicles, that means, focus on less communication time and more response time in order to avoid accidents quickly.
(b) More safety driving depends on the range of IR sensor that we have used.
(c) Wireless transmitters must receive data along the line of sight direction otherwise communication can lead to a failure system.

The proposed architecture must focus on the above important issues to design a fail-safe and robust system.

References

1. Biswas, S., R. Tatchikou, and F. Dion. 2006. Vehicle-to-vehicle wireless communication protocols for enhancing highway traffic safety. *IEEE Communications Magazine* 44: 74–82.
2. Yang, X., J. Liu, F. Zhao, et al. 2004. A vehicle-to-vehicle communication protocol for cooperative collision warning. In *Proceedings of the 1st Annual International Conference on Mobile and Ubiquitous Systems: Networking and Services*, 114–123.
3. Ganeshkumar, P., and P. Gokulakrishnan. 2015. Emergency situation prediction mechanism: A novel approach for intelligent transportation system using vehicular ad-hoc networks. *The Scientific World Journal* 2. Article ID 218379.
4. Wuhan, P.R. 2015. Trajectory following control based on vehicle to vehicle communication. In *IEEE 3rd International Conference on Transportation Information and Safety*.
5. Chou, Y.H., T.H. Chu, S.Y. Kuo, et al. 2018. An adaptive emergency broadcast strategy for vehicular ad hoc networks. *IEEE Sensor Journal* 18: 4814–4821.
6. Lytrivis, P., G. Thomaidis, M. Tsogis, and A. Ambitis. 2011. An advanced cooperative path prediction algorithm for safety applications in vehicular networks. *IEEE Transactions on Intelligent Transportation Systems* 12: 669–679.

Virtual Conversation with Real-Time Prediction of Body Moments/Gestures on Video Streaming Data

Gopichand Agnihotram, Rajesh Kumar, Pandurang Naik and Rahul Yadav

Abstract The exisitng conversation system where the user interacts with the virtual system with voice and virtual system replies to the user based on what user speaks. In this context whenever user makes some gestures to communicate with the virtual system, the virtual system will miss out those communications. For example, user instead of speaking, may nod head for "yes" or "no" and user can also use hand signals to respond to the virtual system. If these events are not addressed then the conversation is not very interactive and natural human-like interaction will start losing important information. The paper describes how the user body moments/gestures will help effective conversation with the virtual system and virtual conversation system can understand the user misspelled conversation, missed conversation effectively with user gesture/body movements.

Keywords Key point detection · Gesture classification · Events computation · Virtual conversation system · User conversation · Feature extraction · Real-time gesture prediction · Convolutional neural networks (CNN)

1 Introduction

The present paper describes a method and system to predict the human body movements/gestures in images or video streaming data which will help the virtual system missing conversation, misspelled conversation and fusion of voice with the user body

G. Agnihotram · R. Kumar · P. Naik · R. Yadav (✉)
Wipro CTO Office, Wipro Technology Limited, Bangalore, India
e-mail: rahul.yadav36@wipro.com

G. Agnihotram
e-mail: gopichand.agnihotram@wipro.com

R. Kumar
e-mail: rajesh.kumar133@wipro.com

P. Naik
e-mail: pradeep.naik@wipro.com

© Springer Nature Singapore Pte Ltd. 2020   113
D. Swain et al. (eds.), *Machine Learning and Information Processing*,
Advances in Intelligent Systems and Computing 1101,
https://doi.org/10.1007/978-981-15-1884-3_11

movements. This leads to virtual machine conversation more human-like and effectively. The key point-based method using deep learning approaches will be used to predict the human body movement.

Human body movement/gesture detection is an important problem in human–machine interface systems for effective virtual conversation/interaction with the machine and the user. The fundamental constraints of natural way of interaction such as speech, touch, contextual and environmental awareness and immersive 3D experience with a goal of computer which can see, listen, learn and talk as a human [1]. In this way, systems can be used to infer non-verbal communication between user and machines. There are challenges for this task such as identifying body parts in the images/sequence of images and inferring body gestures from them such as thumbs up, fist, victory, head nodding, head rotation, and multiple people sitting. The existing approaches [2, 3] for gesture detection assumes prior conditions like body parts are shown in a predefined area of image and identify the gestures from that area using any of the computer vision algorithms. Other few approaches for predicting the different body movements use different heuristic which can not be easily extended to new movements or gestures. This paper describes a system which does not require any such prior assumptions and jointly performs different body gesture detection using geometry of different body parts and other features paired with different machine learning algorithms. Along with geometry and the system also use key points of the human body in this solution which requires drastically smaller training data and gives more accurate inferences. The prediction of body movements/gestures in real time will help the virtual system more interactive with the user.

To compute the user body movements in real time, the system uses machine learning methods, geometry-based approaches for detection of human body parts, human body movements such as hand detection, head detection and gestures such as hand gesture identification, head movements, person standing and sitting, and human count.

For human body parts' detection, our approach consists of a deep learning-powered method which robustly locates body parts in an image/video frames and robust enough to handle any kind of orientation, position, lighting, background, and other variations. For gesture identification, after detection of body parts in an image or frame, we use machine learning-based classification approach which can be trained for the existing gestures and new gestures. For generalization, the system employing key point-based prediction to identify the left hand/left leg or right hand/right leg and the gestures associated with the hand and legs such as fist, thumbs up, victory, ankle lock and withdraw feet which requires less data for training the model and predicting the same in real time. The prediction of these body movements or gestures will help when voice conversation got missed and the virtual system will able to make the conversation with the user based on the system predicted user gestures to communicate to the user efficiently.

The system will also help when user misspelled in the conversation and the user gestures will help to understand the context of the user and communication will be easy in the virtual environment. In this system, we can augment or fuse the user voice with user movements/gestures to communicate in virtual environment with the user.

The rest of the paper is organized as follows. Section 2 describes the related work. Section 3 discusses the detailed solution approach on human body movement detection for hand gestures and other body parts' detection using key points, geometrical and shape features. The subsection of Sects. 3 also discusses the role of gesture events computation on virtual machine interaction. Section 4 deals with real-time human body movement detection using video streaming application. Section 5 outlines the application scenario in the context of user interviews. Section 6 provides the conclusions and future applications followed by References.

2 Related Work

This section discusses the existing work carried out on human gesture/body movement detection and the role of these movements or gestures in virtual conversation systems.

Wu and Hideki [4] proposed a real-time human motion forecasting system which visualizes the future pose/movements of human using simple RGB camera in virtual reality. The authors used residual neural networks and recurrent neural networks to estimate the human 2D pose. The authors also used residual linear networks to recover 3D pose from predicated 2D pose. The lattice optical flow algorithm is used for joints movement estimation.

The authors Badler [5], Kuffner [6] and Ng [7] described different approaches for estimation of human movements for autonomous action prediction. They also portrayed 3D virtual humans for interactive applications for conversations. They have used mid-level PaT-Net parallel finite state machine controller, low-level motor skills and a high-level conceptual action representation system that can be used to drive virtual humans through complex tasks. These complex tasks offer a deep connection between natural language instructions and animation control.

Nguyen [8] used key point-based approach for identifying the gestures. A probabilistic distribution for each key point indicating the likelihood of producing an observable output at that key point is also derived using this approach. The characteristic data pattern is obtained for the new gestures can be compared with the patterns of previously stored known gestures. These gestures will help us to compute the confusion matrix. The confusion matrix describes possible similarities between new and known gestures. This approach is purely probabilistic matching approach to identify the match of existing and new gestures.

Zhou and Cheng [9] outlined hidden Markov models (HMM) and geometrical feature distributions of hand trajectory of a user to achieve adaptive gesture recognition. The method acquires a sequence of input images of a specific user and recognizes the gesture of specific user from the sequence of images using the trained gestured model and geometrical features extracted from the hand trajectory of the user. The geometrical features are extracted from the gesture model based on the relative positions of state transition points and starting point of the gesture. Again, for detection the authors used sequence images to recognize the gesture.

Huang et al. [10] and Smith [11] describe the dynamic gesture recognition using geometrical classification. In each image, the hand area is segmented from the background and used to estimate parameters of all five fingers. The authors proposed method that classifies the hand images as one of the postures in the predefined database and applies a geometrical classification algorithm to recognize the gesture. The algorithm also combines skin colour model with motion information to achieve real-time hand segmentation performance and considers each dynamic gesture as a multi-dimensional volume and uses a geometrical algorithm to classify each volume. Authors applied this approach on gesture recognition for computer, TV control systems and game control systems.

Kurakin et al. [12], Chen et al. [13] and Marin et al. [14] used depth information for dynamic hand gesture recognition. The authors used sensors and 3D images to capture depth information to recognize gestures. The hand gesture recognition will help to predict the human activity understanding for many of the practical applications. This is a real-time approach for dynamic hand gesture recognition. The approach is fully automatic and robust to variations in speed and style in hand orientations. The approach is based on the action graph which shares similar robust properties with standard HMM and requires less training data by allowing states shared among different gestures. The system is evaluated on a challenging dataset of twelve dynamic American Sign Language (ASL) gestures and obtained promising results.

Shan et al. [15] tried with particle filter and mean shift algorithms for visual tracking. The authors found both the algorithms which have their own respective strengths and weaknesses. The authors integrated the advantages of the two approaches for improved tracking. The authors incorporated the MS optimization into particle filtering to move particle to local peaks in the likelihood, the proposed mean shift embedded particle filter improves the sampling efficiency. The authors applied these algorithms in the context of developing a hand control interface for a robotic wheelchair.

The next section describes detailed solution approach followed by the application in the domain of virtual conversation along with human body movements/gestures.

3   Solution Approach

The proposed approach uses machine learning and geometry-based approach for hand gesture detection. The key point-based deep learning approaches are used for other body part movements' detection such as standing and sitting. The solution approach has three stages. The first stage is model training where the model is trained on huge corpus of data for detecting the hand gestures and other body part movement or gestures. The second stage helps to describe the role of gesture/body part movements associated with the trained model in virtual conversation for effective

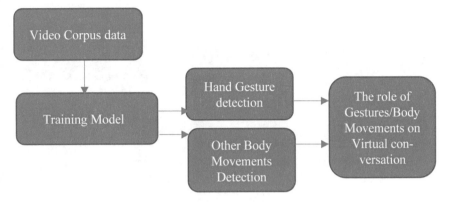

Fig. 1 Training architecture diagram for virtual conversation system

interaction between machine and human. The final stage explains how these trained model and gesture events will help the virtual conversation system with the real-time streaming applications (Fig. 1).

The video corpus data collected from different sources such as YouTube and self-videos are used for training where the hand and other body parts of human are visible. The training model (stage 1) has two parts, one is hand gesture detection model such as victory, fist and thumbs up and body parts' detection model such as standing and sitting. The training models are explained in detail in the next subsections. With the trained model on gesture and body movements' events, how the gap will be filled between machine and human in virtual conversation (stage 2). The details will be explained in the subsequent subsection. In stage 3, the real-time events will be generated on real-time streaming data with the trained model. These events will help in virtual conversation for human and machine interaction in effective way. The details will be given in next Sect. 4.

3.1 Hand Gesture/Movement Detection

As part of training model, the system trains the hand gesture detection for the corpus of training data. The detail steps are shown in Fig. 2, and each step is explained as given below.

Hand Detection—this step locates hand in an image. Input to this module is a raw image, and output is a bounding rectangular box (location of hand) of hand or multiple hands. This module is powered by a deep learning model which is trained for hand detection task using recurrent neural networks (RCNN) or convolutional neural networks (CNN) on hand images. The CNN/RCNN does a binary classification for each pixel in image. The system collects a huge data set of images containing hands (either using simulation and automated labelling or manual image collection and manual

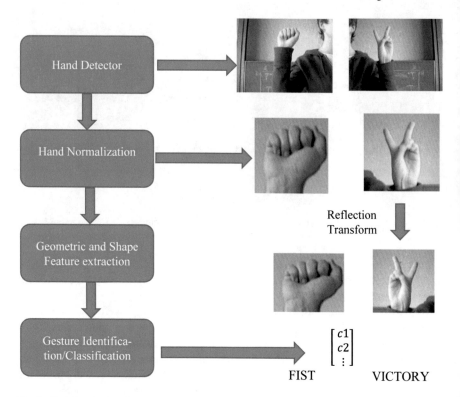

Fig. 2 Training the hand gestures/movements using classifier

labelling). Labelling of these images includes locating the regions (pixels corresponding to hands) of hands/drawing a bounding rectangle around hand. Once we have this labelled data set, we train a CNN (SSD (single-shot detector)/RCNN/Yolo) [16] which takes image as input and gives location of hand in image as output.

Hand Normalization—in this step, resize the detected hand to a standard size (ex-256 × 256) and identify if it is left or right. If it is right hand, then apply reflection transform (rotation along y-axis) and convert it to left hand. For hand and hand side (left or right) detection, use one of the following methods given below.

1. Jointly, predict hand and hand side detection (left or right) by training a labelled data of right and left side using a convolutional neural network (CNN) classifier for hand side prediction
2. Otherwise in the first step, train for hand prediction in the images, and second step, classify which hand side left or right. These two methods require training and extra computation of data. The alternative method using key points for hand and hand side detection is provided in the next step.
3. Key point detection:

a. It is a CNN powered module which takes hand cropped image as input and gives key points as output. To train this CNN, collect a labelled data set which consists of hand cropped image with annotated joint coordinates.

b. For this, use a simulation tool in which human characters of different age, gender, skin colours are simulated with different hand poses. In this simulation, the coordinates of palm and all finger joints are known.

c. After simulating each new frame, capture its image, and save its finger joint coordinates as label. Once large labelled data set is available, train the model using CNN. The trained model provides key points of every input image in real time.

Using key points predicting the hand side (left or right):

- Obtain the key points of the input images for both original and reflection transformed images (explained in above hand normalization step).
- Predict probabilities of key points for both images—original and reflection transformed image.
- For example, what is probability that the key point belongs to thumb in training model—computed using CNN classifier.

$$\begin{bmatrix} 0.005 \\ 0.00067 \\ 0.86 \\ . \\ . \\ . \\ 0.0009 \end{bmatrix} \text{Ex. Probability of Thumb Key point}$$

- Take average of probabilities of all the key points for both images and use that image for further computation which has larger average probability will be hand side (left or right).

Geometric and Shape Feature Extraction—the training model uses geometric features, shape features along with key points as input feature vector for hand classifier. Following features are computed in this step:

1. Curvature of fingers—for each finger, curvature of the finger is computed using following formula
   For ith figure F_i, we have coordinates for all joints (key points)
   $F_i = [(x1, y1), (x2, y2), (x3, y3), (x4, y4)]$; x's and y's—key points
   First, compute mid-point of each finger bones.

$$\overrightarrow{m1} = c((x1 + x2)/2, (y1 + y2)/2) \tag{1}$$

$$\overrightarrow{m2} = ((x2 + x3)/2, (y2 + y3)/2) \tag{2}$$

$$\vec{m3} = ((x3 + x4)/2, (y3 + y4)/2) \tag{3}$$

Now computing angle subtended by (1), (2) on (3) is

$$\text{Curvature} = \text{Cos}^{-1} \frac{\left(\vec{m1} - \vec{m2}\right).\left(\vec{m3} - \vec{m2}\right)}{\left|\vec{m1} - \vec{m2}\right|\left|\vec{m3} - \vec{m2}\right|} \tag{4}$$

2. Angle between fingers—to compute angle between two fingers having edge points $\left(\vec{P1}, \vec{P2}\right)$ and $\left(\vec{P3}, \vec{P4}\right)$ where $\vec{Pi} = (xi, yi)$ at points $\vec{P1}, \vec{P3}$
$\vec{P5} = \vec{P4} + \vec{P1} - \vec{P3}$; Here, $\vec{Pi}$'s are edge points for each $i = 1, 2, 3, 4, 5$

$$\theta = \cos^{-1} \frac{\left(\vec{p2} - \vec{p1}\right).\left(\vec{p5} - \vec{p1}\right)}{\left|\vec{p2} - \vec{p1}\right|\left|\vec{p5} - \vec{p1}\right|} \tag{5}$$

The angle between the edges is shown in Fig. 3. For one hand, compute 4 angles between two consecutive fingers of the five fingers.
3. Direction vector of finger bones—to obtain direction vector of ith finger bone having edge coordinates as $(x1, y1), (x2, y2)$ we use
   direction vector $= \frac{\vec{v}}{|\vec{v}|}$, where $\vec{v} = (x2 - x1, y2 - y1)$
   For each finger, compute 3 direction vectors, which gives 6(=3 * 2) features.
4. Key points of finger joints—coordinates of finger joints (key points), in a reference frame where mid-point of palm is taken as origin, is taken as features. For 21 key points, we get 21 * 2 dimensional features (one point gives two features—one is x-coordinate and other is y-coordinate)
5. Concatenate all these features and normalize them to have norm 1.

   Hand Gesture Identification/Classification: a classifier can be used for gesture identification. Here, a multi-class SVM classifier with RBF kernel will be used for training the hand gesture model. This training will be happening on

Fig. 3 Angle of edges

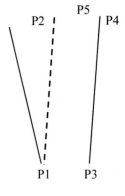

labelled data set where each sample consists of features extracted (geometrical and shape features) from cropped images and gesture type of that images, for example, victory, thumbs up, fist, etc. Here, train the SVM model with the labelled data sets. The model input feature vector is very small and large enough to represent each gesture or body movement type. Here, even a smaller data set is enough for training the classifier. C_i's for $i = 1, 2, \ldots$ is the class label of fist, victory, thumbs up, etc. as shown in Fig. 2.

3.2 Other Body Movements' Detection

For obtaining the other body movements such as person standing, person sitting, number people present in a frame, head nodding for "yes" and "no" which are very important for virtual conversations with machine and human. Based on the human body movements/gestures, models are trained, and trained models are used in real time to derive the gesture events which will help in virtual interaction with the machine and human.

In general, the models are based on key points derived from human body parts as explained using hand gesture classification from above subsection. In addition to the key points, the geometrical features and shape features also used for training the classifiers for different body parts' movements. The key points are derived from the trained convolutional neural networks on different body parts. The geometrical and shape features are obtained from the joint key points of different body parts. The model is trained with a simple SVM classifier using key points, geometrical features and shape features as shown in Fig. 4.

Head rotation and nodding for "yes" and "no": for head, related gestures such as head rotation for "yes" and head nodding for "No" will be computed using the time series-based yaw, roll and pitch feature computation. These features will compute the oscillation and derive the events for head rotation and nodding for "yes" and "no"

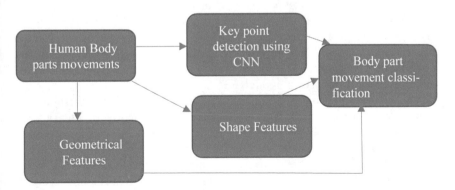

Fig. 4 Human body gesture training model

Fig. 5 Angles of human
head movements

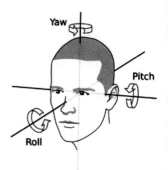

values. Using key points derived from CNN model, the method computes Euclidian angles of roll, yaw and pitch features (please refer Fig. 5) to derive head rotations, head nodding for "yes" and "no" events. The gestures need to be completed within a period, and flags are set whenever there is oscillation. The oscillation provides the head rotation of the user, user opinion for "yes" and "no" events. Here, the gesture/head movements are computed with the key points and geometric features without a need of classifier. These events will help in the conversation for effective interaction of the user in the virtual environment.

3.3   The Role of Gesture/Body Movements on Virtual Conversation

The human gesture/body movements are identified with the trained classifier model as explained in above subsections. Here, the method describes the role of these gesture events in the virtual conversation with the user. These events will help in filling following gaps for effective conversation with the machine and human.

Missed conversation: user instead of speaking, user may nod his head for "yes" or "no". The gesture events will help the virtual system to understand the user intent with these gestures. In another case, user can also use his hand signals such as "Thumb up" and giving responses with finger "1", "2" to response to the virtual system. These gestures will help the effective interaction between human and machine where the user conversation got missed out and virtual system can make out the user conversation with these gestures effectively.

Misspelled/semantic conversation: using the user body movements/gestures, even though the user misspelled the sentences while communicating to the virtual machine the virtual machine can able to figure out the user intention with the misspelled conversation. **For example**: if user saying I want to go room and signed with hand signal. The hand signal will refer to rest room and the user misspelled as room instead of rest room. The virtual system will able to figure out the user intention with the user body movements/gestures, and virtual system will interact to the user accordingly.

Fig. 6 Ankle lock body
movement

Fusion of voice with the gestures: with the fusion of voice and the gesture, the virtual system will able to figure out the user intention and the conversation with the user takes place accordingly. **For example**: the leg and feet gestures as shown in Fig. 6 can able to provide the most accurate clues to someone's mental state. The ankle lock, in a seated position, people sometimes lock their ankles and withdraw their feet below the chair as shown in Fig. 6. The person doing this gesture is holding back a negative reaction. And behind a negative reaction, there is always some negative emotion. So, a person doing this gesture simply has a negative emotion that he/she is not expressing. He/she might be afraid, angry. By fusing this gesture with the voice, the virtual system will able to figure out the intent of the user while interacting.

4 Real-Time Generation of Body Movements for Virtual Machine Understanding

This section describes the real-time generation of body movements for the virtual machine understanding with video streaming application. This method uses the gesture trained model to generate the events in real time. The steps for the real-time body movements' generation are shown in Fig. 7.

The video streaming data will be generated in real time and the frames or sequence of frames will be sent to the gesture training models to generate gestures and those gestures are used by the virtual machine while interacting with the user in real time. These gestures will be handy to the virtual system while interacting to the user where user shows some hand or body movements while communicating to the virtual machine. The body movements' identification makes the machine and human more interactive.

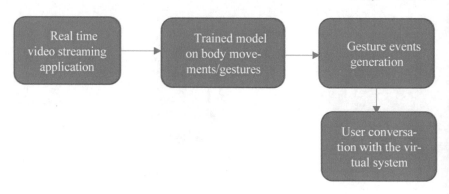

Fig. 7 Real-time gesture generation for virtual conversation

5 Application of Human Body Movements on Virtual Machines

The approach described in above sections can be used in many tasks involving extraction/interpretation of hand movements' recognition in videos/images and action recognition. The solution is deployed in virtual interview platform where user interacts with the virtual system for interview. The user body parts' movements/gestures will help the virtual system for effective communication. The solution is tested on many scenarios with respect to performance and observed that the solution has an accuracy in identifying hand related gestures is more than 90% and more than 80% on other body part movements' detection. Few scenarios listed down where the body movements helped the virtual system for effective interaction with the user.

1. The user interacts with the virtual interviewer where the virtual interviewer will ask the question, Do you know the answer to this particular question. Where user nod his head for "yes"/"no" instead of replying in voice. Hence, the system will able to predict that the user is replying yes/no, and instead waiting for the user voice response, the virtual system will go ahead and proceed for next question or another context.

2. The user also interacts with the virtual interviewer with hand signals. In the interview context, the virtual system will the ask the user multiple choice questions, and the user replies the answer using hand signals such as 1, 2, 3, or 4 with hand fingers. The system will understand the user hand signals and proceed with the next question.

3. In another scenario, the user provides the hand signals such as STOP or waiving to virtual system to provide contextual information rather than voice. The virtual system understands the hand gestures and interacts the user in the same context more effectively. Hence, the user body movements/gestures are foremost important for conversation system where the voice-related information is missing or misspelled while interaction in the virtual environment.

6 Conclusions and Future Work

The solution proposed in this paper discusses the new approach for human body movements or gesture detection and role of these gestures in virtual conversation systems. The method uses key points, geometrical and shape features to train the different human body movements. These features along with key points are used in SVM classifier to train the model instead of using deep learning approaches which needs high computing systems for training, and the solution does not require any huge amount of data to train the model. The geometrical, shape features along with key points will help to train the model for different human body movements' detection. Using the trained model, the real-time events are generated with video streaming application, and these events are consumed by virtual systems for effective communication with the candidate with respect to virtual interview scenario. The candidate body movement events help the interview to go smoothly and more interactive with the virtual machine. The solution is tested on multiple scenarios of interview and observed that the solution able to get different accuracy values for different body part movements. The solution provides 90% of accuracy for hand gestures, and other body parts of the solution can provide average of 80% of accuracy.

In future, this solution will be extended to other applications in the domain of smart health care where patient and machine will interact for diet related, medicine and health-related consultancy. This solution can also be extended to teaching domain and retail domain where virtual machine provides feedback/user query solving.

References

1. Kaushik, Manju, and Rashmi Jain. 2014. Natural user interfaces: Trend in virtual interaction. arXiv preprint arXiv: 1405.0101.
2. Eshed, Ohn-Bar, and Mohan Manubhai Trivedi. 2014. Hand gesture recognition in real time for automotive interfaces: A multimodal vision-based approach and evaluations. *IEEE Transactions on Intelligent Transportation Systems* 15 (6): 2368–2377.
3. Molchanov, Pavlo, Shalini Gupta, Kihwan Kim, and Jan Kautz. 2015. Hand gesture recognition with 3D convolutional neural networks. In *Proceedings of the IEEE Conference on Computer Vision and Pattern Recognition Workshops*, 1–7.
4. Wu, Erwin, and Hideki Koike. 2018. Real-time human motion forecasting using a RGB camera. In *Proceedings of the 24th ACM Symposium on Virtual Reality Software and Technology*. ACM.
5. Badler, I. Norman. 1997. Real-time virtual humans. In *Proceedings of the Fifth Pacific Conference on Computer Graphics and Applications*. IEEE.
6. Kuffner, J.James. 1998. Goal-directed navigation for animated characters using real-time path planning and control. *International Workshop on Capture Techniques for Virtual Environments*, 171–186. Berlin: Springer.
7. Ng, Kia. 2004. Music via motion: Transdomain mapping of motion and sound for interactive performances. *Proceedings of the IEEE* 92 (4): 645–655.
8. Nguyen, H. Katerina. 2001. Method and apparatus for real-time gesture recognition. U.S. Patent No. 6,256,033, 3 July 2001.
9. Zhou Jie, and Pu Cheng. 2016. System and method for gesture recognition. U.S. Patent No. 9,323,337, 26 Apr 2016.

10. Huang, Kuang-Man, Ming-Chang Liu, and Liangyin Yu. 2013. System and method for dynamic gesture recognition using geometric classification. U.S. Patent No. 8,620,024, 31 Dec 2013.
11. Smith Dana, S. 2014. Geometric shape generation using multi-stage gesture recognition. U.S. Patent Application 13/846,469, filed 18 Sept 2014.
12. Kurakin, Alexey, Zhengyou Zhang, and Zicheng Liu. 2012. A real time system for dynamic hand gesture recognition with a depth sensor. In *2012 Proceedings of the 20th European Signal Processing Conference (EUSIPCO)*. IEEE.
13. Chen, L., H. Wei, and J. Ferryman. 2013. A survey of human motion analysis using depth imagery. *Pattern Recognition Letters* 34 (15): 1995–2006.
14. Marin, Giulio, Fabio Dominio, and Pietro Zanuttigh. 2014. Hand gesture recognition with leap motion and kinect devices. In *2014 IEEE International Conference on Image Processing (ICIP)*. IEEE.
15. Shan, Caifeng, Tieniu Tan, and Yucheng Wei. 2007. Real-time hand tracking using a mean shift embedded particle filter. *Pattern Recognition* 40 (7): 1958–1970.
16. Redmon, Joseph, and Ali Farhadi. 2018. Yolov3: An incremental improvement. arXiv preprint arXiv: 1804.02767.

Automatic Diagnosis of Attention Deficit/Hyperactivity Disorder

Sushmita Kaneri, Deepali Joshi and Ranjana Jadhav

Abstract Due to increased exposure to gadgets, social media, competition and other issues, there is an increase in mental health imbalance which leading to mental disorder. Mental health diagnosis is required at early teenage stage so as to prevent further adversities for the patient and the society. Diagnosing of Attention Deficit/Hyperactivity Disorder requires multiple visits to the doctor. Proposed model which is a developed Web application would help diagnosis of Attention Deficit/Hyperactivity Disorder quickly, saving the time of doctor and the patient. Automation of Attention Deficit/Hyperactivity Disorder diagnosis would help to save the time of patient and Consultant, leading to quick treatment. Earlier work states about manual testing, offline questionnaire, Conners test, etc., which requires more time and appointments to be completed. Our model would conduct four different audio–visual tests to check symptoms stated in the Diagnostic and Statistical Manual of Mental Disorders, Fifth Edition (DSM-5). It provides a report based on the tests which doctor would refer to detect Attention Deficit/Hyperactivity Disorder in the patient with less time as compared to existing methodology.

Keywords DSM-5 criteria · LBPH face detection and classification algorithm · Audio processing

1 Introduction

Attention Deficit/Hyperactivity Disorder is Attention Deficit Hyperactivity Disorder which causes a lack of attention and focuses on tasks and routine, leading to departed social life and dropping out of school and staying aloof. For in-person diagnosis of Attention Deficit/Hyperactivity Disorder, a patient has to wait for the appointment date to arrive and several such visits to the doctor. This results in delayed detection of the disorder and its treatment. The proposed model would conduct tests on the

S. Kaneri (✉) · D. Joshi · R. Jadhav
Department of Information Technology, Vishwakarma Institute of Technology, Pune, Maharashtra, India
e-mail: sushmita.kaneri17@vit.edu

© Springer Nature Singapore Pte Ltd. 2020
D. Swain et al. (eds.), *Machine Learning and Information Processing*,
Advances in Intelligent Systems and Computing 1101,
https://doi.org/10.1007/978-981-15-1884-3_12

Fig. 1 Users of the tests

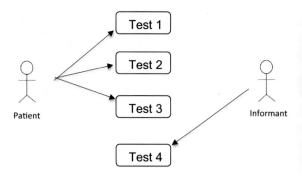

patient prior to the visit and produce a report based on the tests. Tests would check for the symptoms stated in DSM-5. An Attention Deficit/Hyperactivity Disorder patient shows at least 6 out of 9 inattentive and/or hyperactive and impulsivity symptoms for more than 6 months according to DSM-5.

In the proposed system, in order to automate the diagnosis of ADHD, authors have to design four tests which need to be taken by the patient to check for the symptoms, refer Fig. 1. The tests are designed such that it would recheck symptoms multiple times through various scenarios.

Test 1—Completing a task

Patient would be given a task to match the blocks on the screen. While undergoing this test, he/she would be examined for how accurately the task is being completed, attention and patience. This test would look for symptoms like often patient has difficulty in organizing tasks and activities; often he/she avoids or is reluctant to engage in tasks that require sustained mental effort; often does not follow through on instructions and fails to work; often has difficulty sustaining attention in tasks or play activities; often leaves seat in situations when remaining seated is expected; often unable to play or engage in leisure activities quietly.

Test 2—Audio Questionnaire for patient

Patient would be provided with question in audio format, and he/she has to answer through microphone. This would check for hyperactivity and impulsivity. It would be examined for symptoms like often blurts out an answer before a question has been completed; often has difficulty waiting for his or her turn; often interrupts or intrudes on others.

Test 3—Responding to Audio and Visuals

Patient would be given words in audio/image format which he/she has to detect and enter in the text field provided on screen. This test would look for symptoms like often fails to give close attention to details or makes careless mistakes; often does not seem to listen when spoken to directly; often have difficulty sustaining attention in tasks or play activities.

Test 4—Questionnaire for Informants

This test would examine for overall symptoms in different settings like home, school, social. This test would check for the behaviour of patient at home, school and society.

The Web application for these tests has been designed by taking consideration of user prone to Attention Deficit/Hyperactivity Disorder as per the guidelines stated in the paper "Designing for Attention Deficit/Hyperactivity Disorder: in search of guidelines" [1]. The output of this model would be a report containing Attention Deficit/Hyperactivity Disorder symptoms found and a report based on them which can then be taken to the Consultant for further decision.

2 Related Work

Attention Deficit/Hyperactivity Disorder causes a person to lose attention quickly making him/her impulsive or hyperactive. In India, there are moreover one million Attention Deficit/Hyperactivity Disorder cases seen in every year. It is highly observed from 6 to 17 age groups as shown in Fig. 2. Questionnaires answered by informants (parents, teachers, etc.) are an important and efficient part of the diagnostic assessment but cannot be used in isolation to make a diagnosis of Attention Deficit/Hyperactivity Disorder; it has to be validated with other tests and a Consultant. The factors chosen in the model have been approved with the contents of a Seminar in Paediatric Neurology on "Evaluation and Assessment Issues in the Diagnosis of Attention-Deficit Hyperactivity Disorder" [2].

(*x*-axis: age group, *y*-axis: Attention Deficit/Hyperactivity Disorder cases)

Diagnosis by rating scales such as Attention Deficit/Hyperactivity Disorder—RS, Conners CBRS, continuous performance test (CPT) has been practised. They are not efficient if carried out in mutual exclusion. Brain scans—neuroimaging

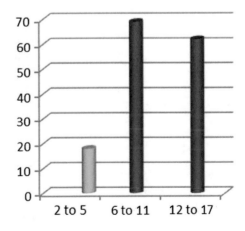

Fig. 2 Attention Deficit/Hyperactivity Disorder age group statistics [Apollo Hospital and others]

procedures, such as positron emission tomography (PET) scans, SPECT scans and magnetic resonance imaging (MRI), have long been used in research studies of Attention Deficit/Hyperactivity Disorder. But their use in diagnosing Attention Deficit/Hyperactivity Disorder has not yet been validated with conclusive scientific research. They have revealed, though, that certain parts of the brain appear different in people who have Attention Deficit/Hyperactivity Disorder than in people who do not have the condition [3].

A doctor from the National Institute of Mental Health and Neurosciences (NIMHANS) has stated that MRI research findings are inconsistent among researches made. Hence, it is difficult to certainly predict Attention Deficit/Hyperactivity Disorder using MRI approach [4].

3   Proposed Model

3.1   Test 1: Completing Task

User would be given a console with few shapes on it which he/she has to match with the impressions besides. While completing this activity, he/she would be monitored for moods the persons goes—whether he/she loses attention by keeping wandering and time required to complete the task. To check first to scenarios, facial expressions would be recognized, and after every defined interval, a photograph would be taken to check whether user is before the screen and his moods (Fig. 3).

Fig. 3 Flow and deliverables of the tests

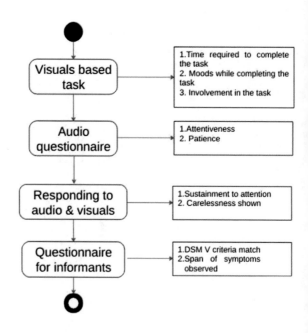

Fig. 4 Flowchart for Test 2—audio questionnaire

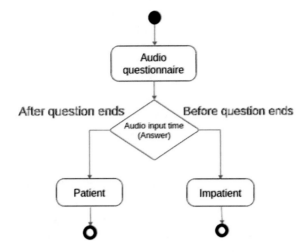

Eigenfaces and Fisherfaces find a mathematical description of the most dominant features of the training set as a whole. LBPH analyses each face in the training set separately and independently. Data set source: Cohn-Kanade (CK and CK+) database.

In LBPH, each image is analysed independently, while the eigenfaces method looks at the data set as a whole. The LBPH method is somewhat simpler, in the sense that we characterize each image in the data set locally; when a new unknown image is provided, we perform the same analysis on it and compare the result to each of the images in the data set. The way in which we analyse the images is by characterizing the local patterns in each location in the image (Fig. 4).

Algorithm for Test 2:

1. Start the task (task is of 2 min).
2. Through the application, pictures of the patient during test would be taken.
3. Check whether face is detected.
4. Analyse and note probability of the facial expressions at that instant using **Local Binary Patterns Histograms (LBPH)** algorithm.
5. Check if task is completed.

Comparison between Eigenfaces Face Recognizer, Fisherfaces Face Recognizer and LBPH was made out of which LBPH had more accuracy and also addressed the light condition issues faced by other two algorithms.

3.2 Test 2: Audio Questionnaire

The user would be provided with audio questionnaire regarding general questions. He/she has to answer the questions in audio format through microphone. Using audio processing (pyAudioAnalysis), this test can be accomplished. The timestamp when the question started playing and the timestamp of audio input (answer) have to be noted. The result whether question was interrupted can be given as:

1. If the difference between Question_Length_time&Response_time is positive, then the user is patient.
2. If the difference between Question_Length_time&Response_time is negative, then the user is impatient.

3.3 Test 3: Responding to Audio and Visuals

The user would be provided with set of words as audios and visuals. These words have to be written in the text field provided. This test would check for accuracy, involvement level and attention level. An Attention Deficit/Hyperactivity Disorder patient tends to neglect things spoken directly or is ignorant of small things. For example, when a word is dictated or shown as "CAT", so the patient may become ignorant to the details and positioning of the letters and end up jumbling them as "CTA" (Fig. 5).

Thus, this test will help Consultant to check for accuracy levels of the patient and his/her detailed observation or ignorant behaviour to look after things approaching them.

3.4 Test 4: Questionnaire for Informants

Diagnostic and Statistical Manual of Mental Disorder (DSM-5) was approved on 1 December 2012 [5] and got published on 18 May 2013, [6] which contains diagnoses and cases for various mental disorders.

As per DSM-5 [7] criteria questions would be drafted which needs to be answered by the patient's parent and teacher. These answers would support the findings found from previous tests. Questions would be asked as set of Group A, Group B and Group C. Group A contains questions related to inattentive symptoms. Group B contains questions related to hyperactive–impulsive symptoms. Group C would be the turning question of the test, depending on Group A and Group B questions which would help predict results ultimately which would be filed in the report submitted to the Consultant.

Fig. 5 Test 3—responding to audio–visuals' flowchart

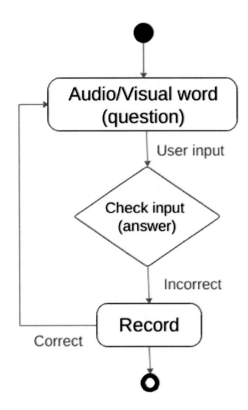

4 Experimentation

In the experimentation of the proposed system, it gave positive result indication for 3 out of 60 sample youngsters from Vishwakarma School and Junior College. Individual test results recorded for positive indication were:

Test 1: Shapes correctly matched—4 out of 10 which show that the user lacks to sustain attention moods observed—angry as well as the user was constantly moving from a place which shows lower engagement level.

Test 2: Incorrect answers—7 out of 10 which show carelessness in following instructions.

Answers before questions complete—6 out of 10 which show impatience, impulsivity and interrupting nature.

Test 3: Incorrect answers—12 out of 20 which show that the person lacks to be focused and tends to make careless mistakes.

Test 4: According to DSM-5, the informants gave positive results by agreeing to observe Attention Deficit/Hyperactivity Disorder symptoms in the child for moreover 6 months. For other 3 out of 57 kids, this result sheet gave negative scores for Attention Deficit/Hyperactivity Disorder while the symptoms were not observed in

Table 1 Confusion matrix

		Actual ADHD		Total
		Positive	Negative	
Predicted	Positive	3	1	4
ADHD	Negative	0	56	56
Total		3	57	60

the kids for more than 6 months. One case was false positive and three were true positive. Following is the confusion matrix of our model. The results were verified by a psychiatrist (Table 1).

Following is the confusion matrix of our tests conducted.

The accuracy of our model is 98% which signifies it is a very effective model to determine the symptoms of Attention Deficit/Hyperactivity Disorder, especially for youngsters.

5 Conclusion and Future Scope

This model will help to solve the problem of the society as it would help Consultant to attend more patients in less time by reducing number of tests to be performed in person and would help briefly to make decision of prevalence to the disorder. It will also help patients to get a quick diagnosis of the disorder that saving time and many scheduled appointments causing overhead. In the future, this model would also help Consultant to suggest treatment based on the severity and span of disorder. Thus, the proposed system saves the diagnosis time by performing a test before the visit to the Consultant.

Acknowledgements Special Thanks to Dr. Amit Kharat MBBS, DM RD, DNB, PhD, FICR Quality Assurance Radiologist and Dr. Makarand Pantoji, Junior Resident, Department of Psychiatry, NIM HANS, for their guidance and support in this research.

References

1. Nass, Ruth. 2006. Evaluation and assessment issues in the diagnosis of attention deficit hyperactivity disorder. *Seminars in Pediatric Neurology* 12: 200–216. https://doi.org/10.1016/j.spen.2005.12.002.
2. McKnight, Lorna. 2010. Designing for ADHD in search of guidelines. In *IDC 2010 Digital Technologies and Marginalized Youth Workshop.*
3. https://www.healthline.com/health/adhd/conners-scale#1.
4. Pantoji, Makarand. Junior Resident, Department of Psychiatry, NIMHANS.

5. Matthews, Marguerite, et al. 2006. Attention deficit hyperactivity disorder. *Current Topics in Behavioral Neurosciences* 16: 235–66. https://doi.org/10.1007/7854_2013_249.
6. https://www.cdc.gov/ncbddd/adhd/data.html.
7. Diagnostic and Statistical Manual of Mental Disorders—Book by American Psychiatric Association.

Classification of Types of Automobile Fractures Using Convolutional Neural Networks

Nikhil Sonavane, Ambarish Moharil, Fagun Shadi, Mrunal Malekar, Sourabh Naik and Shashank Prasad

Abstract Image classification has recently been in serious attention of various researchers as one of the most upcoming fields. For this, various algorithms have been developed and used by researchers. In recent years, convolutional neural networks have gained huge popularity among masses for image classification and feature extraction. In this project, we have used convolutional neural networks for the classification of automobile fractures using their micrographs available on the Internet into their three known types—ductile, fatigue, and brittle. We have used a specific algorithm to extract the best epoch model from the whole model due to loss in the accuracy we encountered.

Keywords Convolution · Deep learning · Max pooling

1 Introduction

It is very important that the components manufactured in the automotive industry be of high quality. In case any automotive component fails, it becomes important to find out the root cause of the failure. Pictures of automotive components which undergo fractures are available which are known as micrographs. These micrographs fall into general three categories known as ductile, brittle, and fatigue. It is nearly impossible to classify these micrographs into their respective categories with human eye. Hence, there has to be diligent classification technique which can accurately classify a micrograph into one of these categories. This is where image classification comes into picture. The most important aspect of any technology related to computer vision is image classification. Even if one does not realize, image classification has a huge impact on day-to-day life. From face detection to security management to traffic tracking, it has a significant involvement in our day-to-day life. Image

N. Sonavane (✉) · A. Moharil · F. Shadi · M. Malekar · S. Naik · S. Prasad
Department of Electronics and Telecommunication Engineering,
Vishwakarma Institute of Technology, Pune, India
e-mail: nikhil.sonavane16@vit.edu

© Springer Nature Singapore Pte Ltd. 2020
D. Swain et al. (eds.), *Machine Learning and Information Processing*,
Advances in Intelligent Systems and Computing 1101,
https://doi.org/10.1007/978-981-15-1884-3_13

classification involves feature extraction and classifies those features on the basis of their contribution.

Traditionally, researchers extract features from different images with label and use classification algorithms such as SVM and random forest on the extracted features for feature classification. The traditional machine learning algorithms require feature extractors to extract features from images. HAARHOG and LBP are some of the known feature extractors. SVM is one of the image classification techniques used in computer vision. It gives good performance with small datasets. But it requires the time-consuming process of initially extracting the features using HAARHOG or LBP. And then after, train the model using these features. Random forest usually requires a lot of data but can give a good robust model. It has been observed that these traditional supervised machine learning algorithms require the data to be structured in feature matrices and labels in a vector of the length equal to the number of observations. Hence, today use of various pre-built algorithms is on the rise and the most popular choice for image classification is convolutional neural networks. Convolutional neural networks fall under the category of deep learning. Many deep learning approaches and algorithms have been discussed and researched upon in the past years. In [1], authors have talked about convolutional neural networks (CNNs),restricted boltzmann machine (RBM), and deep belief network (DBN) in great detail. The authors have also discussed briefly the challenges posed in deep learning and an effective way to tackle them. Convolutional neural networks eliminate this time-consuming process required as it has convolutional layers and pooling layers to take care of feature engineering part. There is no need of manually extracting features in neural networks. Neural networks are considered to be more flexible than the traditional machine learning algorithms. In [1], the authors have discussed how deep learning has highly contributed to computer vision, natural language processing, and artificial intelligence. Also, neural networks perform exceptionally well when the predictor variables are more. Neural networks perform well only when the size of the dataset is large as compared to SVM. But, the size of dataset can be diligently increased by using image augmentation techniques. Thus, it can be observed that neural networks are seen to be more flexible and easier to use than the traditional machine learning algorithms.

2 Convolutional Neural Network

Convolutional neural network [2] is a part of deep learning which has its application in image classification, natural language processing, visual recognition, etc., domains. Convolutional networks are currently the most efficient and accurate deep learning models for classifying pictorial data. CNN is now the efficient model on every image-related problem. Figure 1 shows the representation of a general convolutional neural network.

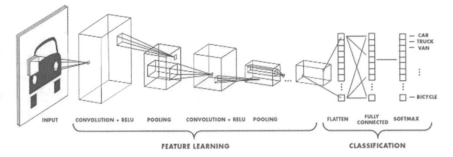

Fig. 1 Architecture of CNN

2.1 Architecture

2.1.1 Convolutional Layer

Convolutional layer is a major part of neural network where the main aim is to extract features from input images. The convolution operation works in a way such that it is able to detect the features such as edges, corners, and sharp edges from a given input image representation [2]. Convolution operation is performed using a kernel/filter and the input image. The kernel is slide over the input image with a stride equal to 1, and a matrix multiplication is performed consequently at every location. The multiplication results in the formation of feature maps (Fig. 2).

Traditionally, researchers extract features from different images with label and use classification algorithms such as SVM and random forest on the extracted features

Fig. 2 Convolution

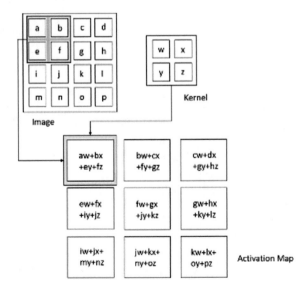

Fig. 3 Pooling operation

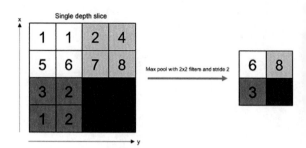

for feature classification. But today the use of various pre-built algorithms is on the rise and the most popular choice for image classification is convolutional neural networks. CNN uses neural networks for classifying images which makes it efficient.

The activation function used in the convolution layer is the RELU which introduces nonlinearity in the images [3].

2.1.2 Pooling Layer

Pooling layer is used after the convolutional layer to reduce the number of parameters and the computation power. It reduces the dimensions of the image but also retains the important information of the image. Pooling can be of three types, namely max pooling, sum pooling, and average pooling. Max pooling is used in this classification process of micrographs for automotive components. Max pooling takes the element with the largest value from a patch in the input image and creates a new pooled feature map. This reduces the size of the image and also prevents overfitting by removing irrelevant information from the input images. The result of the pooling produces a reduced dimension matrix which contains the important information. This is then flattened into a single vector and fed to the hidden layers [4] (Fig. 3).

2.1.3 Fully Connected Layer

The flattened matrix obtained from the pooling layer is given to the fully connected layer. The fully connected layers consist of many neurons. These fully connected layers will help in performing classification based on the learned input features. In a fully connected layer, each neuron is connected with the other neurons from the previous layer, and every connection is assigned a weight [2]. After a certain number of epochs, which are set by the user, the system or the model differentiates the dominating features and the low-level features in the images. Next, these features are classified using softmax classification technique. The output layer is placed at the end of the last fully connected layer.

An activation function is nothing but a mathematical function that is applied in the perceptron; it is basically a mapping of nonlinear complex functions between

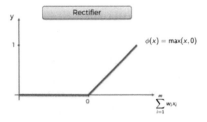
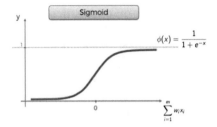

Fig. 4 Rectifier and sigmoid functions

input and output responses. The main purpose of an activation function in a multi-layered artificial neural network (ANN) is to derive an output from an input signal. So, basically in an architecture of ANN with multilayer perceptron, the input signals (X_i) from the input nodes (m) are multiplied to the weights (W_i) of each node and their combined summation is fed into a function $f(x)$ and an output signal is computed which is again fed as an input to the adjoining layer.

The activation functions that we used in our model were rectifier and sigmoid activation functions.

The rectifier activation function was used in the input layer and the hidden layers as it outputs the computation of inputs directly if they are positive and produces the output as zero if the sum of products of weights of the input signals is negative (Fig. 4).

Equation for activation function is given as follows:

$$\mathbf{f(x)} = \sum_{i=1}^{m} W_i X_i$$

For rectifier function,

$$\varnothing(x) = \max(x, 0)$$

For sigmoid function,

$$\varnothing(X) = 1/(1 + e^{-x})$$

The activation function that was used at the output layer was sigmoid activation function. The purpose of using sigmoid is that it gives the accurate value between 0 and 1, just what we get in logistic regression. So, the output can now be measured in probabilistic terms and it helps in analyzing the results especially when we are targeting to solve a classification problem.

2.2   Model Description and CNN Training

The most important part of this model was the data, and it was difficult to fetch it from a single source. We created a dynamic Web scraper to parse the images from various sources [5]. After careful observation, filtering and manual sorting, we arrived at 420 (Four Hundred and Twenty) images which were suitable to train the model. Although 420 images indicate a very less dataset for a deep network to train on, the model was built and constructed successfully after numerous alterations and changes.

The images were sorted into three categories as brittle, ductile, and fatigue, and the aim was to train the CNN on this available data and predict the type of failure or fracture accurately.

To build this algorithm, TensorFlow and Theano were taken into consideration, but due to the ease offered by Keras and due to its rejection toward low-level computation, Keras was used to script the algorithm and also due to the fact that it runs on top of TensorFlow as well as Theano.

The dataset we fetched was a set of 3D images with RGB matrix values and had to be converted into a one-dimensional set of images for the ease of training.

With 1D images, it was easy to perform image augmentation procedures.

To increase the accuracy of the model, the first and most important factor was to increase the dataset, and instead of a quantitative approach, we went for the qualitative one. Several layers of image augmentation techniques like rotation, shearing, blurring, grayscale conversion, adding salt-and-pepper noise, scaling, translation, flipping, nonlinearizing, perspective transformation, and alteration of lightning conditions were performed using the Keras API.

So, this dataset of 420 images was expanded to a final set of 3890 images.

The total hidden layers or multilayer perceptrons used in the tail ANN were 6.

The optimizer used for the model was 'Adams' optimizer, and the loss function used was 'categorical_crossentropy'. While training the ANN, the model reaches inconsistent values of accuracies. It becomes very difficult to fetch the model with low loss and high validation accuracy, when the total batch size was getting trained for the fixed number of 500 epoch cycles.

2.2.1   Ensemble Learning

Ensemble learning is a process that is practiced quite often in machine learning. Various models like decision trees and random forests work on the algorithm of ensemble learning. Ensemble learning can be described as combining the output of multiple models into a single output; i.e., various models are cascaded in a parallel connection, and the output of these models is converted into a singular output combining several predictive models into one [6].

Initially, many classification models were created using different splits of the same training dataset. The first classifier (convolutional neural network in this case)

Fig. 5 Representation of
ensemble learning

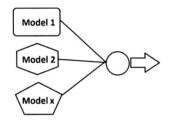

was constructed. The shape of the input image was (64, 3, 3), and 64 feature maps
were obtained at the end of convolution. Then, these 64 feature maps were further
subsampled by the fourth layer or the max pooling layer into 128 output matrices.
Four convolutional and max pooling layers were used. The max pooling layer as
explained in Sect. 2.1.2 extracted the impact features and then gave them as the
output to the flattening layer for flattening these matrices to feed them into the input
layer of the full connection. The full connection through majority voting of the
multilayer perceptrons classified the input image as ductile, brittle, or fatigue.

The second classifier (convolutional neural network in this case) was constructed.
The shape of the input image was (32, 3, 3), and 32 feature maps were obtained after
the first convolutional layer operation. Then, these 32 feature maps were further
subsampled by the fourth layer or the max pooling layer into 64 output matrices. In
the second classifier too, around four convolutional and pooling layers were used.
The max pooling layer as explained in Sect. 2.1.2 extracted the impact features and
then gave them as the output to the flattening layer for flattening these matrices to
feed them into the input layer of the full connection. The full connection through
majority voting of the multilayer perceptrons classified the input image as ductile,
brittle, or fatigue.

The third classifier (convolutional neural network in this case) was constructed.
The shape of the input image was (128, 3, 3), and 128 feature maps were obtained
as the output. Then, these 128 feature maps were further subsampled by the fourth
layer or the max pooling layer into 256 output matrices. In the third classifier, four
convolutional and max pooling layers were used. The max pooling layer as explained
in Sect. 2.1.2 extracted the impact features and then gave them as the output to the
flattening layer for flattening these matrices to feed them into the input layer of
the full connection. The full connection through majority voting of the multilayer
perceptrons classified the input image as ductile, brittle, or fatigue.

All these three models were combined into a single predictive model, and the
weighted average of all these three models was obtained as the final output (Fig. 5).

2.2.2 The Callback Method

The model tended to reach max and suitable values in between and while having
these peaks extracting the end epochs with high loss and low validation accuracy
was a backdrop, but this was a major backdrop and we successfully overcame this

backdrop by using the callback method [7]. The callback method is a very probabilistic approach of extracting the most accurate epoch. A limit of 500 epoch cycles was decided, and a patience value of 47 epoch cycles was given in the 'callback method' (the value 47 was decided after numerous training iterations and cycles). Now, when the model reached a peak value, i.e., high validation accuracy and low loss, the model would carefully observe the next 47 cycles after this peak was attained, if the accuracy value did not reach as high as the peak; the training iterations would stop, and the peak model was saved making it the best model. Also while the training process went on after the peak was attained, if the model reached higher accuracy values and lower difference between training and validation loss, the model now gave up the initial peak value and considered this new epoch as the new peak value and observed for next 47 cycles until it reached the maximum value of 500 epoch cycles. This process saved the best models while training and helped to achieve higher validation accuracy.

The classes of the callback method were built in a way to stop the training process if the algorithm was unable to find an epoch cycle with accuracy higher than that of the most frequent peak in the defined patience window.

3   Experimental Analysis and Results

In this section, we explained the working and basic structure of our model along with the discussion of output.

3.1   Dataset

The automobile fracture dataset consisted of 420 images extracted from internet using Web scraping. The images were micrographs from various sources and were labeled properly before using it in the model.

3.2   Architecture

For getting the best model among various iterations, we used a callback function which stops the models from further iterations if there is no progress in the accuracy of model. Thus, it helps in extracting the best model (Figs. 6, 7 and 8).

Finally, we achieved a highest validation accuracy score of 92.43% with training and validation loss as 0.9068 and 0.6866, respectively.

125/500 [==============================] - 4 s 153 ms/step - loss: 0.2987 - acc: 0.9068 - val_loss: 0.6866 - val_acc: 0.9243

Figure 9 shows the best model epoch.

Fig. 6 val_acc versus time

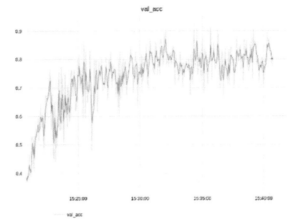

Fig. 7 Training accuracy versus time

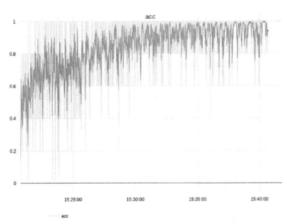

Fig. 8 Training loss versus time

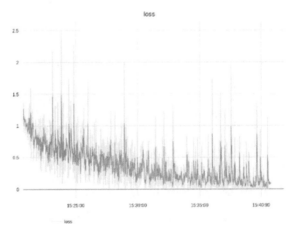

Fig. 9 Val_loss versus time

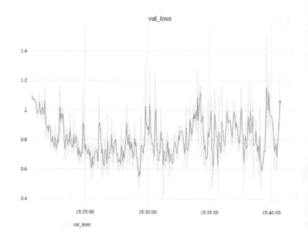

4 Conclusion

In this approach discussed in the paper, we have trained a model to classify automobile fracture types in categories: brittle, fatigue, and ductile on the basis of the micrographs of these fractures. The classification is done using convolution neural networks, and it is trained on 420 images. With tuning the convolutional layers, pooling layers, batch size, and many such parameters, we can see a drastic change in the efficiency of the model. And by using the callback function, we can stop the training if the accuracy of the model is not increasing anymore.

References

1. Jia, Xin. 2017. Image recognition method based on deep learning. In *29th Chinese Control and Decision Conference (CCDC)*.
2. Guo, Tianmei, Jiwen Dong, Henjian Li, Yunxing Gao. 2017. Simple convolutional neural network on image classification. In *2017 IEEE 2nd International Conference on Big Data Analysis*.
3. https://www.datascience.com/blog/convolutional-neuralnetwork.
4. Yim, Junho, Jeongwoo Ju, Heechul Jung, and Junmo Kim. 2015. Image classification using convolutional neural networks with multi-stage feature. In *Robot Intelligence Technology and Applications 3*. Advances in Intelligent Systems and Computing, ed. J.-H. Kim et al., vol. 345, 587. Springer International Publishing Switzerland.
5. Tajbakhsh, N., et al. 2016. Convolutional neural networks for medical image analysis: Full training or fine tuning? *IEEE Transactions on Medical Imaging* 35 (5).
6. Krizhevsky, A., I. Sutskever, and G.E. Hinton. 2012. ImageNet classification with deep convolutional neural networks. In *Proceedings of Advances in neural information processing systems*.
7. He, K., X. Zhang, S. Ren, and J. Sun. 2016. Deep residual learning for image recognition. In *Proceedings of IEEE Conference on* Computer *Vision and Pattern Recognition*, June.
8. Boureau, Y.L., F. Bach, Y. LeCun, and J. Ponce. 2010. Learning midlevel features for recognition. In *CVPR*.

9. https://towardsdatascience.com/a-comprehensive-guide-to-convolutional-neural-networks-the-eli5-way-3bd2b1164a53.
10. Jmour, Nadia, Sehla Zayen, and Afef Abdelkrim. 2018. Convolutional neural networks for image classification. 978-1-5386-4449-2/18/$31.00 ©2018 IEEE.
11. Wang, Tao, David J. Wu, Adam Coates, and Andrew Y. Ng. End-to-end text recognition with convolutional neural networks.
12. Vo, An Tien, Hai Son Tran, Thai Hoang Le. 2017. Advertisement image classification using convolutional neural network. In *2017 9th International Conference on Knowledge and Systems Engineering (KSE)*.

A Subjectivity Detection-Based Approach to Sentiment Analysis

Nilanjana Das and Santwana Sagnika

Abstract With the rise of Web 2.0 where loads of complex data are generated every day, effective subjectivity classification has become a difficult task in these days. Subjectivity classification refers to classifying information into subjective (expressing feelings) or objective (expressing facts). In this paper, we use Yelp reviews dataset. Our aim is to prove that a dataset with the objective sentences removed from each review gives better results than the dataset containing both subjective and objective sentences. To achieve this, we have used two approaches, each divided into two phases. The first phase of both the approaches is mainly the subjectivity classification phase where we filter out the objective sentences and keep the subjective sentences in the reviews, thus creating a new dataset with purely subjective reviews. The second phase of the first approach uses CountVectorizer which creates word vectors, and we fit the model to the classifiers. The second phase of first approach is repeated for both the datasets, and we get better results for the newly created dataset which contains purely subjective reviews. The second phase of the second approach uses Word2Vec, an implementation of neural network which creates distributed word vectors. We fit this Word2Vec model to the classifier, and we analyze the results. Again, the newly created dataset gives better results after we repeat this phase of the second approach for both the datasets.

Keywords Sentiment analysis · Subjectivity detection · Opinion mining · Natural language processing

N. Das (✉)
TATA Consultancy Services, Gitanjali Park, Kolkata, India
e-mail: nilanjanadas010@gmail.com

S. Sagnika
School of Computer Engineering, Kalinga Institute of Industrial Technology (Deemed to be University), Bhubaneswar, Odisha, India
e-mail: santwana.sagnika@gmail.com

© Springer Nature Singapore Pte Ltd. 2020
D. Swain et al. (eds.), *Machine Learning and Information Processing*,
Advances in Intelligent Systems and Computing 1101,
https://doi.org/10.1007/978-981-15-1884-3_14

1 Introduction

Subjectivity classification is the categorization of information into either subjective (expressing sentiments) or objective (expressing facts). Sentiment analysis is the processing of natural language to categorize subjective information into positive, neutral, or negative. Subjectivity detection is a step before sentiment detection. A text is first classified into subjective or objective. Then, the detected subjective sentences are classified into positive, neutral, or negative. For example, the sentence 'This place is one of the best places I have ever visited' is first analyzed for subjectivity. Once it is classified to be non-neutral, sentiment detection is carried out which then finds out that the sentence contains a positive sentiment [1, 2].

With the onset of Web 2.0 loads of data is being generated every minute. Now, Web 2.0 has numerous platforms where people can share their opinions expressing subjectivity or objectivity [3]. These opinions can be used to analyze and predict their behavior toward a particular topic or product. Analysis and categorization of such opinions have become difficult because people have started using special characters or complex words, phrases to express their opinions. For example, text messages like email id 'example@gmail.com,' URLs, complex tweets, and many more can produce difficulty in detecting subjectivity. Consequently, accuracy of the model used to train the dataset decreases. Accuracy is significantly higher for complex datasets with proper preprocessing [4, 5]. In this paper, we will compare the accuracies for both, a dataset in its actual form and the same dataset with the objective sentences removed from each review in the dataset. Through our experimentation, we will try out new techniques to prove that it is more precise to train a model with subjective sentences.

In [6], three metaheuristic methods and subjectivity lexicons are used. However, we are going to explore more on naive Bayes, support vector machines, and random forest classifier. In [7], each word of a review is classified into subjective or objective first, and then, the probability of the sentence is predicted to be subjective or objective using a supervised machine learning technique. We follow a similar approach where using TextBlob, we first take out the subjectivity score of each sentence of a review, and then, highly objective sentences have to undergo a part-of-speech (POS) tagging process where each word in the objective sentence is checked with their respective POS tags, and if it belongs to any of the categories such as adjective, verb, adverb or noun, it is assumed to be subjective, and a separate dataset is created for only subjective reviews. For example, 'What a wonderful place' gives a subjectivity score of 1.0; therefore, it is subjective, and no POS tagging process is required for this sentence. Another example 'I like this place' gives a subjectivity score of 0.0, but we know the sentence has a positive sentiment in it, so the sentence is first tokenized into words like 'I,' 'like,' 'this,' 'place.' Then, each word is checked for POS tags. We see that the word 'like' has the tag name as 'VBP,' i.e., verb, so the whole sentence is assumed to be subjective irrespective of its score. Also, we use two approaches, namely CountVectorizer and Word2Vec, for comparing the classification reports of the Yelp and the newly created dataset.

We have discussed some of the past work in this field in Sect. 2. Section 3 discusses the approaches we follow. Section 4 discusses the experimental setup and the results of the experimentation.

2 Related Work

In this section, we will study some of the related works done previously on subjectivity classification. In [1], different concepts and methodologies on sentiment analysis are discussed by Pawar et al., where sentiment classification can be done both at the sentence level and document level. Bravo-Marquez et al. [2] use a combination of meta-level features for subjectivity and sentiment classification. In [8], three different categories of methods are used for subjectivity detection by Chaturvedi et al. The categories used are hand-crafted, automatic, and multi-modal.

SemEval 2016 stance detection task dataset was used by Dey et al. [9] where the labels were positive, negative, and neutral stances. A feature-driven model was developed, where in the first phase the tweets were classified into either neutral or non-neutral; i.e., subjectivity classification was carried out. In the second phase, the non-neutral tweets were classified into either positive or negative; i.e., sentiment classification was carried out.

Rashid et al. [10] is a survey paper by Rashid et al. which discusses the various supervised, unsupervised, case-based reasoning, and machine learning techniques used for computational treatment of sentiments in opinion mining. A semi-supervised machine learning technique is used in [11] by Esuli and Sebastiani to determine whether a term has a positive, neutral, or negative connotation. Zhuang et al. [12] use a multi-knowledge-based approach which is the integration of statistical analysis, WordNet, and movie knowledge to extract features from opinion, thus classifying them into positive or negative.

Kim and Hovy [13] propose a system that could align the pros and cons to their respective sentences in each review. Then, they train a maximum entropy model on the resulting dataset and use it to extract pros and cons from review datasets which are available online and are not labeled explicitly.

Xuan et al. [14] use the concept of features extracted from the text. They use a movie review dataset proposed by Pang and Lee earlier. They proposed 22 syntax-based patterns to extract the linguistic features to classify a sentence into subjective or objective using the maximum entropy model (MaxEnt). Rustamov in [15] uses different supervised machine learning techniques like hidden adaptive neuro-fuzzy inference system (ANFIS), markov model (HMM), and fuzzy control system (FCS) for subjectivity detection.

Keshavarz and Abadeh in [6] propose three metaheuristic methods, namely genetic algorithm, simulated annealing, and asexual reproduction optimization, to create subjectivity lexicons in which the words are classified into either subjective or objective. It concluded that the proposed approaches perform better by comparing the accuracy and f-measure to the baseline.

Kamal [7] proposes a supervised machine learning technique to classify sentences from customer reviews into either subjective or objective. Then, it uses a rule-based method to extract feature–opinion pairs from the subjective sentences obtained earlier, by applying linguistic and semantic analyses of texts.

3 Proposed Approach

We aim to prove that a dataset which has its objective sentences from each review removed will train a classifier more efficiently. We have followed two approaches to gain the desired results. For achieving this, we use Yelp reviews dataset. The dataset had 10 columns, out of which we used only 2 columns which served our purpose. The columns used were customer reviews and the star ratings given by the customers. For our purpose, we considered star ratings 1 and 5. We had divided our overall work into two phases to achieve our aim. Let us study it phase-wise.

Phase I—Preprocessing

Phase I is common for both the approaches. All the reviews in the Yelp reviews dataset have a star rating from 1 to 5. Each review however may contain some objective sentences which express mere facts and no opinion or emotions toward the place. Our aim is to remove these objective sentences and create a new dataset with only the subjective sentences present in each review. This process of finding out subjectivity instances is known as subjectivity detection. To achieve this, we follow a series of steps.

(1) **Subjectivity Detection**: At first, we take each review and tokenize it into a list of sentences. We then find the subjectivity score of each sentence using an available Python library TextBlob. The TextBlob library provides a subjectivity score for sentences using an available function. The subjectivity score ranges from 0 to 1. We consider 0 as an extremely objective score but that does not mean it does not contain any trace of subjectivity. Tokenized sentences of a particular review with subjectivity score higher than 0 are concatenated directly, whereas those sentences with score equal to 0 have to undergo a POS tagging process, discussed in the next step.

(2) **POS**: Part-of-speech (POS) is very effective in subjectivity detection. Most researches have concluded that adjectives play a very important role in expressing opinions or emotions and are a good indicator of subjectivity. In [7], it is discussed that apart from adjectives, nouns like great, brilliant and delicious, and verbs like hate and amaze can also indicate subjectivity in a sentence. As sentences with subjectivity score equal to 0 may have some trace of subjectivity, and we filter out those sentences by checking the POS tags such as adjectives, nouns, and verbs for each word in each sentence of a review. If any word is found to possess any tag from among these three, then the sentence is assumed to be subjective. The sentences which were filtered out are concatenated, and we get the final review with subjective sentences.

(3) **Data Cleaning**: The review which we finally obtain has to undergo a cleaning process. Using the Beautiful Soup library, we remove any HTML markup and tags present in the review. This reduces the noise present in the review. After this, any non-alphabet character, for example, digits, '!' and ')', is removed from the reviews. Finally, all the words are converted into lowercase and we get the clean review. The cleaned subjective review is then updated in the new dataset. These steps are carried out for each review present in the Yelp dataset. Also, we do not remove the stopwords as it may lead to loss of information. Therefore, it is better to keep the stopwords. The new dataset is used in phase two where we work out different methods to prove that new dataset more efficiently trains a classifier, as compared to the older one. Next, we discuss phase II.

Phase II—Feature Representation

Phase II is different for both the approaches, and we will discuss them one by one. This phase is applied to Yelp and the newly created dataset to compare the results.

In the first approach, we use a simple bag-of-words model. As we know that Yelp dataset and the newly created dataset from Yelp have star ratings from 1 to 5 and have only 10,000 rows of data, we consider only those rows with star ratings 1 and 5 for phase II of the first approach of each dataset. Thus, training the model becomes more efficient. We use CountVectorizer for phase II of our first approach of each dataset one by one. It creates a feature vector for each of the reviews in a dataset. Once the feature vector gets created, we fit this model to train our classifier. In this approach, we have used multinomial naive Bayes, random forest, and support vector machine classifiers as they suit our purpose.

In phase II of our second approach, we use the concept of distributed word vectors which is created by the Word2Vec algorithm. The Word2Vec model was initially published by Google in the year 2013. Word2Vec is an implementation of neural networks that creates distributed word vectors. One of the advantages of Word2Vec is that it performs well on unlabeled datasets. It uses the concept of placing words with similar meaning in a single cluster. To implement Word2Vec, we need to have gensim package installed. Although Word2Vec performs well on unlabeled dataset, we still use the same labeled Yelp dataset and the newly created dataset to compare our final results in the second approach. In this approach too, we consider the rows with star ratings 1 and 5. Since Word2Vec model works on single sentences, each one as a list of words; ultimately, what we pass on to the Word2Vec model is a list of lists. For this purpose, we process each review one by one. One by one, each review is tokenized into sentences and passed on for data cleaning similar to what is already discussed in phase I. The same steps are carried out for each sentence it receives except for one difference. Stopwords are not removed again so that we get high quality of word vectors. Each sentence is then returned as a list of words. The above process continues for each sentence in each review, and finally, we get a list of sentences as a list of words. There is only one difference between the data cleaning function of the two approaches; that is, the first approach returns sentence as a whole, whereas the second approach returns sentences as a list of words. Finally, the list of parsed sentences is used to train the Word2Vec model. Word2vec model

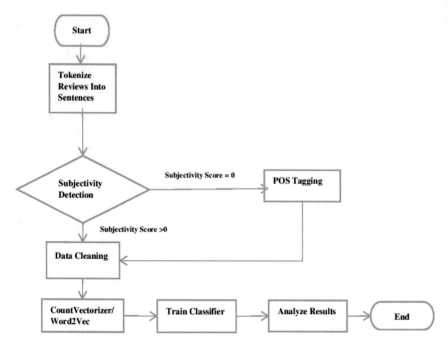

Fig. 1 Process flow diagram

trained creates a feature vector for each vocabulary word. The number of rows in the vocabulary represents the number of words contained in the word vocabulary, and the number of columns represents the feature vector size. Next, we proceed to vector averaging. Since our Yelp reviews dataset and the newly created dataset contain reviews of variable length, we work out a method that will take each word vector and convert it into a feature set that will have the same length for each review present in the datasets. For this, we first send the training and testing reviews to the data cleaning method again. This method now returns each review as a list of words unlike the previous one, where each sentence was returned as a list of words. Then, this list of reviews as list of words is processed for vector averaging. Finally, we try to create the average of word vectors for each review. We pass on the average of word vectors to train our random forest or support vector machine classifier (Fig. 1).

4 Experiments

This is the section where we discuss our experiments and their results. We compare the classification reports of both the datasets for each of the approaches.

Data Description

As we have also discussed earlier, we used Yelp reviews dataset in this paper. Yelp reviews are a predefined dataset which is available in Kaggle. It had 10 columns out of which only 2 columns, namely the customer reviews and the star ratings given by them, were used in this paper. Since the size of the dataset is small (10,000 rows) and star ratings were on a range of 1–5, therefore we considered only star ratings 1 and 5 so that we get better results; otherwise, it was ineffective trying to classify reviews using 5 star ratings on such a small dataset. Our aim in this paper is to prove that a dataset containing purely subjective reviews when trained using a classifier gives better results when compared to a dataset with mixed type of reviews. We used subjectivity detection of each sentence of a review at a time using a Python library and TextBlob. This gives a score between 0 and 1. Sentences with score above 0 were assumed to be subjective, whereas sentences with score equal to 0 were considered objective. Then, further processing of objective sentences was done. They had to undergo a POS tagging method. Each word in the sentence was checked for tags like adjective, noun, verb, and adverb which if found in any of the words, then the sentences were assumed to be subjective. The final subjective sentence that we got undergoes a data cleaning method. There we remove all the HTML markup and tags using Beautiful Soup, an available Python package. Another library 're' was used to remove any non-alphabet characters from the review. Then using the lower function in Python, all words were converted into lowercase. Stopwords are not removed anywhere in this paper. The above processes are done for each review in the dataset, and we finally update the subjective reviews in a new dataset. After this, first we use CountVectorizer which creates word vectors for each of the reviews of both the datasets. The model is then fit into the classifier. In this case, we use three classifiers support vector machine, random forest, and naive Bayes and compare all three outputs for both the datasets one by one. Both the datasets using CountVectorizer are trained on three classifiers one at a time and show how accurately the classifiers predict the reviews into 1 and 5 star ratings. Their results are then analyzed. After CountVectorizer another approach, we tried out Word2Vec which creates feature vector for each vocabulary word. We then fit this model to the classifier. In this approach, random forest and support vector machine classifiers are used for both the datasets to again predict the reviews into 1 and 5 star ratings. The final classification report is compared for both the datasets.

Evaluation

Our original dataset is the Yelp reviews dataset, and the subjective dataset is the new dataset. Here, all the values that we mention are an average we get by running the code for five times for each type of dataset for each classifier in each approach. Thus, we were able to avoid overfitting and underfitting. In the first approach where we use CountVectorizer, three classifiers, namely support vector machine, random forest, and naive Bayes, are used for both the datasets. Naive Bayes gives an accuracy of 94.13% on the new dataset and 93.89% on the Yelp dataset using CountVectorizer. Support vector machine gives an accuracy of 92.91% on the new dataset and 92.79% on the Yelp dataset using CountVectorizer. Random forest classifier gives an accuracy of 86.41% on the new dataset and 86.18% on the Yelp dataset using CountVectorizer.

On the new dataset using naive Bayes algorithm, we get precision of 0.93, recall of 0.74, and $f1$-score of 0.82 for 1 star rating. For 5 star rating, precision is 0.94, recall is 0.99, and $f1$-score is 0.96. On the new dataset using support vector machine algorithm, we get precision of 0.86, recall of 0.73, and $f1$-score of 0.79 for 1 star rating. For 5 star rating, precision is 0.94, recall is 0.97, and $f1$-score is 0.96. On the new dataset using random forest, we get precision of 0.97, recall of 0.27, and $f1$-score of 0.42 for 1 star rating. For 5 star rating, precision is 0.86, recall is 1.00, and $f1$-score is 0.92.

On the Yelp dataset using naive Bayes algorithm, we get precision of 0.92, recall of 0.73, and $f1$-score of 0.81 for 1 star rating. For 5 star rating, precision is 0.94, recall is 0.99, and $f1$-score is 0.96. On the Yelp dataset using support vector machine algorithm, we get precision of 0.85, recall of 0.74 and $f1$-score of 0.79 for 1 star rating. For 5 star rating, precision is 0.94, recall is 0.97, and $f1$-score is 0.96. On the Yelp dataset using random forest, we get precision of 0.97, recall of 0.25, and $f1$-score of 0.40 for 1 star rating. For 5 star rating, precision is 0.86, recall is 1.00, and $f1$-score is 0.92. Accuracy is higher for naive Bayes, random forest, and support vector machine for the new dataset. However considering $f1$-scores, for new dataset using naive Bayes it is $0.82 - 0.81 = 0.01$ higher for 1 star rating. $F1$-score for new dataset using random forest classifier is $0.42 - 0.40 = 0.02$ higher for 1 star rating. Other $f1$-scores for the first approach are the same as can be inferred from Table 1.

Now coming to the second approach which uses Word2Vec, let us analyze its results. On the new dataset using random forest classifier, we get precision of 0.76, recall of 0.48, and $f1$-score of 0.59 for 1 star rating. For 5 star rating, precision is 0.89, recall is 0.97, and $f1$-score is 0.93. On the new dataset using support vector machine, we get precision of 0.85, recall of 0.46, and $f1$-score of 0.59 for 1 star rating. For 5 star rating, precision is 0.89, recall is 0.98, and $f1$-score is 0.93. On the Yelp dataset using random forest classifier, we get precision of 0.80, recall of 0.44 and $f1$-score of 0.56 for 1 star rating. For 5 star rating, precision is 0.88, recall is 0.97, and $f1$-score is 0.93. On the Yelp dataset using support vector machine, we get precision of 0.84, recall of 0.45, and $f1$-score of 0.58 for 1 star rating. For 5 star rating, precision is 0.89, recall is 0.98, and $f1$-score is 0.93. $F1$-score for the new dataset using random forest classifier is $0.59 - 0.56 = 0.03$ higher for 1 star rating. $F1$-score for new dataset using support vector machine is $0.59 - 0.58 = 0.01$ higher for 1 star rating. Rest $f1$-scores are same for the second approach as can be inferred from Table 2.

5   Conclusion and Future Scope

This paper mainly aims to prove that a dataset which has objective sentences removed from each review gives better accuracy than its original dataset. Here, in the first phase of both the approaches, we took out the subjectivity score for each sentence in a review and the objective sentences had to undergo a POS tagging process where each sentence was tokenized into words and checked for POS tags. If any word in the

Table 1 Classification report using CountVectorizer

Classifier	Star ratings	Using CountVectorizer								
		Original data				Subjectivity-filtered data				
		Precision	Recall	F1-score	Accuracy	Precision	Recall	F1-score	Accuracy
Naive Bayes	1	0.92	0.73	0.81	93.89	0.93	0.74	0.82	94.13
	5	0.94	0.99	0.96		0.94	0.99	0.96	
Support vector machine	1	0.85	0.74	0.79	92.79	0.86	0.73	0.79	92.91
	5	0.94	0.97	0.96		0.94	0.97	0.96	
Random forest	1	0.97	0.25	0.40	86.18	0.97	0.27	0.42	86.41
	5	0.86	1.00	0.92		0.86	1.00	0.92	

Table 2 Classification report using Word2Vec

Classifier	Star ratings	Using Word2Vec embeddings									
		Original data				Subjectivity-filtered data				
		Precision	Recall	F1-score	Accuracy	Precision	Recall	F1-score	Accuracy
Random forest classifier	1	0.80	0.44	0.56	87.65	0.76	0.48	0.59	87.70
	5	0.88	0.97	0.93		0.89	0.97	0.93	
Support vector machine	1	0.84	0.45	0.58	88.29	0.85	0.46	0.59	88.53
	5	0.89	0.98	0.93		0.89	0.98	0.93	

sentence had certain POS tag, then the sentence was assumed to subjective. The final subjectivity-rich reviews had to go through a data cleaning method. In the second phase of the first approach, we used CountVectorizer model with three classifiers naive Bayes, random forest, and support vector machines. We saw that accuracies for the new dataset were greater than those for Yelp dataset. The accuracies for naive Bayes, random forest, and support vector machines for new dataset were 94.13%, 86.41%, and 92.91% and for Yelp dataset were 93.89%, 86.18%, and 92.79%, respectively. In the second phase of the second approach, we used Word2Vec model with random forest classifier and support vector machine. We saw that accuracies for the new dataset were greater than those for Yelp dataset. The accuracies for random forest and support vector machine for new dataset were 87.70% and 88.53% and for Yelp dataset were 87.65% and 88.29%, respectively. We aim to further improve the subjectivity detection by applying several other techniques as discussed in subjectivity/objectivity analyzer in paper [7]. More the reviews are enriched with subjectivity, more is the accuracy.

References

1. Pawar, A.B., M.A. Jawale, and D.N. Kyatanavar. 2016. Fundamentals of sentiment analysis: Concepts and methodology. *Sentiment analysis and ontology engineering*, 25–48. Cham: Springer.
2. Bravo-Marquez, F., M. Mendoza, and B. Poblete. 2014. Meta-level sentiment models for big social data analysis. *Knowledge-Based Systems* 69: 86–99.
3. Liu, B. 2012. Sentiment analysis and opinion mining. *Synthesis Lectures on Human Language Technologies* 5 (1): 1–167.
4. Pandey, S., S. Sagnika, and B.S.P. Mishra. 2018. A technique to handle negation in sentiment analysis on movie reviews. In:*2018 IEEE international conference on communication and signal processing (ICCSP)*, 0737–0743.
5. Baldonado, M., C.-C.K. Chang, L. Gravano, and A. Paepcke. 1997. The stanford digital library metadata architecture. *International Journal on Digital Libraries* 1: 108–121.
6. Keshavarz, H.R., and M. Saniee Abadeh. 2018. MHSubLex: Using metaheuristic methods for subjectivity classification of microblogs. *Journal of AI and Data Mining* 6 (2): 341–353.
7. Kamal, A. 2013. Subjectivity classification using machine learning techniques for mining feature-opinion pairs from web opinion sources. arXiv preprint arXiv:1312.6962.
8. Chaturvedi, I., E. Cambria, R.E. Welsch, and F. Herrera. 2018. Distinguishing between facts and opinions for sentiment analysis: Survey and challenges. *Information Fusion* 44: 65–77.
9. Dey, K., R. Shrivastava, and S. Kaushik. 2017. Twitter stance detection—A subjectivity and sentiment polarity inspired two-phase approach. In *2017 IEEE international conference on data mining workshops (ICDMW)*, pp 365–372.
10. Rashid, A., N. Anwer, M. Iqbal, and M. Sher. 2013. A survey paper: areas, techniques and challenges of opinion mining. *International Journal of Computer Science Issues (IJCSI)* 10 (6): 18–31.
11. Esuli, A., and F. Sebastiani. 2006. Determining term subjectivity and term orientation for opinion mining. In *11th Conference of the European chapter of the association for computational linguistics*.
12. Zhuang, L., F. Jing, and X.Y. Zhu. 2006. Movie review mining and summarization. In *Proceedings of the 15th ACM international conference on Information and knowledge management*, 43–50.

13. Kim, S.M., and E. Hovy. 2006. Automatic identification of pro and con reasons in online reviews. In *Proceedings of the COLING/ACL on main conference poster sessions*. Association for Computational Linguistics, 483–490.
14. Xuan, H.N.T., A.C. Le, and L.M. Nguyen. 2012. Linguistic features for subjectivity classification. In *2012 IEEE international conference on asian language processing*, 17–20.
15. Rustamov, S. 2018. A hybrid system for subjectivity analysis. In *Advances in fuzzy systems*.

Image Processing and Machine Learning Techniques to Detect and Classify Paddy Leaf Diseases: A Review

Jay Prakash Singh, Chittaranjan Pradhan and Srikanta Charana Das

Abstract In Asian countries, paddy is one of the major staple foods, and the agricultural sector is the largest employer in India's economy. Hence, an effective mechanism should be adopted for food security and its production. Due to infections caused by pests like virus, fungus and bacteria, there is a huge loss in quality and quantity of the rice. This results in a huge loss to the farmers. This survey presents the different image processing techniques for paddy disease identification and further classification. The challenges involved in each step of diseases detection and classification are analyzed and discussed. Image preprocessing, segmentation, feature extraction and classification are the main stages involved in paddy disease identification and classification. All these four stages are well addressed here. This survey gives a complete overview along with the challenges of each stage and also focuses the research challenge and research gap. This survey concludes that automatic paddy leaf disease detection, and classification requires much more advancement. There is a need for more advanced techniques to automate the system effectively and efficiently.

Keywords Image processing · Image segmentation · Feature extraction · Classification

J. P. Singh (✉) · C. Pradhan
School of Computer Engineering, Kalinga Institute of Industrial Technology,
Bhubaneswar, India
e-mail: jaykiit.research@gmail.com

C. Pradhan
e-mail: chitaprakash@gmail.com

S. C. Das
School of Management, Kalinga Institute of Industrial Technology, Bhubaneswar, India
e-mail: srikant@kiit.ac.in

© Springer Nature Singapore Pte Ltd. 2020
D. Swain et al. (eds.), *Machine Learning and Information Processing*,
Advances in Intelligent Systems and Computing 1101,
https://doi.org/10.1007/978-981-15-1884-3_15

1   Introduction

Agriculture has always played a vital role in the economy of the most of the developing countries, mainly in South Asian countries. Farming is not only meant to feed the increasing population, but at the same time, it helps to handle global warming problems. Agricultural production is greatly affected by the crop diseases. The amount of crops that are damaged by adverse climatic condition and by invasion of pathogens can never be neglected. There are many plant diseases that present in the crop field due to soil fertilizer, mineral deficiency, environmental agents and other various factors. It is very important to monitor plants/crops from early stage in order to detect and avoid the diseases. It includes so many tasks like soil preparation, seeding, using manure/fertilizers, irrigation, using required amount of pesticides, timely harvesting and storage. Improvement in crop quality can only be achieved by the proper automation techniques of the disease detection [1]. The majority of population depends on rice as their staple food. A vast land area in the India is mainly used to cultivate rice crops. Farmers incur huge loss in rice production due to infection of rice crops by diseases caused by various pathogens like fungi, bacteria and viruses. It is very important to detect and classify such diseases at the early stage so that proper measure can be taken. Main four paddy crop diseases are rice blast disease (RBD), brown spot disease (BSD), narrow brown spot disease (NBSD), bacterial leaf blight disease (BLBD), etc. Normal leaf image is shown in Fig. 1a.

Fig. 1 Image of healthy leaf **a**, **b–e** diseased leaves of rice blast, brown spot, narrow brown spot, bacterial leaf blight, respectively

Rice blast disease is a fungal disease (fungus—Pyricularia grisea). It is the main disease in rice plants. It is widely distributed in more than 80 countries [2]. RBD affected leaf is shown in Fig. 1b. Symptoms are:

- At initial stage, looks like small bluish green specks soaked in water.
- Affected leaf portions are mainly football shaped (elliptical) with brown to brown-red margins and gray-white centers.

Brown spot disease is also a fungal disease (fungus—Helminthosporium oryzae). Initially, it changes from small oval to circular brown spots. It appears on leaves and glumes of maturing plants. BSD affected leaf is shown in Fig. 1c.

Narrow brown spot disease is also a fungal disease (fungus—Cercospora jansean). The symptoms are similar to that of brown spot disease. This disease becomes more active as the rice attains ripening stage. It causes prematurity damage to leaves and leaf sheaths. This disease is one of the causes of chalkiness in the rice grains. Leaf is shown in Fig. 1d.

Bacterial leaf blight disease is bacterial diseases (bacteria—Xanthomonas oryzae pv. oryzae). The lesions soaked in water gradually moves from tip toward the edges of leaves and it slowly changes to yellowish and stripes are straw colored with wavy margins. The leaf is shown in Fig. 1e.

Manually, disease detection requires a vast experience, and it does not work fine always. It is also very time consuming. It also lacks accuracy level, and it becomes very complex in large-scale farming [3]. Thus, it is very important to detect and classify diseases precisely and accurately by using some advanced automated technology. It has been observed that when paddy crops are infected by diseases their leaf shape, size, color and texture changes. Image processing techniques play an important role in the detection and classification of diseases bases on shape, size, texture and color of the paddy leaves. The texture represents homogeneity, contrast, energy and cluster influence which can be identified by the use of proper classifiers like support vector machine [4] to categorize the diseases.

Owning to the importance of automation of paddy leaves' disease detection and classification, we present a detailed survey on preprocessing, segmentation, feature extraction, classification, along with their challenges, benefits and limitations. Also, described the research gaps among the existing techniques. The overall view of our survey is presented in Fig. 2.

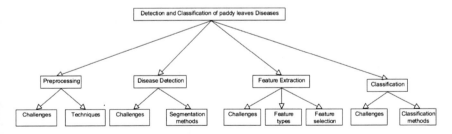

Fig. 2 Overview of the survey

The rest of this paper is categorized into three sections: Sect. 2 presents review of different approaches to paddy disease detection and further classification of the diseases. Section 3 presents the discussion on research gap and focuses on relevant techniques that are feasible for the existing problems. Lastly, Sect. 4 gives the conclusion of the survey.

2   Disease Detection and Classification

The complete overview of image processing techniques that are used for the paddy disease detection, and further classification is presented in this section. Image processing technology has been widely used and studied in agriculture to detect and classify leaf diseases. It has shown a good potential to identify the most of the diseases that invade the paddy crops besides some serious challenges that will be discussed below. The four main steps (preprocessing, segmentation, feature extraction and classification) of image processing techniques have their own challenges. In Fig. 3, the block diagram of our survey is presented. In the survey, our main point of focus is about the limitations of each step along with the existing methods

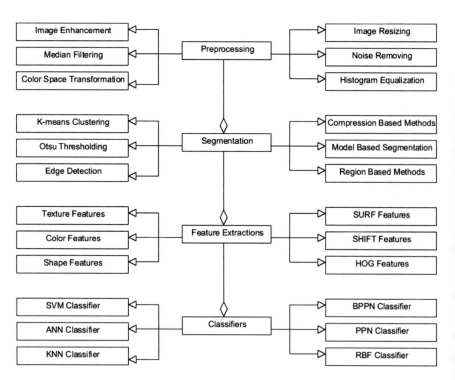

Fig. 3 Leaves' disease detection and classification techniques

and developments in the recent approaches. Our focus is also on the advantages and disadvantages of the recent approaches.

2.1 Preprocessing-Based Approaches

It is important as it involves image enhancement, noise removal. The image smoothing is also done in this step. Various image enhancement techniques like gray scaling, shadow removal image resizing, green pixel matching, color spaces, filtering (mainly median filter), etc. are used in this step. To achieve effective results, the selected images would be cropped properly [5]. To enhance image quality, image filters are used [6].

2.1.1 Challenges

The preprocessing step makes it easier to detect the diseased spot of the leaf as the diseased portion is more visible now. The main tasks of preprocessing step are:
(a) Noise affected background; (b) variation of lighting; and (c) low intensity image. These directly influence the accuracy of the segmentation process. There are various preprocessing methods like median filtering, green plane extraction, image binarization, color spaces, etc.

2.2 Segmentation-Based Approaches

The process of segmenting an image into several segments based on the similarity of pixels is image segmentation. The main motive behind it is to find the infected region in the image. Various segmentation approaches are available for detection of infected portion in the image. There are certain challenges associated with the segmentation.

2.2.1 Challenges

Segmentation one of the most challenging step. In paddy leaf disease detection, its major aim is to segregate the diseased part and background from the images. Challenges of this stage are (a) color change of disease, (b) illumination problem, (c) morphological changes in disease portion and (d) affected portion texture. These affect the accuracy level of the disease detection and performance degrade. Some of the techniques of segmentation are: K-means, 8-connected component analysis, region growing segmentation technique, color-based segmentation, thresholding and edge detection. Summary of segmentation-based techniques is presented in Table 1.

Table 1 Summarization of segmentation-based methods

Methods	Advantages	Limitations
K-means clustering	Easy to implement and high computational performance, suitable for large database	Performance is poor in the worst cases
Edge detection	High efficiency for high contrast disease images	Tough to implement in multiple edges environment
Otsu thresholding	It is appropriate for execution of two classes like foreground and background.	Computationally very fast
Histogram matching	Calculation complexity is low	Spatial information is ignored, and segment regions may not be adjacent
Region-based technique	Correctly separate the regions that have the same properties we define	Computationally expensive and sensitive to noise

2.3 Feature Extraction-Based Approaches

The main objective is to extract the meaningful/useful components. The features (shape, size, color, texture) of the image can be extracted from an image to fulfill the purpose. Diseases can be classified based on shape, size, color and texture of the affected leaf. Each feature has its own challenges. Out of these all features, color feature is the most important one as the most of the diseases have distinct disease color.

2.3.1 Challenges

The leaf features (color, shape, size and texture) are extracted from paddy leaf to detect the diseases. Each one has its own advantages and disadvantages. The color features are affected by outside lights and shape, and size of different diseases may be similar. Some of the techniques of feature extraction are: principal component analysis, color intensity methods, wavelet transform, color space transformation, etc.

2.4 Classifier-Based Approaches

Based on the feature extraction, the classifiers are utilized for the classification of various diseases of the paddy leaf. There are various classification techniques such as SVM [4], ANN [7], PNN [8], BPN [9], KNN [10], MDC [11], etc. Each one

Table 2 Summarization of classifiers

Classifiers	Advantages	Drawbacks
SVM	Fit to both linear and nonlinear classification Easy to implement and high accuracy	Training process is time consuming Its algorithm structure is complex It works in only two classes
ANN	It is self-adaptive Noisy data managed with higher efficiency	Issue of computing process High volume of training image needed Slow training process
KNN	Easy to implement No need of training for small data	Attracted by irrelevant data Testing step is very slow
BPN	Easy implementation Covers large number of issues	Slow learning phase Tough to predict neuron layers

has their specific benefits and drawbacks. The different classification techniques are summarized in Table 2 along with their benefits and drawbacks.

2.5 Application of Techniques in Research Work

The techniques involved in the image preprocessing are mainly image enhancement, color-based transformation and noise reduction and resizing. Green plane extraction approach based on median filtering and binarization was utilized by researchers [12]. This approach is utilized for image preprocessing phase, and features are extracted by extracting the green color components in order to highlight the affected area of the leaf. The intensity difference of the original gray scale image and green value is calculated in order to detect the infected area irrespective of the brightness and leaf age. They used median filter in order to preserve edges, to remove unnecessary spots and for proper smoothing.

Another researcher Devi et al. [13] used CIELUV color space and median filter for the preprocessing of the paddy leaves to detect the diseases in the leaves. Color space is applied to manage the color specification in the specified standard. In this work, they have used the liner mapping property of the LUV model.

Narmadha et al. [14] used histogram equalization and contrast adjustment technique for image enhancement. From the input image, occurrence matrix is generated based on the probability distribution. To remove the noise and unwanted spots, K-means technique is applied.

Another technique is based on image normalization. Firstly, R, G, B components are extracted, and mean, max (mean) is calculated. Now scaling factor for each color component is calculated by dividing each extracted mean by the maximum value. Normalized image is obtained by multiplying the R, G, B component values with the scaling factors. For segmentation, normalized RGB image is converted to YCbCr space [15].

Skeletonization technique is used to extract the shape of the infected spot. For classification of diseases, MDC and KNN classifiers are used.

Kurniawati et al. [16] used local entropy (H) threshold method and Otsu threshold method to convert RGB image to the binary image. An occurrence matrix was generated from the input image based on probability distribution (p_i) required for entropy measures. The local entropy is defined as:

$$H = -\sum_{i=1}^{n} p_i \log p_i \tag{1}$$

$$p_i = \frac{a_i}{a_1 + a_2 + \cdots + a_n} \tag{2}$$

$a_1 + a_2 + \cdots + a_n$ are the brightness levels in the central pixel windows.

Ghosh et al. [16] used color-based transformation methods to enhance the image. The techniques used to detect the diseases are color space transformation, image filtering and segmentation (K-means). They used these techniques to analyze the severity and lesions of plant leaf for feature extraction and classification. The main focus of this approach was extraction of leaf features using these mechanisms and to design the classification mechanism for disease classification.

Singh et al. [17] designed a leaf color chart (LCC) in such a way that all the shades of green color could lie in the range. They used LCC as a reference to carry out their research. They designed LCC in MATLAB based on information furnished in the brochure [18]. The shades are marked as Lightest Green, Lighter Green, Average Green, Dark Green, Darker Green and Darkest Green. Test image was converted to HSI from RGB in order to avoid any validation problem.

After conversion from RGB to HSI, it is assumed that RGB values are normalized between 0 and 1, and the angle of conversion is measured with respect to the red axis of HSI color space. Hue is normalized to [0, 1] by dividing all the values by 360°. Region of interest is found by leaf clipping. For cropping a hybrid model, 2D masking is done and then shifting to the cropping of RGB/HSI image done.

A system is developed by Chaudhary et al. [19] based on preprocessing approaches for paddy leaf disease detection. The proposed system performs in four steps: Firstly, the image is captured, and then, color transformation (HSI color space conversion), intensity adjustment and histogram analysis are performed on the captured image. Secondly, the region of interest is extracted. Thirdly, features (shape, size and color) of the segmented portion are obtained, and finally, the classification of the diseases is done by the artificial neural network (ANN) classifier.

Al Bashish et al. [20] proposed a system based on preprocessing techniques to detect and classify the leaves' lesions. The whole process is divided into four working steps. Firstly, HSI color space transformation applied on RGB leaves. Secondly, for segmentation of affected regions K-means method is used. Thirdly, a texture feature of segmented leaves is extracted, and finally, classification of the features is done by the use of a neural network classifier.

3   Comparative Analysis

In this chapter, we discussed various approaches to detect the paddy leaf diseases and based on the symptoms/features, and these diseases are classified by the use of efficient classifiers. The preprocessing step is one of the most important steps in the disease detection. Filtering, image enhancement, color space transformation and image cropping are the main components of this step. Various segmentation techniques like K-means clustering, thresholding and edge detection are used for the segmentation process. The segmentation error generally occurs due to uneven illumination of the captured images. The Otsu method is often affected by this problem. The main issue with the feature extraction is that it often extracts inaccurately the type of lesions. Based on the features of the diseases, they are classified using various classifiers like SVM, ANN, KNN/PNN classifiers.

The comparison of various approaches is presented in Table 3, based on which we can conclude that histogram equalization, median filtering, Euclidean distance-based approaches, color-based approaches, green plane extraction-based techniques and gray scaling-based approaches perform well on all major paddy diseases. It has been found that K-means clustering and area thresholding performs well if the input image is not a complex and affected region is quite visible. The feature extraction is an important factor based on which diseases can be classified. It has been observed that support vector machine (SVM) and neural network plays an important role in disease classification with maximum accuracy level. It has also been observed that feature reduction and feature selection methods are not utilized which are very much important for the outperformance of the classification techniques.

In order to minimize the size of the extracted features also, there is need to have an effective feature reduction and feature selection techniques. Also, in segmentation step to identify infected portions, simple techniques like K-means, thresholding are used that cannot handle when an image is complex or if it has a high level of brightness.

4   Conclusion

Through this survey, we have discussed various image processing and machine learning techniques to detect and classify paddy diseases. All the four steps that are involved to detect and classify diseases are well addressed along with the challenges associated with each step. Comparison of the techniques is done based on the accuracy level. We conclude that the accuracy of segmentation stage is directly influenced by the preprocessing stage. Also, it has been found that the most suitable technique for segmentation of infected leaves is K-means. Also, it is concluded that SVM and NN utilize texture features. These techniques are very much effective for the detection and classification of paddy leaf diseases. The above-mentioned techniques need to be implemented and validated with existing datasets. A new paddy leaf disease

Table 3 Comparison of various approaches

References	Year	Diseases	Techniques	Efficiency measures	Results (%)
Suman et al. [12]	2015	Brown spot, narrow brown spot, bacterial leaf blight, rice blast	Green plane extraction, median filtering threshold, 8-connected component analysis, principal component analysis, SVM classifier, confusion matrix	Accuracy	70
Devi et al. [13]	2014	Brown spot, narrow brown spot, bacterial leaf blight, rice blast	LUV color space conversion, median filtering, Region growing segmentation, mean shift segmentation,	Structural component, Normalized cross correlation, Peak signal-to-noise ratio	
Narmadha et al. [14]	2017	Blast, brown spot, narrow brown spot	Histogram equalization, Occurrence matrix, K-mean technique, ANN, fuzzy logic, SVM	Accuracy	94.7
Joshi et al. [15]	2016	Rice bacterial blight, rice blast, rice brown spot, Rice sheath rot	Color space conversion, Euclidean distance technique, color moments, skeletonization technique, MDC and KNN classifiers, Manhattan distance	Accuracy	89.23

(continued)

Table 3 (continued)

References	Year	Diseases	Techniques	Efficiency measures	Results (%)
Ghosh and Dubey [16]	2013	Anthracnose	Color space conversion, clustering (K-means), gray-level co-occurrence matrix, SVM	Accuracy	96
Singh et al. [17]	2015		Leaf color chart technique, color space conversion, hybrid model of cropping, histogram, Chi-square goodness of fit test,	Accuracy	100
Chaudhary et al. [19]	2012	Citrus canker	Color transformation, edge detection, ANN		
Al Bashish et al. [20]	2011	Frogeye leaf spot, brown spot, blueberry leaf spot	Median filter, color space conversion, Otsu method		

detection technique can be proposed including effective feature reduction and feature selection techniques.

References

1. Gutte, V.S., and M.A. Gitte. 2016. A survey on recognition of plant disease with help of algorithm. *International Journal of Engineering Science* 7100.
2. Scardaci, S.C., R.K. Webster, C.A. Greer, J.E. Hill, J.F. Williams, R.G. Mutters, and J.J. Oster. 1997. Rice blast: A new disease in California. *Agronomy Fact Sheet Series* 2: 1–2.
3. Kobayashi, T., E. Kanda, K. Kitada, K. Ishiguro, and Y. Torigoe. 2001. Detection of rice panicle blast with multispectral radiometer and the potential of using airborne multispectral scanners. *Phytopathology* 91 (3): 316–323.
4. Hearst, M.A., S.T. Dumais, E. Osuna, J. Platt, and B. Scholkopf. 1998. Support vector machines. *IEEE Intelligent Systems and their applications* 13 (4): 18–28.
5. Gonzalez, C.R., E.R. Woods, and L.S. Eddins (eds.). 2001. *Digital image processing*. Pearson Education.
6. Zhang, M. 2009. *Bilateral filter in image processing*. Master's Thesis, Louisiana State University, USA.

7. Schalkoff, R.J. 1997. *Artificial neural networks*, vol. 1. New York: McGraw-Hill.
8. Specht, D.F. 1990. Probabilistic neural networks. *Neural networks* 3 (1): 109–118.
9. Buscema, M. 1998. Back propagation neural networks. *Substance Use and Misuse* 33 (2): 233–270.
10. Mucherino, A., P.J. Papajorgji, and P.M. Pardalos. 2009. K-nearest neighbor classification. *Data Mining in Agriculture*, 83–106. New York: Springer.
11. Hodgson, M.E. 1988. Reducing the computational requirements of the minimum-distance classifier. *Remote Sensing of Environment* 25 (1): 117–128.
12. Suman, T., and T. Dhruvakumar. 2015. Classification of paddy leaf diseases using shape and color features. *International Journal of Electronics and Electrical Engineering* 7 (01): 239–250.
13. Devi, D.A., and K. Muthukannan. 2014. Analysis of segmentation scheme for diseased rice leaves. In *International conference on advanced communication control and computing technologies (ICACCCT)*, May 2014, 1374–1378. IEEE.
14. Narmadha, R.P., and G. Arulvadivu. 2017. Detection and measurement of paddy leaf disease symptoms using image processing. In *International conference on computer communication and informatics (ICCCI)*, January 2017, 1–4. IEEE.
15. Joshi, A.A., and B.D. Jadhav. 2016. Monitoring and controlling rice diseases using image processing techniques. In *International conference on computing, analytics and security trends (CAST)*, December 2016, 471–476. IEEE.
16. Ghosh, S., and S.K. Dubey. 2013. Comparative analysis of k-means and fuzzy c-means algorithms. *International Journal of Advanced Computer Science and Applications* 4(4).
17. Singh, A., and M.L. Singh. 2015. Automated color prediction of paddy crop leaf using image processing. In *International conference on technological innovation in ICT for Agriculture and Rural Development (TIAR)*, July 2015, 24–32. IEEE.
18. Kular, J.S. 2014. *Package of practices for Kharif Crops of Punjab*. Punjab Agriculture University Ludhiana. https://www.pau.edu/. Accessed December 25, 2014.
19. Chaudhary, P., A.K. Chaudhari, A.N. Cheeran, and S. Godara. 2012. Color transform based approach for disease spot detection on plant leaf. *International Journal of Computer Science and Telecommunications* 3 (6): 65–70.
20. Al Bashish, D., M. Braik, and S. Bani Ahmad. 2011. Detection and classification of leaf diseases using K-means-based segmentation and Neural- networks-based Classification. *Information Technology Journal* 10 (2): 267–275.

A Comparative Study of Classifiers for Extractive Text Summarization

Anshuman Pattanaik, Sanjeevani Subhadra Mishra and Madhabananda Das

Abstract Automatic text summarization (ATS) is a widely used approach. Through the years, various techniques have been implemented to produce the summary. An extractive summary is a traditional mechanism for information extraction, where important sentences are selected which refers to the basic concepts of the article. In this paper, extractive summarization has been considered as a classification problem. Machine learning techniques have been implemented for classification problems in various domains. To solve the summarization problem in this paper, machine learning is taken into consideration, and KNN, random forest, support vector machine, multilayer perceptron, decision tree and logistic regression algorithm have been implemented on Newsroom dataset.

Keywords Text summarization · Extractive · Sentence scoring · Machine learning

1 Introduction

A compact version of the original text which produces the same concept as the original document is known as summary. In 1958, Luhn introduces the concept of abstract generation out of text data. That gives boost to the idea of automatic text summarization [1, 2] (ATS). In recent years, high availability of text data helps in growth of natural language processing, especially in the field of text summarization. Summarization can be of two categories, such as abstractive and extractive. Extractive text summarization is one of the oldest and widely used approaches among researchers. The idea behind extractive summarization is to extract the sentences

A. Pattanaik (✉) · S. S. Mishra · M. Das
School of Computer Engineering, Kalinga Institute of Industrial Technology (Deemed-to-be University), Bhubaneswar, India
e-mail: anshumanpattanaik21@gmail.com

S. S. Mishra
e-mail: sanjeevani321@gmail.com

M. Das
e-mail: mndas_prof@kiit.ac.in

© Springer Nature Singapore Pte Ltd. 2020
D. Swain et al. (eds.), *Machine Learning and Information Processing*,
Advances in Intelligent Systems and Computing 1101,
https://doi.org/10.1007/978-981-15-1884-3_16

173

from the document in a way that those sentences will represent the core idea of the document [2]. Extractive summary is the basic summarization technique. Sentences are selected in the basis of some scores and ranks. Scoring and ranking of sentences are done by feature mapping and selection. Features can be of different types such as frequency-based and prediction-based. Frequency-based features are more widely used in extractive summarization. Summarized data can be helpful in various fields. A summarized document helps in understanding the whole document in less amount of time. One can find relevant document from a query search more faster by going through summaries. In case of scientific data or medical data, one can easily produce a report through the summaries. Summaries can help in creating proper indexing of multiple documents in much lesser time. Several methods have been implemented over the years for creating improved summaries. Machine learning is one of the best suited methods for resolving classification and clustering-based problem nowadays according to the state-of-the-art approaches on different fields. This work presents a technique for extractive summarization as a classification problem instead of an information extraction problem. Extractive summary has been considered as a two-class problem where each sentences can be considered as either 0 or 1. If the sentence is not selected for the summary, then it is classified as 0 else 1. Different machine learning techniques have been implemented on the same dataset, and comparisons have been made.

Rest of the paper is organized as follows: Sect. 2 represents related works, Sect. 3 explains the proposed method, Sect. 4 presents experimental setup and result analysis. Finally, we have concluded our work in Sect. 5.

2   Related Works

Automatic text summarization (ATS) is a technique to generate summary from a provided text data. Statistical algorithms, graph-based algorithms, etc., are used by ATS to generate the desired summary. These multiple algorithms use specified mathematical models and computational devices. ATS has a wide range of applications, larger diversification and is quite reliable in generating the requisite summary. Hence, it has attracted the attention in the field of research and development. Foreseeing the advantages, the researchers have invested a great effort in modifying and developing ATS techniques.

In extractive summarization, sentence selection is the main criteria. Frequency-based methods and predictive methods are available for sentence selection. In 2014, Meena and Gopalani [3] gave an analysis on different frequency-based features such as term frequency, TF-IDF, sentence location, title similarity, proper noun, word co-occurrence, numerical values in sentences and sentence length. In their work, they have concluded that in most cases TF-IDF, word co-occurrence, sentence length and location give better performance together. Researchers can consider this in case of extractive summarization as a combined parameter. However, more number of different combinations can be taken into consideration.

Pattanaik et al. [4] considered the extractive summarization as an optimization problem. The objectives are to find an optimized solution for the given text considering high coverage of context and lower redundancies between the output sentences. BAT algorithm outperforms the existing model in their experiment. They have taken TF-IDF and sentence similarity as their sentences selection features.

Machine learning algorithms have been amazing when it comes to prediction or classification of any data. Naïve Bayes classifier, support vector machine, kernel support vector machine, decision tree, logistic regression, etc., techniques are widely used in different fields for prediction.

Joachims [5], in 1998, introduced a classification mechanism to text categorization. Text categorization is a standard classification problem where input data is text articles, and output is a set of categories out of which text data will fall into either one or many categories. So, this dataset is not linearly separable. Author explained that the concept of high dimensionality of feature vector is the factor for which general machine learning mechanism fails to achieve the desired output. Support vector machine transcends the state-of-the-art techniques.

3 Proposed Method

In this work, extractive summarization is treated as a two-class classification problem, where each sentence of the document either falls under 0 or 1 class. If a sentence is not selected in the summary, it is considered as class 0 and as class 1 if selected. Each sentence of the document goes through the classifier for prediction. In this work, different machine learning classification models are tested over the same dataset. The proposed model work flow is explained in Fig. 1.

The input document goes through preprocessing. In this phase, highly unstructured text data goes through data cleaning (stemming, tokenizing, stop word removal). After that, the cleaned data is converted to numerical form with different frequency-based scoring. Numerically presented data flows through the machine learning model for training and testing. TF-IDF [6, 7], keywords [8] and sentence length are the parameters taken into consideration for the feature selection. TF-IDF is one of the sentence scoring methodologies where it focuses on the rare terms that appear in any document and have some more weight regarding the query. For single document, it calculates from term frequency inverse sentence frequency (TF-ISF). For multiple document, it uses term frequency inverse document frequency (TF-ID).

$$TF - IDF(\omega) = \frac{tf}{(tf + 1)} \log\left(\frac{N}{df}\right) \qquad (1)$$

where tf is term frequency, N is for number of total document, and df stands for document frequency. Higher the tf/idf value rarer the term in the document and has more weight for term occurrence in summary section. In the ML modeling phase, every sentence of the document is first trained with the model. The labeled data,

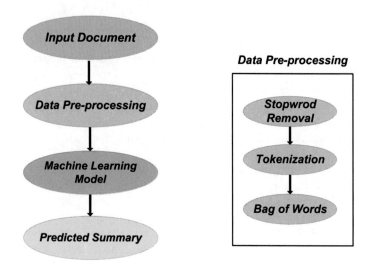

Fig. 1 Workflow of proposed model

i.e., the human generated extractive summaries are the extracted sentences from the original text. It can be considered as two class problem with class value 0 and 1. The value 0 refers to the sentences those are not in the summary where as, value 1 presents sentences available in human generated summary.

In case of machine learning modeling, K-nearest neighbor (KNN), random forest, support vector machine, multilayer perceptron, decision tree and logistic regression models have been implemented. KNN algorithm works on grouping approach by considering the k-nearest majority voting. In case of support vector machines, it handles large dimensions more precisely and also removes irrelevant features easily as explained by Joachims [5] in 1998. Decision tree classifies categorical values more accurately. Multilayer perceptron is a feed forward network which trains through back propagation.

4 Experimental Setup and Result Analysis

Generating extractive summary from the text data is a high computing task when it involves machine learning algorithm. Table 1 elaborates about every specification that author have taken into consideration. Newsroom [9] is one of the widely used dataset for text summarization. This dataset consists 1.3 million news articles and associated summaries by various authors and editors. Extractive summary articles are taken into account, and experiment has been done on these data as explained in Sect. 3. In this dataset, extractive, abstractive and mixed categories of summaries are available in different volumes. In this work, only extractive summaries are taken into account. K-nearest neighbor, random forest, support vector machine, multilayer

Table 1 Experimental setup

Hardware specification	Programming language used	Software specification	Dataset	Sentence selection feature	Machine learning classifier
Intel-core i5 7th generation processor 2.71 GHz clock cycle 32 GB RAM 1 TB HDD	Python 3.7	Windows 10 64 bit Anaconda Distribution 64 bit	NewsRoom [9]	TF-IDF Keyword Sentence length	KNN Random forest SVM with RBF MLP Decision tree Logistic regression

perceptron, decision tree and logistic regression classifiers have been implemented. In Table 1 experimental setup, software used and models used are given for better understanding of the work.

In K-nearest neighbor, ten neighbors are taken into consideration with uniform weights. In random forest 1000, 10,000 n_estimators are taken into consideration with 'gini' criterion for splitting quality of the data is used. In decision tree, "entropy" is used as criterion function. In support vector machine, radial basis function is used as kernel function which focuses on worst case class splitting. Multilayer perceptron classifier is used with 100 hidden layers, having RELU activation function with stochastic gradient-based optimizer for the back propagation and weight updating solver. A classification problem analyzes by its classification report [10] which includes precision, recall and F1-score values. Precision is the score which indicates what percent of your prediction was correct. Recall indicates how much positive prediction is done by the classifier. F1-score can be calculated by taking weighted harmonic mean of recall and precision. Table 2 illustrates classification scores for each class. 0 indicates the sentences are not in the summary, and 1 indicates sentences are in the summary.

True Negative (TN): originally negative and predicted negative
True Positive (TP): originally positive and predicted positive
False Negative (FN): originally positive but predicted negative
False Positive (FP): originally negative but predicted positive

$$\text{Precision} = \frac{TP}{(TP + FP)} \tag{2}$$

$$\text{Recall} = \frac{TP}{(TP + FN)} \tag{3}$$

$$F1\text{-Score} = \frac{2 * \text{Precision} * \text{Recall}}{(\text{Precision} + \text{Recall})} \tag{4}$$

Table 2 Classification report for K-nearest neighbor, random forest, support vector machine, multilayer perceptron, decision tree and logistic regression classifiers

		0	1
K-nearest neighbor (KNN)	Precision	0.78	0.75
	Recall	0.92	0.49
	F1-score	0.84	0.59
Random forest (RF)	Precision	0.80	0.61
	Recall	0.82	0.58
	F1-score	0.81	0.60
Support vector machine (SVM)	Precision	0.78	0.74
	Recall	0.92	0.58
	F1-score	0.84	0.58
Multilayer perceptron (MLP)	Precision	0.80	0.75
	Recall	0.91	0.54
	F1-score	0.85	0.63
Decision tree (DT)	Precision	0.78	0.57
	Recall	0.78	0.57
	F1-score	0.78	0.57
Logistic regression (LR)	Precision	0.78	0.74
	Recall	0.92	0.47
	F1-score	0.84	0.58

In this paper, six different types of classifiers were being examined. Multilayer perceptron classifier seems to perform better in terms of precision, recall and f1-score for class 1. Multilayer perceptron has the advantage of hidden layer which provides some meta-data to the classifier, and the back propagation mechanism keeps the classifier more accurate. In the following tables, i.e., from Tables 3, 4, 5, 6, 7 and 8, confusion matrix values for all the six classifiers. Confusion matrix gives the brief idea about how much data predicted correctly.

Receiver operating characteristic (ROC) [11] curve is a widely used curve to analyze the performance of classifiers. ROC is a curve between true positive rate and false positive rate of any classifier. The curve having more area under the curve

Table 3 Confusion matrix: KNN

	Predicted 0	Predicted 1
Actual 0	402	36
Actual 1	112	107

Table 4 Confusion matrix: RF

	Predicted 0	Predicted 1
Actual 0	358	80
Actual 1	93	126

Table 5 Confusion matrix: SVM

	Predicted 0	Predicted 1
Actual 0	402	36
Actual 1	114	105

Table 6 Confusion matrix: MLP

	Predicted 0	Predicted 1
Actual 0	399	39
Actual 1	101	118

Table 7 Confusion matrix: DT

	Predicted 0	Predicted 1
Actual 0	343	95
Actual 1	94	125

Table 8 Confusion matrix: LR

	Predicted 0	Predicted 1
Actual 0	402	36
Actual 1	115	104

(AUC) is better than others. In Fig. 2, ROC curve of different classifiers is given. Multilayer perceptron classifiers cover more area under the curve as compared to other classifiers. KNN, random forest, logistic regression and support vector machine are having same AUC value.

5 Conclusion and Future Work

In this paper, the authors considered the extractive text summarization problem as a classification problem. A document is classified under a two-class problem. Class 0 indicates that the sentences in the document are not considered as summary sentences. Class 1 is the class of summary sentences. TF-IDF, keywords and sentences length are taken into consideration for evaluation. K-nearest neighbor, random forest, support vector machine, multilayer perceptron, decision tree and logistic regression classifiers have been implemented on Newsroom dataset. The experimental analysis of the algorithms with precision, recall, f1-score and confusion matrix of all the classifier is mentioned in Tables 2, 3, 4, 5, 6, 7 and 8. Confusion matrix is calculated over the test data, which indicates the polarity of data according to the classifiers. The ROC curve is also plotted. All the analyses state that although above classifiers do not satisfy the goal more accurately, and out of them multilayer perceptron classifier gives better result. MLP has hidden layers and back propagation principle which

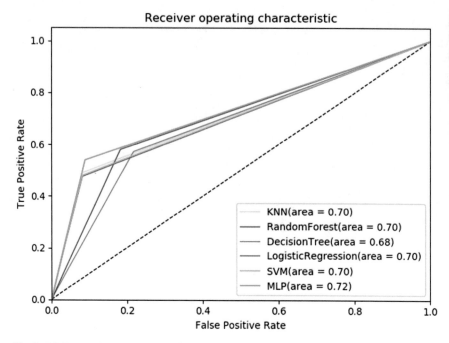

Fig. 2 ROC curve for different classifiers

enhances the quality of the classifier. In this work, MLP provides 72% accuracy. This work can be considered for future work by modifying different parameters and adding more features to the input vectors.

References

1. Luhn, H.P. 1958. The automatic creation of literature abstracts. *IBM Journal of Research and Development* 2 (2): 159–165.
2. Gambhir, M., and V. Gupta. 2017. Recent automatic text summarization techniques: A survey. *Artificial Intelligence Review* 47 (1): 1–66.
3. Meena, Y.K., and D. Gopalani. 2014. Analysis of sentence scoring methods for extractive automatic text summarization. In *Proceedings of the 2014 international conference on information and communication technology for competitive strategies*, November 2014, 53. ACM.
4. Pattanaik, A., S. Sagnika, M. Das, and B.S.P. Mishra. 2019. Extractive summary: An optimization approach using bat algorithm. *Ambient communications and computer systems*, 175–186. Singapore: Springer.
5. Joachims, T. 1998. Text categorization with support vector machines: Learning with many relevant features. In *European conference on machine learning*, April 1998, 137–142. Springer, Berlin, Heidelberg.
6. Nobata, C., S. Sekine, M. Murata, K. Uchimoto, M. Utiyama, H., and Isahara. 2001. Sentence extraction system assembling multiple evidence. In *NTCIR*.

7. Jafari, M., J. Wang, Y. Qin, M. Gheisari, A.S. Shahabi, and X. Tao. 2016. Automatic text summarization using fuzzy inference. In *22nd International conference on automation and computing (ICAC)*, September 2016, 256–260. IEEE.

8. Matsuo, Y., and M. Ishizuka. 2004. Keyword extraction from a single document using word co-occurrence statistical information. *International Journal on Artificial Intelligence Tools* 13 (01): 157–169.

9. NewsRoom Dataset Available (2017) Cornell Newsroom. https://summari.es. 2017.

10. Powers, D.M. 2011. *Evaluation: From precision, recall and F-measure to ROC, informedness, markedness and correlation.*

11. Davis, J., and M. Goadrich. 2006. The relationship between precision-recall and ROC curves. In *Proceedings of the 23rd international conference on machine learning*, June 2006, 233–240. ACM.

Capabilities of Chatbots and Its Performance Enhancements in Machine Learning

Mahendra Prasad Nath and Santwana Sagnika

Abstract To keep in pace with ever-increasing customer demands, providing instant and useful responses is a prominent need of service providers. Latest technical developments have led to the advent of a faster, easier solution: chatbot. It is an artificial intelligence-based simulation of human conversation to automate customer interactions, thereby reducing manual effort. Not necessarily limited to answering simple product-related queries, chatbots can provide complex predictive analysis within limited response time, by the help of machine learning. Creating a chatbot has become simplistic enough, even for any non-technical person. It can be configured and integrated into any messenger within an organization or social network. This paper discusses the capabilities of chatbots and the enhancements in their performance by the contribution of machine learning.

Keywords Chatbot · Artificial intelligence (AI) · Natural language processing (NLP) · Machine learning (ML) · Service registry

1 Introduction

An interested customer wants to know more about a certain company products. The customer care will be called by him, or a few sites will be surfed by him to get the information, which is very tedious, and all the required information may not be gotten by him which is looking by him. Another scenario can be a situation where an internal business person is trying to reach IT through any channel, for example, by raising an incident/ticket for any technical issue that he might be facing. Since all this is manual

M. P. Nath (✉)
Cognizant Technology Solutions India Pvt. Ltd., Manyata Embassy Business Park,
Bangalore, Karnataka, India
e-mail: mahendranath272@gmail.com

S. Sagnika
School of Computer Engineering, Kalinga Institute of Industrial Technology (Deemed to be University), Bhubaneswar, Odisha, India
e-mail: santwana.sagnika@gmail.com

© Springer Nature Singapore Pte Ltd. 2020   183
D. Swain et al. (eds.), *Machine Learning and Information Processing*,
Advances in Intelligent Systems and Computing 1101,
https://doi.org/10.1007/978-981-15-1884-3_17

effort by customer/customer facing business team and the resolution/response is also dependent on human interaction and reaction, the turnaround time is subject to conditions. In this fast-paced world, customer might get impatient with the wastage of time. An instant response tool which simulates human interaction by a computer program can be a time savior and a helping hand named chatbot, which is a chatting mechanism like chatting with human. A chatbot is nothing more than a computer program which helps to maintain a simple conversation with a user/person/client. It helps us save time and allows focusing on those parts of our jobs that are really of benefit: understanding the business issues, designing powerful consumer insights, and bringing those to life [1, 2].

Implementation and Architecture

Architecture and implementation is made easy by the available open-source natural language processing (NLP) and messengers that are available in the industry. NLP is the main component which analyzes text allowing machines to understand how human' speak. In our implementation, we have used API.AI as the NLP and Skype for Business as the messenger [3, 4]. A messaging server musts interact with the API.AI (available open-source NLPs) to decide what action must be taken. Once the action is selected, it calls respective Java Service from the registry which in turn calls the desired application to get the response. NLP helps define a machine's capability to accept user's input, breaks it down to required format for comprehending its meaning, decides appropriate action required, and then responds to the user with appropriate language that user will able to understand. NLP is the part of machine learning layer in chatbot [4] (Fig. 1).

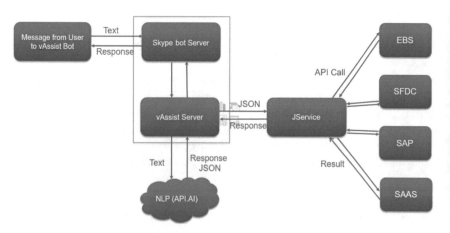

Fig. 1 Technical architecture diagram

2 Components of Chatbot Architecture

i.   Messenger (Client)
ii. API.AI (NLP)
iii. Java Service
iv. APIs from respective applications.

In the above technical architecture diagram, vAssist acts as service registry. Service registry simply works as a phonebook. It lets clients/end users search for different services using their logical names. There are a number of service registries available these days. But out of all those service registries, Netflix's Eureka, HashiCorp Consul, and Apache Zookeeper are more popular. Spring Cloud's DiscoveryClient abstraction offers a handy client-side API to work with service registries. A major benefit of using a service registry is load balancing on the client side. This factor allows clients to select from among the registered instances of a service. Ribbon is an efficient load balancer for the client side by Netflix. Spring Cloud integrates Ribbon and applies it automatically at various layers of the framework, be it is the RestTemplate, declarative REST clients which are powered by Netflix's Feign, or the Zuul microproxy.

Messenger (client):

Messenger can be any company wide messenger or a social messenger like Skype for Business, Facebook, Slack, etc. [4, 5].

Components of Messenger (Skype for Business):

We are using Skype for Business as messenger, and below is the high-level architecture of the same.

i.   Skype server
ii. Application server
iii. Windows service (Fig. 2).

Skype server:

A trusted end point has to be created on Skype server, and it can reside on multiple nodes.

Application server:

A trusted application has to be created and mounted on application server. A trusted channel is to be established between end point and application to keep the data in sync.

Windows service:

Windows service comprises UCMA and collaboration platform.

1. UCMA is a Skype plug-in to enable chatbot.
2. Collaboration platform is used to maintain the status of the application user as online always.

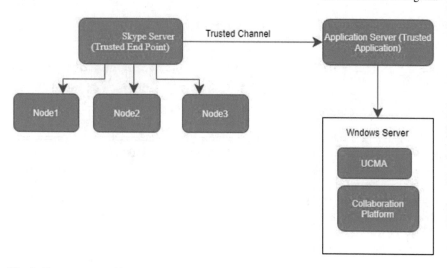

Fig. 2 Skype server architecture diagram

Links given in references section can be used for Skype for Business implementation.

API.AI (Dialogueflow.com)

API.AI is a platform to build interfaces that facilitate conversation between bots, devices, and applications. This provides input and output methods and can react to actionable data. An optional web hook can be implemented for connection to web services. The web service becomes able to call external APIs, performs business logic, or accesses data stores [3, 4].

API.AI accepts input data in the form of a query. A query is in the form of either an event name or text in natural language. API.AI converts the query text into actionable data and then returns a JSON response object as output data. This process of transforming natural language into actionable data is called natural language understanding (NLU).

Terminology in API.AI (Dialogueflow.com)

a. **Intents**:

Intent is a representation of a mapping between the statement of a user and action needed to be taken by the software. Standard intent interfaces consist of the following sections:

• User says
• Action
• Response
• Contexts.

b. **Action**:

An action refers to what the application will do after identifying the intent based on user input.

c. **Parameters**:

Parameters can be identified as the input values extracted from the user text, for the action.

d. **Context**:

Context is used to make system remember topic of conversation in order to link multiple intents. Each intent can be linked with other using input and output contexts.

e. **Agents**:

Grouping of different context and intents. Allows you to set language, time zone, theme, and other settings for your bot.

f. **Talent Manager (vAssist)**:

Talent Manager talks to API.AI to get the output JSON that is generated with all the information related to the question user has asked along with intent, context, parameters and transforms into more readable format. It maintains the registry with a list of Java Services for various applications. After reading the intent, it calls appropriate Java Services according to the mapping for further action. This also receives the output from the Java Service and passes it to the messenger as a response to the user via API.AI. The Talent Manager setup can be skipped, and messenger can directly interact with API.AI to call the Java Service; however, having the repository of java services in internal layer ensures data integrity.

g. **Java Service**:

The Java Service determines which application to call based on the intent and the inputs that are received from API.AI. Each application can have a different JS and can be configured to be called accordingly. It reads the received JSON file and parses the values to route it to the appropriate API. This also reads the output that the application API sends and returns it to the messenger.

Application Layer:

Code residing on the application layer performs the task and returns a response to the java service (Fig. 3).

3 Machine Learning and Chatbots

The most well-known thing in this day and age is that you will almost certainly hear an alternate prologue to the specialized field on regular routine. The greater part of this innovation is encompassing information as it were. It is an extensive field. It

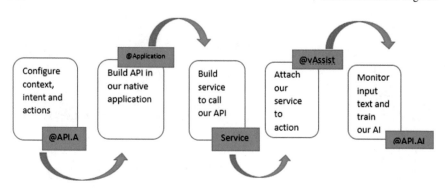

Fig. 3 Steps for implementation

is very difficult to leave it without implementing computer knowledge. One such pertinent expansion is machine learning.

Connection of machine learning with chatbot

Prior to finding any alternate solution, we ought to appreciate the possibility of chatbot. Well-known products are Cortana, Alexa, and Siri. Many are using these to set an alert call and send a text message.

Individuals are utilizing it for their daily schedule errands without really looking through anything or playing out a manual assignment. Only an order and your work are finished. In spite of this, it is hard to compel the word out of these effusive operators about the mutual or sensible subjects.

Everything considered, we get to know how AI is used nowadays. We can use the concept of natural language processing and machine learning in improving the functionality of chatbots [6, 7].

It is necessary to collaborate artificial intelligence with different features of machine learning to a particular language which we can use comprehension and translation with the existing and upcoming frameworks. This area is now popularly known as spoken discourse frameworks, exchange frameworks, or chatbots. The aim of our study is to give a concrete answer, discuss the scenario, and use human behavior perfectly.

It is not certain to ignore the strength and diverging capabilities of robots. Currently, in software world, we have two kinds of chatbot.

1. General conversation-based chatbots (e.g., Microsoft Taybot)
2. Objective situation-based chatbots (e.g., Cortana, Alexa, and Siri).

The objective organized is made to deal with the issues of clients using standard language, while across the board discourse is utilized to exchange with individuals on a tremendous stage with two or three subjects. The discussion structure really relies on the neural systems with many moved ways. Through this paper, we will take a few to get back some composure tight how AI is completed to make better chatbot models [8, 9].

1. Procreative and selective models

Generally, the conversational model is divided into two distinct sorts, out of which, one is selective and the other is generative models. Be that as it may, at times the hybrid model is, moreover, used for improving the chatbot. A primary concern is that these frameworks consolidated into the regular inquiry individuals may pose and their answers. It is fundamentally a considered method for putting away discourse setting and just anticipating the proper responses in the individual circumstance.

Basically, the words embedded are experienced the framework instead of the IDs which are commonly delineated as a gathering content eaten up by the framework or the words are given up to the RNN [10].

2. Dialogue data depiction

By and by we may confounded in transit information is genuinely tended to and what kind of dataset is utilized. In any case, it is incredibly basic; a huge amount of information is framed that contains setting and answer. The setting is any condition or a development of request that needs an answer quickly by the framework. Regardless, the requests are a structure of tokens that are utilized to structure its language.

It is a course of action of a regular talk held between two individuals. A social occasion is framed that shows the sentences and their answers, for example,

- Hi! How are you?
- Hi! I am great. What's more, you?
- I am great as well. What is your age?
- I am twenty years of age. What's more, you?
- Me as well!

<eos> token is utilized near the culmination of every line that suggests "part of the bargain." This is utilized to make it fundamental for the neural structure to get a handle on the sentences obstacles and associates in studying inside state brilliantly. Be that as it may, it is conceivable that metadata is utilized for some data in the models, for example, feelings, sexual orientation, speaker ID, and so forth [11].

Since we think about the key, then we can continue forward to a generative model, generally called neural conversational model. An exchange is conveyed as arrangement to succession structure that is effectively finished with the assistance of machine interpretation documented that aids in the simple adjustment of issues. This model incorporates two distinctive recurrent neural networks (RNNs) with totally various arrangements of criterion. One family is called as encoders, and another is called as decoders. The primary work of encoders is to grasp the setting grouping tokens individually. Exactly when the entire strategy is done, then the secured state is passed on that aide in the direct hardening of the setting to make answers. Regardless, the decoder functions for the most part take the delineation of the setting gone by the encoder, and a brief span later makes the sensible answer. It includes a layer of softmax present with the language that aids in dealing with RNN. This works by accepting the concealed state and equipping out the distributive response. Here are experiences in which the basic age happens [10, 11].

In any case, this framework is a touch of the model interface, yet there is an other way like model arranging part that is utilized to pick every single movement. We ought to just use right word, i.e., y_t. The basic spotlight is on getting the benefit next word each time. We can even imagine the join word by giving a prefix. After a short time, nothing is impeccable even in the headway world. The thing has some hindrance that makes human to continue refreshing and altering them. The basic issues in the generative models are as referenced underneath:

- It gives a standard reaction. For example, the likelihood of getting "yes", "no", "okay", and so on at a wide stage is conceivable. In any case, there are individuals that are attempting to change this. The fundamental issue to regulate is to change the target furthest reaches of the model interface. Something else is that it is essential to display the fake metric that will be utilized as a reward.
- Something different that can be seen is a conflicting reaction to the solidified point of view. The originators are attempting to overhaul the model with a further made encoding state that can cause an answer and that to can no ifs, ands, or buts do speaker embeddings.

After we are finished with the generative model, now we have to comprehend what selective neural conversational model is and in what manner may it work. The specific model works on clear breaking point, for example, answer, setting, and word (w). There are, on an exceptionally fundamental level, two towers each distributive to "setting" and "answer." These two have specific structures. The working is fundamental, one zenith takes in the information and the count is done in the vector space. The fitting response that sounds coherently immaculate as shown by the setting is the picked answer.

This all functions with the triplet hardship, for example, setting, reply_correct, and reply_wrong. The fitting reaction wrong is basically utilized for the negative model which is only some answer in the most phenomenal likelihood answer. Through this, we get the arranging of cutoff that is not much educational. The best possible response with the best score will be picked as the significant answer.

Intent- versus Flow-Based Chatbot Communication

Intent-Based Bot

An intent-based communication implies that the bot will comprehend the client questions by consolidating two information sources—state and context. State alludes to assessing the visit history, while context alludes to investigating contributions from outer information focuses. For example, if a client says "Book a Dr's. appointment," the bot can take logical data like area and state data like talk history in regard to conceivable sickness conditions to recommend fitting specialists.

Intent-based bots unravel client inquiries with respect to a coordinated premise. With each addressed inquiry, it adjusts to the client conduct. The degree of insight increments as these bots gets more information. Mainstream instances of intent-based bots are Google Assistant, Siri, Amazon Alexa, and so forth [7].

Flow-Based Bot

A flow-based bot centers with respect to settling client questions such that it moves them bit by bit to an answer. The stream is plainly characterized remembering the conceivable client concerns. For example, a stream-based bot of a fabric retailer would move this way:

- Ask for gender preference and event (formal, easygoing, or partywear).
- Ask the client about what he is searching for—shirt or pants.
- Showcase a few examples—in the event that he/she picks something, at that point move to checkout, else proceed with more examples.
- If even after rehashed cycles there is no decision made, at that point give the client the alternative to continue to the versatile application.

This is an extremely precise way to deal with bot advancement as it requires negligible improvement endeavors and is simpler to shape. A lion's share of e-commerce chatbots are stream based like H&M Bot, Wholesome Bot, and so on. In the event that the stream is little and essential, you can even decide on a no code choice.

4 Benefits of Chatbots

1. Chatbots once created will available to customers or end users 24*7.
2. It helps to keep up with the modern trends as it uses simple messaging platform. Nowadays, smart phone users do not download any new app in a month as users have their core apps like Facebook, Instagram, WhatsApp, etc.
3. Using chatbots, customer service is improved a lot. These are always available for customer support. Chatbots interact with customers proactively.
4. Chatbots make the engagement of customers more interactive with products by great sense of humor.
5. Chatbots are great tools to communicate with customers. With the feedback collected with simple questions, the service can be improved.
6. Chatbots make thing easier to approach global market.
7. Chatbots help in cost saving. Creating, implementing a fully functional chatbot is cheaper than creating cross-platform apps and having employees to support those apps.

5 Drawbacks of Chatbots

1. Chatbots are not suitable for all business. These are suitable for predictable and stereotyped conversation such as pizza delivery service, taxi retails, and travel companies. But in Business-to-Business (B2B) companies, direct client interaction needed; there chatbots do not work with 100% efficiency.
2. Chatbots have limited audience.

3. Sometimes chatbots are confused with user's request. As they are not human beings, they cannot improvise with information when it is needed.

6   Conclusion

Artificial intelligence and machine learning are two confounding terms. Man-made brainpower is the investigation of preparing machine to impersonate or recreate human assignment. A researcher can utilize distinctive techniques to prepare a machine. Toward the start of the AI's ages, software engineers composed hardcoded programs, that is, type each coherent probability the machine can face and how to react. At the point when a framework develops complex, it ends up hard to deal with the tenets. To conquer this issue, the machine can utilize information to figure out how to deal with every one of the circumstances from a given situation. The most vital highlights to have a great AI are to have enough information with extensive heterogeneity. For instance, a machine can learn diverse dialects as long as it has enough words to gain from.

References

1. Dialogflow. https://docs.api.ai/. Accessed August 10, 2019.
2. https://www.slideshare.net/DoYouDreamUP/chatbots-use-benefits-and-key-success-factors. Accessed July 15, 2019.
3. Nath, M.P., S. Sagnika, M., Das, and P. Pandey. 2017. Object recognition using cat swarm optimization. *International Journal of Research and Scientific Innovation (IJRSI)* IV(VIIS): 47–51.
4. Nath, M.P., K. Goyal, J. Prasad, and B. Kallur. 2018. Chat Bot—an edge to customer insight. *International Journal of Research and Scientific Innovation (IJRSI)* V(V): 29–32.
5. Skype for Business bot using UCWA.F5. https://ankitbko.github.io/2017/02/Sykpe-For-Business-Bot-Using-UCWA/. Accessed August 3, 2019.
6. Das, K., and R.N. Behera. 2017. A survey on machine learning: Concept, algorithms and applications. *International Journal of Innovative Research in Computer and Communication Engineering* 5 (2): 1301–1309.
7. Witten, I.H., and E. Frank. 2005. *Data mining: Practical machine learning tools and techniques*, 2nd ed. San Francisco: Morgan Kaufmann.
8. Karp, N.B., and D. Song. 2006. Paragraph: Thwarting signature learning by training maliciously. In *Recent advances in intrusion detection (RAID), (LNCS 4219)*, ed. D. Zamboni and C. Kruegel, 81–105. Berlin: Springer.
9. Mitchell, T., et al. 2014. Never-Ending Learning. *Communications of the ACM* 61 (5): 103–115.
10. Mitchell, T.M. 1997. Does machine learning really work? *AI Magazine* 18 (3): 11–11.
11. Carlson, A., J. Betteridge, B. Kisiel, B. Settles, E.R. Hruschka, and T.M. Mitchell. 2010. Toward an architecture for never-ending language learning. In *Twenty-Fourth AAAI conference on artificial intelligence*.
12. Chatbot on Skype for Business. http://acuvate.com/blog/chatbots-on-skype-for-business/. Accessed July 31, 2019.

A Deep Learning-Based Approach for Predicting the Outcome of H-1B Visa Application

Anay Dombe, Rahul Rewale and Debabrata Swain

Abstract The H-1B is a visa that allows US employers to employ foreign workers in specialty occupations. The number of H-1B visa applicants is growing drastically. Due to a heavy increment in the number of applications, the lottery system has been introduced, since only a certain number of visas can be issued every year. But, before a Labor Condition Application (LCA) enters the lottery pool, it has to be approved by the US Department of Labor (DOL). The approval or denial of this visa application depends on a number of factors such as salary, work location, full-time employment, etc. The purpose of this research is to predict the outcome of an applicant's H-1B visa application using artificial neural networks and to compare the results with other machine learning approaches.

Keywords H-1B · Labor condition application · Deep learning · Artificial neural networks

1 Introduction

The H-1B is a work visa granted by the USA which allows an individual to temporarily work at an employer in a specialty occupation. According to the Immigration Act of 1990, a total of 65,000 foreign nationals may be issued a visa each fiscal year. However, an additional 20,000 H-1Bs are available to foreign nationals holding a master's or a higher degree from US universities. An individual can stay in the USA on an H-1B visa for a period of 3 years which is extendable up to six years. Although the final outcome of the visa status (approved or denied) is based on the lottery system, the LCA is scrutinized by the US DOL prior to entering the lottery pool. A number of factors such as the salary, job location, and job designation are taken into account by the DOL to determine the approval or denial of the application. The application is forwarded to the lottery pool once it has been approved by the DOL. If denied, the applicant has to reapply before his/her current H-1B visa term expires.

A. Dombe · R. Rewale · D. Swain (✉)
Vishwakarma Institute of Technology, Pune, India
e-mail: debabrata.swain@vit.edu

© Springer Nature Singapore Pte Ltd. 2020
D. Swain et al. (eds.), *Machine Learning and Information Processing*,
Advances in Intelligent Systems and Computing 1101,
https://doi.org/10.1007/978-981-15-1884-3_18

193

At present, there is no such model which predicts the outcome of H-1B visa application. The applicant applies through a standard process and waits for the decision. If the visa is denied, then the applicant has to reapply, which is in a way inefficient for the applicant. It is therefore essential to have an automated tool that could give insights to the applicant prior to applying.

Machine learning (ML)-based approaches are becoming popular for prediction problems in different domains including health care, banking, security, computer vision, and natural language processing. Artificial neural networks (ANN), which are inspired by human brain, are a state-of-the-art ML technique to find patterns in data and make predictions. Here, an automated system is proposed using ANN to predict the outcome of H-1B visa application by taking into account multiple factors such as salary, work location, and job designation. This will help the applicant to determine his/her chances of approval. This model has been compared with other ML algorithms to understand the overall efficiency.

2 Literature Review

A large number of changes are taking place while recruiting new employees due to the restrictions imposed by H-1B visa. For instance, Monica [1] has presented the changes in recruiting tracts because of the changes in H-1B policies. H-1B visa has played a key role in hiring international talent. Dreher et al. [2] have showed how the USA has managed to improve its post-secondary education by hiring top faculty from overseas. This shows how important an H-1B visa is to the employers.

The approval of H-1B visa application is decided on several parameters. In the current scenario, the salary parameter is extremely important due to some of the changes, which have been proposed by the administration, and as a result, it is crucial to understand the salary trends of the H-1B workers. Interesting salary versus case status trends have been presented by Lin et al. [3]. Lin et al. have used decision tree and K-means clustering to analyze the attributes of foreign workers such as salary and work location. They have presented a model which shows a picture of different approval rates compared with different conditions based on the data. It is evident how salary changes from one state to another from the salary versus job location trends presented by Renchi Liu et al. These trends can be useful not just from the application point of view but also prior to finding a job. For example, a trend shows that people working in California are earning quite higher than any other state. This trend can be helpful for people who are looking for the highest paid job. **Interesting trends presented by Lin et al. [3] helped us in identifying the effect of parameters across the dataset.**

3 Problem

3.1 Problem Statement

Given the attributes of an applicant as input, determine the likelihood of H-1B visa approval.

3.2 Approach

The system has been designed in two phases

Phase 1: Data pre-processing
Phase 2: Prediction of the H-1B visa application (approved or denied).

4 Dataset and Data Pre-processing

The dataset was acquired through Kaggle [4]. The dataset has about 3 million entries. However, we truncated this dataset to a size of 1 million for testing our prototype. The name and description of each column are given below:

1. EMPLOYER NAME: Name of the employer
2. YEAR: H-1B visa petition year
3. SOC NAME: Name associated with the SOC CODE, which is an occupational code, associated as the job has been granted temporary labor condition; this is required for Standard Occupational Classification (SOC) System
4. JOB TITLE: Title of the job
5. PREVAILING WAGE: Temporary labor condition is requested for the job and to set prevailing wages. The wage is scaled and listed in USD. The definitions on prevailing wages are as follows: the average wage paid to similar employed workers in the requested occupation in intended employment. It is an indication of employer's minimum requirements at the company for the specified job title
6. FULL TIME POSITION: Y = Full-time position; N = Part-time position
7. WORKSITE: Information of the applicant's city and state intended for work in USA
8. CASE STATUS: It states if an applicant has been granted a visa or not (certified implies that a visa has been granted)
9. LAT/LON: It is the latitude and longitude of the worksite of the employer.

4.1 Chosen Attributes

The dataset we have used had data for the year 2016 only. Therefore, we decided to remove the YEAR attribute. We are trying to predict the likelihood of approval based on the geographical area the individual works in. The exact location (longitude and latitude) may adversely affect the generality of the algorithm. Therefore, we decided to remove the same and keep WORKSITE only in the dataset. The resulting dataset used for training and testing had the following attributes: EMPLOYER_NAME, SOC_NAME, JOB_TITLE, FULL_TIME_POSITION, PREVAILING_WAGE, WORKSITE, and CASE_STATUS.

4.2 Removing Outliers

The only numeric attribute in the dataset was PREVAILING WAGE. We used box-plot to detect the outliers and removed the same using Tukey's 1.5*IQR method [5].

4.3 Balancing Dataset

The original dataset had four possible values for the attribute CASE_STATUS, which were CERTIFIED, DENIED, WITHDRAWN, and CERTIFIED-WITHDRAWN.

We removed WITHDRAWN and CERTIFIED-WITHDRAWN rows from the dataset because it solely depends on the applicant. In addition, the original dataset was highly skewed. The number of certified rows was very less compared to that of denied. Therefore, we randomly selected rows from the dataset such that the resulting dataset had equal number of certified and denied rows to avoid bias.

4.4 One-Hot Encoding

Except PREVAILING WAGE, all other attributes were in textual format. To feed them to the neural networks, it was necessary to convert them into numeric format. FULL_TIME_POSITION column had only two possible values: Y and N. We replaced Y with 1 and N with 0. CASE_STATUS column also had only two possible values after removing withdrawn cases in the previous step. So we replaced CERTIFIED with 1 and DENIED with 0. The remaining four columns EMPLOYER_NAME, SOC_NAME, JOB_TITLE, and WORKSITE were textual. To get the numeric representation of these columns, we used one-hot encoding technique implemented in Keras library. The pictorial representation of the one-hot

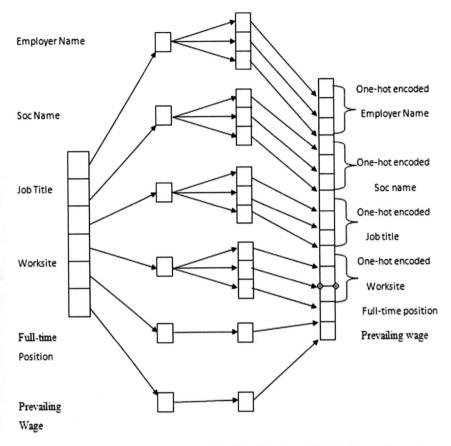

Fig. 1 Attributes chosen as input are on the left side of the figure. Each of these attributes is pre-processed separately. Some of them are one-hot encoded and others are kept as they were. Finally, these are clubbed together to get the final attributes which will be fed to the model

encoding is shown in Fig. 1. Left side of the figure shows six columns. Four out of the six columns are then one-hot encoded individually. All the resultant one-hot encoded columns are merged to produce the final feature set of size 16,978.

4.5 Normalization

After one-hot encoding, all attributes, except prevailing wage, had values 0 or 1. Therefore, we normalized prevailing wage using min-max normalization to scale its value in the range [0, 1]

$$y = \frac{x - \min\{X\}}{\max\{X\} - \min\{X\}}$$

x An attribute value
X An array of all possible values for attribute x
y Normalized value of attribute x

After pre-processing, we got a dataset having 26,962 samples. This resulting dataset was used for training and testing.

5 Experiments

Artificial Neural Networks
Artificial neural networks, which are inspired by human brain, are a state-of-the-art technique to find patterns in data and make predictions. An ANN contains neurons connected to each other in different ways. A neuron basically represents an activation function, which transforms an input into output.

A linear activation function is easy to train but cannot be used for learning complex patterns. Nonlinear functions, like sigmoid and hyperbolic tangent, can learn complex patterns in data but are sensitive to input only when it is in the middle range and they can saturate. This makes ANN containing these nonlinear units difficult to train. Also, when the number of layers increases, it may result in vanishing gradient problem. Rectified linear function is a piecewise linear function that outputs the input as it is if it is positive and outputs zero if it is negative. The graph of the function is shown in Fig. 2. The rectified linear activation function and its derivative are easy to calculate, and this makes ANN containing rectified linear unit (ReLU) easy to

Fig. 2 Graph of rectified linear function for ReLU

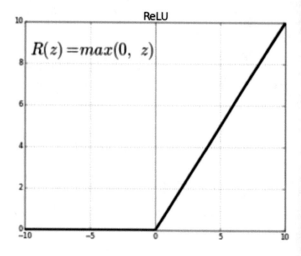

train. The fact that it outputs 0 for negative values leads to sparsity, which is often desirable in representation learning. Because of the piecewise linearity, there is no gradient vanishing effect [6].

We applied multi-layer perceptrons (MLPs) to our problem to predict the case status of a visa application. The ANN architecture we used is shown in Fig. 3. The input to the neural network is the one-hot encoded data generated after pre-processing. The input layer is of size 16,978. It has four hidden layers having neurons 512, 128, 64, and 32, respectively, of ReLU type and an output layer containing single neuron of sigmoid type. Visa status prediction being a binary classification problem, sigmoid was the suitable choice for the output layer. The output neuron gives a probability value of the case status. The output probability is then rounded off to get the final output, i.e., if probability is 0.5 or <0.5 then it becomes 0, which means DENIED and if probability is >0.5, then it becomes 1, which means CERTIFIED. Since this is a binary classification problem, we used binary cross-entropy function.

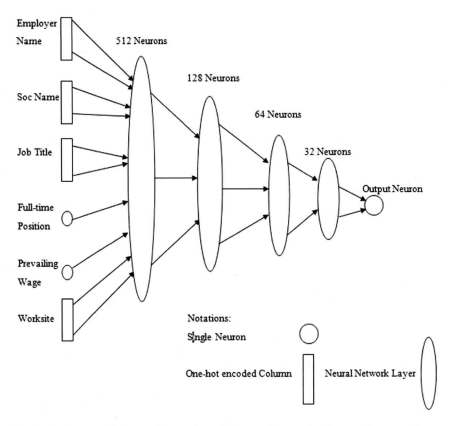

Fig. 3 Architecture of the neural network used. The one-hot encoded input attributes are shown on the left side of the figure. Hidden layers of 512, 128, 64, and 32 neurons are used followed by an output layer consisting of single neuron which gives the prediction

Fig. 4 Accuracy obtained
using different machine
learning algorithm

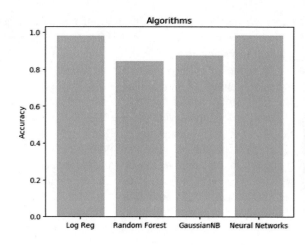

Table 1 Accuracy for H-1B
visa prediction task for
different machine learning
algorithms

Algorithm	Accuracy (%)
Logistic regression	96.27
Random forest classifier	84.30
Gaussian NB	87.50
Neural networks	98.3

For training the ANN model, we experimented with Stochastic Gradient Descent
and Adam [7] optimizers implemented in Keras library and chose Adam since it was
converging faster.

Other ML Algorithms:

We compared the outcome of ANN with other standard ML algorithms like logistic
regression, Gaussian Naïve Bayes, and random forest and found the prediction from
ANN to be more accurate. The results are shown in Fig. 4 and Table 1.

5.1 Training and Testing

The entire dataset was split into training and testing in the ratio 80:20. For ANN,
20% of the training dataset was separated and used as validation data during training.
The ANN model was trained on the training dataset for multiple epochs. After each
epoch, the validation dataset was fed to the model to validate if the model is being
trained properly or not. The training was stopped when the accuracy on the validation
dataset stopped improving considerably. The trained ANN model was then used on
the testing dataset for verification.

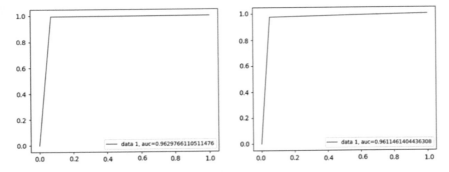

Fig. 5 ROC curve for ANN (on left) and for logistic regression (on right)

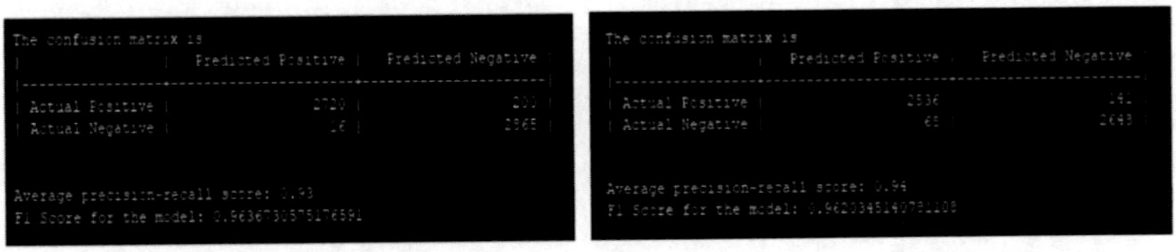

Fig. 6 Confusion matrix and F1 score for ANN (on left) and for logistic regression (on right)

5.2 Results

Figure 4 shows accuracies obtained for different algorithms. The ROC plots for the logistic regression and ANN models are shown in Fig. 5. The area under the curve (AUC) value is 0.9629 for ANN and 0.9611 for logistic regression model. Figure 6 shows the confusion matrix, precision–recall score, and F1 score. It is clear from Fig. 6 that the false-positive value is 16 for ANN and 68 for logistic regression model.

6 Future Scope

In this paper, we are just predicting whether the applicant's H-1B visa application will be approved or not. Also, we trained the model on a truncated dataset to test our prototype. However, in future we will be using the entire dataset and will be creating a recommendation system that will guide the applicant in the right direction. The recommendation system will generate recommendations if the prediction of the outcome of H-1B visa for an applicant is DENIED and will suggest improvisations in the applicant's profile to increase the chances of H-1B visa approval.

References

1. Boyd, Monica. 2014. Recruiting high skill labour in North America: Policies, outcomes and futures. *International Migration* 52 (3): 40–54.
2. Dreher, Axel, and Panu Poutvaara. 2005. *Student flows and migration: An empirical analysis.*
3. Jing-Lin. H-1B Visa Data Analysis and Prediction by using K-means Clustering and Decision Tree Algorithms. [Online] Available: https://github.com/Jinglin-LI/H1B-Visa-Prediction-by-Machine-Learning-Algorithm/blob/master/H1B%20Prediction%20Research%20Report.pdf.
4. H-1B Visa Petitions 2011–2016—Kaggle. [Online]. Available: https://www.kaggle.com/nsharan/h-1b-visa/data. Accessed October 20, 2017.
5. Seo, Songwon. 2006. *A review and comparison of methods for detecting outliers in univariate data sets.* Master's Thesis, University of Pittsburgh (Unpublished).
6. Glorot, Xavier, Antoine Bordes and Y. Bengio. 2010. Deep sparse rectifier neural networks. *Journal of Machine Learning Research* 15.
7. Kingma, Diederik P., Ba Adam Jimmy. *A method for stochastic optimization.*

Data Leakage Detection and Prevention: Review and Research Directions

Suvendu Kumar Nayak and Ananta Charan Ojha

Abstract Disclosure of confidential data to an unauthorized person, internal or external to the organization is termed as data leakage. It may happen inadvertently or deliberately by a person. Data leakage inflicts huge financial and nonfinancial losses to the organization. Whereas data is a critical asset for an organization, recurrent data leakage incidents create growing concern. This paper defines data leakage detection and prevention system and characterizes it based on different states of data, deployment points and leakage detection approaches. Further, this paper follows a systematic literature review considering a decade of the existing research efforts and makes a critical analysis thereof to highlight the issues and research gaps therein. The paper then proposes important research directions in the field of data leakage detection and prevention. This review helps fellow researchers and interested readers understand the research problem, appreciate the state-of-the-art techniques addressing the research problem, draw attention toward the research challenges and derive motivation for further research in this promising field.

Keywords Data leakage detection and prevention · Machine learning · Natural language processing · Information processing · Security and privacy

1 Introduction

In the present digital era, organizations are becoming more and more data-driven. Data and information are one of the most important assets for any data-driven enterprise. Data-driven decision making helps organizations manage the business operations well while keeping the customers and other stakeholders satisfied. Effective

S. K. Nayak (✉) · A. C. Ojha
Department of CSE, School of Engineering and Technology, Centurion University of Technology and Management, Bhubaneswar, Odisha, India
e-mail: suvendu.sonu@gmail.com

A. C. Ojha
e-mail: acojha2002@yahoo.co.in

© Springer Nature Singapore Pte Ltd. 2020
D. Swain et al. (eds.), *Machine Learning and Information Processing*,
Advances in Intelligent Systems and Computing 1101,
https://doi.org/10.1007/978-981-15-1884-3_19

management practices in terms of data capture, processing, storage, usage and protection of data are indispensable for such organizations. In particular, the protection of data from loss and leakage is paramount to organizational policies and practices in order to stay ahead of business competition.

Data leakage is defined as the accidental or intentional distribution and disclosure of confidential or sensitive data to an unauthorized party. It may be caused by internal or external entities to the organization. Sensitive data of an organization may include financial data, corporate strategy information, trade secrets, intellectual property, data about future projects, personal data of customers and employees such as patient records, credit card data, biometric data and many such data depending upon the business and industry. Similarly, sensitive data for a government may involve data about internal security and law enforcement, military and defense secrets, relationships and transactions with political parties, confidential diplomatic engagements, etc.

While distribution and sharing of data are a necessary requirement for business operations, leakage of sensitive or confidential data results serious consequences such as heavy financial loss, damage of reputation and credibility, regulatory penalties, decrease of company share price and likes. According to Data Breach QuickView Report [1], the year 2015 reported all-time high 3930 incidents of data breach exposing 736 million records. The highest number of incidents accounted for the business sector was 47.2%, followed by education (13.9%), government (12.2%) and medical (6.8%). All other sectors combined were only 19.9% of the reported incidents. Several other reports and studies reveal that incidence of data leakage is frequent and significant in several organizations [2]. As per IBM Security's 2019 Data Breach Report [3], the global average cost of data breach is estimated to USD 3.92 million, and the most expansive sector health care accounts for USD 6.45 million. India accounts for USD 1.8 million as the total average cost of data breach with the highest cost in industrial sector estimated at USD 2.7 million.

Data leakage is a serious and growing security concern for every organization which creates a pressing need for leakage detection and prevention of sensitive data and information. Consequently, it drives ample research attentions toward development of effective solutions from both academia and industry. Although a plethora of research proposals and data leakage prevention systems available in the literature, there is a pressing need to find an effective approach to this problem of security and privacy of sensitive data and information [4, 5]. Thus, it remains an active field of research.

This survey paper provides a comprehensive understanding of the field of study and presents state-of-the-art approaches for data leakage detection and prevention problem. The paper is structured as follows. Section 2 discusses on phases of a data leakage detection and prevention system and presents a taxonomy to characterize data leakage detection and prevention systems. Section 3 provides a review of select research proposals in the field of study. Section 4 lists an array of research challenges and provides pointers for future research. Section 5 concludes the paper.

2 Data Leakage Detection and Prevention

Although a number of data leakage prevention solutions are available from software vendors, there is less clarity on the exact definition of a data leakage prevention system. While there is no commonly agreed definition of what exactly a data leakage detection and prevention (DLDP) system should be, an attempt is made here to avoid the ambiguity. It provides an understanding of the DLDP system and its characterization.

2.1 Phases of DLDP System

Data leakage detection and prevention systems aim at identifying sensitive data and information, monitoring its usages and movement inside and out of the organization and taking action to prevent unintentional or deliberate disclosure of it. As shown in Fig. 1, identifying the sensitive data is the first important step in the DLDP systems. What to be included and what not to be included within the scope of sensitive data are a very crucial decision to be taken by the organization. The DLDP system must be configured correctly; otherwise, the system may result in false negative, and the leakage of potentially sensitive data will go unnoticed. The sensitive data must be kept in a readable digital format so that the DLDP system can monitor its usage and movement through multiple channels such as email, instant messaging, file transfer, access using HTTP, blogs and copying to USB. While it is rather challenging to develop a system to monitor all possible channels of data leakage, most systems focus on critical channels such as email, instant messaging, file transfer, HTTP and USB copy. Several techniques such as content matching, image recognition, fingerprinting, and statistical analysis can be used by DLDP systems to detect sensitive data leakage during channel monitoring. Once the leakage is detected, the system may perform one or more of the following actions. It may simply log the incident for later analysis and investigation, notify the risky behavior of the individual involved in the leakage to designated authorities and block the data being transmitted.

2.2 Characterizing DLDP Systems

DLDP systems can be characterized mainly based on which data state they handle, where they are deployed and what approach they employ to detect leakage. The classification is shown in Fig. 2. Since techniques for monitoring data are different

Fig. 1 Core phases of DLDP system (adopted from [21])

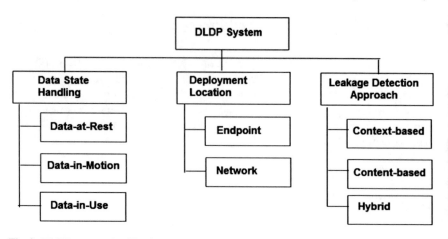

Fig. 2 DLDP system classification (adopted from [4])

for different data states, DLDP systems may be characterized based on three states of data such as Data-at-Rest, Data-in-Motion and Data-in-Use. Data-at-Rest can be defined as data in storage such as in file systems, databases, servers, desktop and laptops. Data-at-Rest can be local or remote to the DLDP system. A DLDP system may monitor Data-at-Rest at regular intervals or on demand. It checks if the sensitive data is encrypted or not. It checks if the access control policies are violated or not. Data-in-Motion can be defined as data that is passing through a network such as the Internet or local area network. A DLDP system usually monitors the network traffic by inspecting data packets to identify sensitive data traveling over the network using variety of protocols such as SMTP, FTP and HTTP. Data-in-Use can be defined as data that is being processed on endpoint devices. A DLDP system monitors the sensitive data and checks if it is being transmitted from one endpoint to another. Usually, the system monitors operations such as copy and paste between applications, screen capture, print, download to portable storage device such as USB drives and CD/DVD.

DLDP systems can be characterized based on their deployment location. A DLDP system can be deployed directly on endpoint devices or on network level. The DLDP system that is deployed on an endpoint device monitors and controls access to sensitive data, while a central server is responsible for policy creation and management, policy distribution and violation management, system administration and creation logs for incidents. On the other hand, a DLDP system deployed at network level inspects traffic based on leakage detection policy at multiple monitoring points in the network and blocks suspicious packets. DLDP systems can be further characterized based on the approaches employed to detect leakage of data. They can be context-based, content-based and hybrid. In a context-based approach, a DLDP system analyzes contextual metadata such as source, destination, sender, recipients, header, file type and format. A content-based system analyzes the data content using several techniques such as natural language processing, statistical analysis, fingerprint and keyword matching. A DLDP system using both content-based and context-based analysis is a hybrid one.

3 Literature Review

Several research efforts have been found in academic research literature to address the data leakage problem. Most of the works have been made to detect and prevent data leakage considering different states of data. These proposals mostly used either content analysis or context analysis for leakage detection. A few of them used the hybrid approach. The analysis techniques used in the works differ from one to another. This paper considers select research works carried out in the past decade based on their relevance. These are summarized below and their characterization is shown in Table 1.

Zilberman et al. [6] proposed an approach to detect email leakage based on topic identification and sender-recipient analysis in a group communication. A person in an organization belongs to several topic groups. The recipient of an email is classified as legal or illegal based on his/her association with the topic of the email being exchanged. The email content is analyzed to establish its topic using k-means clustering. The proposal was evaluated on a standard data set and provides 85% of

Table 1 Characterization of surveyed approaches

Proposal, Year	Leakage detection approach	Data state handling	Deployment location
Zilberman et al., 2011 [6]	Hybrid	Data-in-Motion	Endpoint
Shapira et al., 2013 [7]	Content-based	Data-in-Use	Endpoint
Shu et al., 2015 [8]	Content-based	Data-at-Rest, Data-in-Motion	Network
Costante et al., 2016 [9]	Hybrid	Data-at-Rest	Endpoint
Papadimitriou et al., 2011 [10]	Context-based	Data-in-Use, Data-in-Motion	Network
Alneyadi et al., 2016 [11]	Hybrid	Data-in-Motion	Endpoint
Shu et al., 2016 [12]	Content-based	Data-in-Use, Data-in-Motion	Network
Katz et al., 2014 [13]	Context-based	Data-in-Motion	Endpoint
Gomez-Hidalgo et al., 2010 [14]	Content-based	Data-in-Motion	Endpoint
Trieu et al., 2017 [15]	Content-based	Data-in-Motion, Data-in-Use	Endpoint
Lu et al., 2018 [16]	Hybrid	Data-in-Motion, Data-at-Rest	Network
Chhabra et al., 2016 [17]	Context-based	Data-in-Use	Network

accuracy. However, it suffers from a serious problem when the trained model does not have any information about a recipient.

Shapira et al. [7] proposed a content-based approach that uses well-known fingerprinting methods. The proposal uses k-skip n-gram technique and extracts fingerprints from confidential contents while skipping nonconfidential contents of a document to address the problem of traditional fingerprinting methods. It also takes care of rephrasing of contents and makes it possible to detect intentional leakage incidents. While it outperforms the traditional fingerprinting methods, it requires extensive data indexing when comes to implementation.

Shu et al. [8] proposed fuzzy fingerprint technique to enhance data privacy during leakage detection. The approach enables the data owner to safely delegate the content inspection task to a leak detection provider without disclosing the content. Although the network-based approach is efficient and provides satisfactory results under various data-leak scenarios, it suffers from computational complexity and realization difficulty.

Costante et al. [9] proposed a hybrid approach that combines signature-based and anomaly-based techniques to detect and prevent data leakage. It identifies insider threats by monitoring the activities of users and detecting anomalous behavior. Once the malicious behavior is detected, it is flagged up and the transaction is blocked. An attack signature is created and recorded to prevent such type of activities in the future. It uses a rule base which is updated automatically when a new anomaly is detected. While it attempts to combine both detection and prevention of leakage attacks using a rule-based technique, the system is flooded with false-positive alerts in the initial stage of system operation.

Papadimitriou et al. [10] studied guilty agent identification. The authors proposed data allocation strategies using which a distributer discloses sensitive data to a set of supposedly trusted agents. The distributer then assesses the likelihood that an agent is a guilty agent and responsible for the leakage when data is leaked. The approach is robust in case the released data is altered. However, the study does not capture adequate leakage scenarios.

Alneyadi et al. [11] proposed a hybrid method to detect potential data leakage in email communications of an organization. The method first performs context analysis using five contextual parameters in an email to measure RAI. The mail that scores high RAI is subjected to semantic analysis of its content in order to detect data leakage. The proposal showed encouraging results in detecting data leakage in emails. However, it suffers from implementation issues and consequently poor performance when it is not possible to capture all five contextual parameters in an organization.

Shu et al. [12] proposed a content-based approach that detects data leakage in transformed content. When the leaked data is modified, it becomes difficult to detect using usual n-gram technique. The problem is addressed using an alignment method in which a pair of algorithms, one sampling algorithm and another alignment algorithm, is used to compute a similarity score between the sensitive data sequence and the content sequence under inspection. The approach is efficient and results in high specificity; i.e., the percentage of true positive is very high than false-positive cases. However, it suffers from computational complexity.

Katz et al. [13] proposed a context-based model called CoBAn to detect data leakage in a controlled communication channel of an organization. During training phase, the model considers both types of documents, confidential and nonconfidential, and identifies various subjects dealt in these documents. Using k-mean clustering, it develops clusters of documents representing subject or context approximation. The model generates a context-based confidential term graph for each cluster. In detection phase, documents are analyzed and matched with one or more term graphs to estimate their confidentiality score. A document is detected confidential if its score crosses a predefined threshold. Although the model is able to find confidential data hidden in a document, the time complexity of the model is very high. Further, the model results in high rate of false positive which may not be acceptable in a real-world system.

Gomez-Hidalgo et al. [14] proposed a context-based approach that uses named entity recognition (NER) technique to detect data leakage of individuals and companies. Using Twitter data, experiments were conducted on a developed prototype which demonstrated encouraging accuracy level. However, the attributes considered in the NER technique are mostly homogenous.

Trieu et al. [15] proposed a method that used semantic and content analysis of documents to detect sensitive data leakage. The model uses document embedding to generate vector representation of a document or a text fragment. This vector representation is evaluated using a sensitivity corpus to find the sensitivity score of the document or text fragment. Experimental results show very high detection accuracy.

Lu et al. [16] proposed an approach for collaborative data leakage detection over distributed big data sets. The approach performs a privacy-preserving collaborative training on each owner's data which eventually trains and optimizes a global model. It uses a graph masking technique on the local weighted graphs representing local data and develops a trained template graph that represents global graph space. Then, the document to be tested is converted to graph and matched with the trained template graph to compute the sensitivity score of the test document. Although the approach can handle leakage detection with efficiency, the computational complexity remains high.

Chhabra et al. [17] studied data leakage detection in MapReduce computation in cloud computing environment. They used s-max algorithm to the reduced data in order to identify the guilty agent when any data leakage happens. They conducted a simulation for parallel processing of weather forecasting data using Hadoop framework and cloud analyst tool. However, the probability of indentifying the guilty agent is not very significant, and it reduces with increase in number of agents.

4   Challenges and Research Directions

DLDP systems face several challenges while preventing leakage of sensitive data [5]. One of them is abundant leaking channels such as email, social media, USB, printer, fax, smart phone and laptop. It is difficult to manage and secure all the channels. It is also very complex to model all possible activities of a person with sensitive data as well as leakage channels while developing a DLDP system. Another major challenge is the transformed data which is very hard to detect. When sensitive data is modified, its identity and patterns are changed making leakage detection very challenging.

The advent of big data and cloud computing has amplified the challenges of a DLDP system [18]. The system should be scalable to process massive data using parallel processing in a distributed environment. Real-time leakage detection is a requirement but a huge challenge for DLDP systems while dealing with big data. Anonymization and privacy preservation in big data are rather challenging to protect sensitive information [19]. Since data is kept in a remote location away from the data owner in a cloud environment, security and privacy remain a major concern which adds to the leakage detection problem. Multi-tenancy model in cloud computing offers threats of data leakage due to vulnerabilities in inter-tenant isolation [20]. Successful DLDP systems must be able to deal with the above said challenges in order to address growing concern of data leakage.

Additionally, there exist several areas with research opportunities which require further efforts from the researcher community. Deep learning has been successfully applied in various domains. The efficacy of deep learning can be exploited in both context and content analysis to detect data leakage and identify the insider threat with higher accuracy while achieving timely protection of sensitive data. In particular, deep learning-based leakage detection may be investigated in transformed data, wherein sensitive information is hidden in the exposed content. Cloud computing offers a new avenue for DLDP systems. Leakage detection and prevention can be offered as Software-as-a-Service (SaaS). Consequently, privacy preservation of the sensitive data becomes a major concern if DLDP system is offered as SaaS. Further, data leakage detection in MapReduce computation is a key research direction in cloud computing.

5   Conclusions

Data leakage is a persistent problem in organizations and inflicts grave consequences. It requires constant research efforts to mitigate the problem. The paper has reviewed several research contributions published in the recent past in order to portray the state-of-the-art techniques in the field of data leakage detection and prevention. The objective of the paper has been to provide a comprehensive reference to the field and attract research attention of fellow researchers toward it. The review reveals that the existing solutions are not satisfactory enough to tackle the perils of data leakage. In particular, the new challenges are thrown by big data and cloud computing invite further investigation. Nevertheless, organizational policies play a very effective role in curbing the menace of data leakage. Organizations should have concise policies to identify their critical data, and its handling since successful DLDP solutions heavily rely on classified information. Organizations should have policies to monitor access and activities on sensitive data at network level as well as on various endpoints. Additionally, a bare-minimum policy of encrypting sensitive data should be universally implemented across the organization.

References

1. Data breach quick view. 2015. Data breach trends. Available at https://www.riskbasedsecurity.com/2015-data-breach-quickview/. Accessed on 5 Sept 2019.
2. Data leakage news. Available at https://infowatch.com/analytics/leaks_monitoring. Accessed on 5 Sept 2019.
3. IBM security's cost of a data breach report 2019. Available at https://www.ibm.com/security/data-breach. Accessed on 5 Sept 2019.
4. Asaf, Shabtai, Yuval Elovici, and Lior Rokach. 2012. A survey of data leakage detection and prevention solutions, 1st ed. Springer: New York Heidelberg Dordrecht London. https://doi.org/10.1007/978-1-4614-2053-8.
5. Alneyadi, S., E. Sithirasenan, and V. Muthukkumarasamy. 2016. A survey on data leakage prevention systems. *Journal of Network and Computer Applications* 62: 137–152.
6. Zilberman, P., S. Dolev, G. Katz, Y. Elovici, and A. Shabtai. 2011. Analyzing group communication for preventing data leakage via email. In *Proceedings of 2011 IEEE international conference on intelligence and security informatics*, 37-41, 10–12 July 2011. Beijing, China: IEEE.
7. Shapira, Y., B. Shapira, and A. Shabtai. 2013. Content-based data leakage detection using extended fingerprinting. arXiv preprint arXiv:1302.2028.
8. Shu, Xiaokui, Danfeng Yao, and Elisa Bertino. 2015. Privacy-preserving detection of sensitive data exposure. *IEEE Transactions on Information Forensics and Security* 10 (5): 1092–1103.
9. Costante, E., D. Fauri, S. Etalle, J.D. Hartog, and N. Zannone. 2016. A hybrid framework for data loss prevention and detection. In *Proceedings of 2016 IEEE security and privacy workshops*, 324–333. IEEE Computer Society.
10. Papadimitriou, P., and H. Garcia-Molina. 2011. Data leakage detection. *IEEE Transactions on Knowledge and Data Engineering* 23 (1): 51–63.
11. Alneyadi, S., E. Sithirasenan, and V. Muthukkumarasamy. 2016. Discovery of potential data leaks in email communications. In *Proceedings of the 10th international conference on signal processing and communication systems (ICSPCS)*, 1–10. Gold Coast, Australia: IEEE.
12. Shu, Xiaokui, Jing Zhang, Danfeng Daphne Yao, and Wu-chun Feng. 2016. Fast detection of transformed data leaks. *IEEE Transactions on Information Forensics and Security* 11 (3): 1–16.
13. Katz, G., Y. Elovici, and B. Shapira. 2014. CoBAn: A context based model for data leakage prevention. *Information Sciences* 262: 137–158.
14. Gomez-Hidalgo, J.M., J.M. Martin-Abreu, J. Nieves, I. Santos, F. Brezo, and P.G. Bringas. 2010. Data leak prevention through named entity recognition. In *Proceedings of IEEE 2nd international conference on social computing*, 29–34, Minneapolis, USA.
15. Trieu, Lap Q., Trung-Nguyen Tran, Mai-Khiem Tran, and Minh-Triet Tran. 2010. Document sensitivity classification for data leakage prevention with twitter-based document embedding and query expansion. In *Proceedings of 13th International Conference on Computational Intelligence and Security*, 537–543, 15–18 Dec 2017. Hong Kong, China: IEEE.

16. Lu, Yunlong, Xiaohong Huang, Dandan Li, and Yan Zhang. 2018. Collaborative graph-based mechanism for distributed big data leakage prevention. In *2018 IEEE Global Communications Conference (GLOBECOM)*, 9–13, Abu Dhabi, UAE.
17. Chhabra, S., and A.K. Singh. 2016. Dynamic data leakage detection model based approach for MapReduce computational security in cloud. In *Proceedings of fifth international conference on eco-friendly computing and communication systems (ICECCS-2016)*, 13–19. IEEE.
18. Cheng, L., F. Liu, and D. Yao. 2017. Enterprise data breach: causes, challenges, prevention, and future directions. *WIREs Data Mining and Knowledge Discovery* 7: e1211. https://doi.org/10.1002/widm.1211. Wiley & Sons, pp. 1–14.
19. Basso, T., R. Matsunaga, R. Moraes, and N. Antunes. 2016. Challenges on anonymity, privacy and big data. In *Proceedings of seventh Latin-American symposium on dependable computing*, 164–171, Cali, Colombia, 19–21 October. IEEE Computer Society.
20. Priebe, C., D. Muthukumaran, D. O'Keeffe, D. Eyers, B. Shand, R. Kapitza, and P. Pietzuch. 2014. CloudSafetyNet: detecting data leakage between cloud tenants. In *Proceedings of the 6th edition of the ACM workshop on cloud computing security*, 117–128, Scottsdale, Arizona, USA, November 7–7, 2014. ACM.
21. Data Leakage Prevention (DLP)-ISF Briefing Paper. Available to https://www.securityforum.org/research/data-leakage-prevention-briefing-paper/. Accessed on 5 Sept 2019.

A Blockchain Based Model to Eliminate Drug Counterfeiting

Monalisa Sahoo, Sunil Samanta Singhar and Sony Snigdha Sahoo

Abstract The drug counterfeit problem has become global and so huge that it has drawn significant attention from everyone. The fake drug industry is worth $10B per year as estimated by survey. Also estimated by WHO, about a million die per year due to fake drugs. The distribution of fake drugs is a crucial issue. One of main reasons behind drug counterfeiting is imperfect supply chain. There are many loopholes in our drug supply chain. In the present scenario of supply chain, either the information is not at all shared between the parties during the hand-off process or a little or irrelevant information is shared, which has led to counterfeiting. The counterfeit drug not only affects the health condition of patients but also results in the financial loss of genuine manufacturer. In this paper, the details of the drug counterfeit have been explained along with its impact. Also, various measures to fight the counterfeits using blockchain technology have been discussed. Using blockchain technology traceability, visibility and security can also be incorporated into the drug supply chain. This proposed system will track the drugs from its origin, the manufacturer to the end, the consumer.

Keywords Distributed architecture · Drugs supply chain · Security · Traceability · Blockchain · Drug counterfeit

M. Sahoo · S. S. Singhar
Department of Computer Science, Utkal University, Bhubaneswar, India
e-mail: monalisasahoo.ln@gmail.com

S. S. Singhar
e-mail: sunilsinghar@gmail.com

S. S. Sahoo (✉)
DDCE, Utkal University, Bhubaneswar, India
e-mail: sony28788@gmail.com

© Springer Nature Singapore Pte Ltd. 2020   213
D. Swain et al. (eds.), *Machine Learning and Information Processing*,
Advances in Intelligent Systems and Computing 1101,
https://doi.org/10.1007/978-981-15-1884-3_20

1 Introduction

Imagine a situation, where someone dies because of fake medicine. It is not a distant possibility, and this has become a reality for most developing countries even for some developed countries. Fake medicines have turned to be multi-billion-dollar problem on a global level. The shape, size, color of the pharmaceuticals and even the packaging exactly look like the original. Small amounts of the active ingredients can be found in these bogus products or sometimes none at all or may be even worse like some fatal ingredients. It is upsetting, but fact is, India is the epicenter of manufacturing of fake medicines. It is beyond doubt that fake medicines are an increasing threat to consumers and the pharmacy industry. So, the real-time visibility of drug production and management is necessary. Blockchain is the answer for dealing with counterfeit drugs. It can be accessed freely because the transactions are stored in digital ledger format without compromising on the security and privacy of the users [1].

Blockchain technology concept came from bitcoin cryptocurrency [2] where no third party is required for transaction. In blockchain, transaction is being carried out in a distributed peer-to-peer system. It has been a common thought that blockchain can only be used for financial sector as it is based on the idea of bitcoin. Only recently, the true potential of blockchain has been realized by the researcher community. This decentralized technology has many useful applications [3] such as health care, logistics, Internet of Things (IoT), reputation system, public service, supply chain which are beyond the financial sector. If only health care sector [4] is considered, it may further be divided into many parts like medical record management, health care data access control, clinical trial, drug supply management, insurance claim adjudication, etc. (Fig. 1).

The remaining of the paper is organized as follows. Section 2 provides a detailed background study. Section 3 explains about fake drugs and how they enter into the

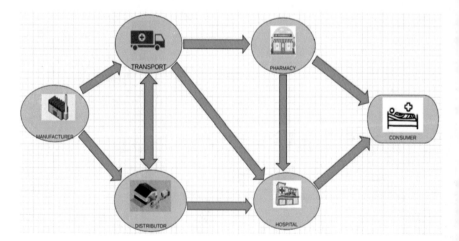

Fig. 1 Distributed architecture diagram of drugs supply chain

supply chain. Sections 3.1 and 3.2 list down the challenges posed by fake drugs and those by the drugs supply chain. Section 4 describes the blockchain solution, and Sect. 5 demonstrates the proposed solution. Finally, Sect. 6 concludes this research work.

2 Background

The primary objective of pharmaceutical industry is to discover, develop, produce and market drugs. The drugs can be used for many purposes, such as medications to cure patients, vaccination and alleviate any kind of symptoms. Pharmaceutical companies usually deal with two types of medicine, generic and brand medications. They also at times deal with medical devices to some extent.

The history of Indian pharmacy dates back to Gupta period which was approximately from 320 to 550 CE. The Ayurveda signature of Indian medication is based on "Charak Samhita" and "Sushruta Samhita," two great texts named after two legendary persons Charak (Father of Medicine) and Sushruta (Father of Surgery). Before the Britishers came to India, Indians were usually dependent only upon the indigenous medicines. They were totally unaware of any other type of medications. Allopathic medication was introduced by British rulers in India. But such medicines were not produced in India. Raw materials were imported from India, foreign countries used their units to produce final products, and then, those medicines are exported back to India.

At the end of nineteenth century, some of the scientists of India like P. C. Ray, T. K. Gajjr and A. S. Kotibhaskar established a foundation for a pharmaceutical industry. Bangal Chemical was the first Indian pharmaceutical industry established by Acharya P. C. Ray in Calcutta in 1901. Later on, many Indian entrepreneurs started various pharmaceutical industries. Only 13% of country's medicinal requirement could be fulfilled by the initial achievements of Indian drug industry. The Second World War (1939–1945) caused a massive fall in the supply of medicines from non-native companies. So, there arose an urgent need and opportunity to fulfill the requirement of people, and thus, many pharmaceutical companies originated in India. These are Unichem, Chemo Pharmaceuticals, Zandu Pharmaceutical Work, Calcutta Chemicals, Standard Chemicals, Chemical Industrial and Pharmaceutical Laboratories (Cipla), East India Pharmaceutical Works, etc. By the time of independence, with the help of newly established pharmaceutical industries, nearly 70% of India's need could be met.

If statistics be considered, over 20,000 registered pharmaceutical industries in India have sold nine billion dollars worth of formulations and considerably huge amount of drugs in 2002. Around 85% of these drugs were sold in India, and more than 60% of the formulations were exported, mainly to the USA and Russia. Most of the participants in the market are small-to-medium enterprises. Two hundred and fifty of the largest companies' control 70% of the Indian drug market. It is appreciated that because of the 1970 Patent Act, multinationals represent only 35% of the Indian

Table 1 Top ten countries importing drugs from India in 2017–18 [5]

Rank	Country	Value (US$)	Share (%)
1	USA	3.8 Billion	32.9
2	South Africa	461.1 Million	3.9
3	Russia	447.9 Million	3.8
4	UK	444.9 Million	3.8
5	Nigeria	385.4 Million	3.3
6	Kenya	233.9 Million	2
7	Tanzania	225.2 Million	1.9
8	Brazil	212.7 Million	1.8
9	Australia	182.1 Million	1.6
10	Germany	178.8 Million	1.5

market, down from 70% thirty years ago. Exports of pharmaceuticals products from India have also risen from US 6.23 billion dollars in 2006–07 to US 8.7 billion dollars in 2008–09. India has sold 11.7 billion dollars worth of pharmaceutical products in 2014. Pharmaceutical export from India stood at US 17.27 billion dollars in 2017–18 and is estimated to increase by 30% to reach US 20 billion dollars by the year 2020 (Table 1).

3 Fake Drugs and Their Entry into the Supply Chain

A counterfeit drug is one which is composed of the wrong active ingredients, the proper active constituents at the wrong dose or none of the prescribed active ingredient. Counterfeit drugs have found a place in the mainstream of drug supply chain. More than one third of spurious drugs being produced globally have originated from India, which erodes the reputation of our country in international market. How the fake drugs find such a big chunk of share in the market? There are multiple points of contact which are not secure and unsubstantiated, any of which could have been an entry point of counterfeit drugs.

A number of probable reasons have been listed:

– The global supply chain of drug market is so complex that it is very difficult to decide on the drugs or drug ingredients.
– Fragmented supply chain. (Before the arrival at their final destination, sometimes the drugs pass through various trading companies who usually do not check for quality)
– One or several members in the supply chain may be corrupt: Adulterating, Replacing or Mislabeling products in the supply chain.
– Uncertified source of raw material. (A trusted manufacturer might be unaware that the ingredients possibly came from an unverified source).

- Fake drug manufacturers or the fake drug brands successfully pose as legitimate pharmaceutical companies and reputed drug brands with the help of false paperwork.
- Lack of consumer awareness and high price of genuine medicine.
- Weak enforcement law of legislation toward corruption and advanced method used in counterfeiting.

3.1 Challenges Posed by Fake Drugs

In drug supply chain system, lots of challenges have to be faced for overcoming the fake drugs supply like:

- Loss of Business Possibilities: The existence of counterfeit drugs can cause the loss of market share and business possibilities for producer of authentic pharmaceutical products. It has been predicted that fake drugs have given rise to a 46 billion dollars loss yearly to pharmaceutical industries globally.
- Affecting the authorization of Generics: Estimation is that 90% of the cost of India's medicine market is influenced by generic trademarks. In order to lessen the health care overheads, many governments are recommending the use of low-cost generic medications, but the presence of counterfeits is serving as a barrier.
- Accumulating the Financial and Societal Difficulties: The consumption of spurious drugs has caused a significant rise in expense of the health care system due to the requirement for further interventions for undesired reactions and/or advanced disease progression. This is a particular issue for developing countries like India, where drug spending is already high at almost 70% of the income, and affordability levels are low.
- Resourcing: To deal with the problem of counterfeits, the Government of India has employed various anti-counterfeiting programs, but with inadequate effects, largely due to India's Central Drugs Standard Control Organization, the country's drug regulator, having only 323 employees in 2014, about two percentage the size of the FDA of USA. This under-resourcing is likely to influence the success of any subsequent approaches.

3.2 Challenges in the Drug Supply Chain

With multiple entry points and so many loopholes and absence of a transparent and robust security system, tracking the drug supply chain has become a herculean task. The hindrances to the present supply chain system are as follows:

- Absence of unified and inter-operable labeling and identifications standard.

- Soiled, fractured and opaque supply chain infrastructure. There is no easy and clear way to track down the product's journey in the supply chain that can unveil the real origin and hand-off points.
- Improper monitoring and management of cold chain supply.
- Hand-offs between various parties, including packager and transport agencies, as the drugs are transferred between multiple stages, which could be the possible points of entry of spurious drugs.
- Different systems have been adopted by different manufacturers. This leads to a compatible problem where the distributors and transport agencies to keep different type of solutions along their supply chain systems. Which can cause a confusion in the information or delivery of the drugs and eventually risk to consumer.

4   Blockchain: The Solution

In this context [6], the issue of counterfeit medicine is an increasingly acute worldwide issue and demands immediate attention. With\$10B lost in the war against counterfeit drugs annually, crumbling patient safety issues, a chain-of-custody log that blockchain could potentially implement, holds major promise. This type of blockchain [7] system could ensure a standard log of tracking at each step in the supply chain at the specific drug or product classification level.

Blockchain technology [8] allows large amounts of correlated data to be distributed but not copied. Although these types of digitally distributed ledgers have been around for years, blockchain uses new technology, like specialized algorithms, to validate and authenticate transactions within a decentralized architecture while progressing in real time through protected layers of digital encryption.

The decentralized framework of blockchain [9] in addition to distributed digital ledger that records and transfers data in a secure, transparent and fast manner may be put to use. A drug supply chain based on blockchain depends on a trusted network geared with blockchain and IoT. The required entities or nodes of the blockchain network are the manufacturer, distributors, the transport agencies and pharmacist or hospital. A supply chain is to be created including all the stakeholders such as suppliers, vendors, distributors and pharmacies, hospitals. Drug bottles are to be labeled with serial numbers and unique fingerprint of its manufacturer. The drug bottles packaged for transportation and the packages must also contain the serial numbers of the drugs inside. The serial numbers are to be scanned at every point of its journey from the manufacturer to the consumer, and information recorded in the blockchain. The whole journey of the drug is to be made visible to all entities in the blockchain [10]. The manufacturing inputs like chemical ingredients and other parameters can be updated and linked to the serial no of the products. So, when the consumer finally buys the drug by just scanning the label by a smartphone, he/she will be able to know everything about the drug.

5 Proposed Model

Basically, we can divide the whole drug supply chain from the manufacturers to the consumers into six parties.

1. Manufacturer
2. Distributor
3. Transporter
4. Pharmacy
5. Hospital
6. Consumer.

The model can be described as follows. The manufacturer sends the drugs to the distributor or to the transport agency according to their requirement. The distributor can directly collect drugs from the manufacturer or order it through a transport agency. The distributor sends drugs to hospitals and pharmacies again through the transport agencies. If the hospital's requirement is very large, it can directly order from the manufacturer through transport agency, or it can collect drugs from the distributor directly. The smaller hospitals can order drugs from their nearby pharmacies. At last, the consumer or patient will get the drugs either from the hospital or from the pharmacy store.

Excluding the last party (the consumer), a blockchain framework can be created among the first five parties. It is quite obvious question that the last party, which is the consumer, has to be kept out of the blockchain. This is because every node in the blockchain framework participates in the transaction verifying process. The consumer community is very diverse from illiterate to highly educated, from high school kid to senior citizens, from a person who lives in a remote area to the person who lives in a city, so they all cannot participate in the transaction verifying procedure. And also one more thing is that if every consumer is added to the blockchain framework, then the system will be heavily loaded as there are billions of consumers in our nation. Managing such a huge data will be expensive, and the complexity will also increase. But there is a platform where every consumer can check the authenticity of the drug, which has been discussed later in this section.

As our traditional system consists of some loopholes, so a decentralized blockchain framework is to be created with a shared distributed ledger with the five entities. Each entity should be verified before adding them into the blockchain framework (Fig. 2).

First, all the drug manufacturers must have to be certified by the Central Drugs Standard Control Organization (CDSCO), India. Each manufacturer should have a unique id. The drug manufacturer should be a trusted entity. It has to make sure that all the registered manufacturers produce genuine drugs, and the Government of India must take responsibility of the genuineness. Once a drug is produced, it has to be labeled by the unique id of the manufacturers, the drug name, the composition of the drug and the other details. The drug information is added to the shared ledger with all the details of the drug. Now, the manufacturer will transfer the drugs to the distributor.

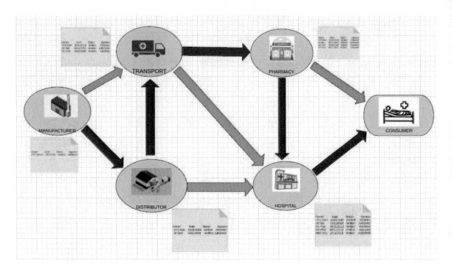

Fig. 2 Blockchain-based drugs supply chain architecture

All the distributors registered in the blockchain should also be assigned with a unique id. During the hand-off from manufacturers to distributor, the both parties should digitally sign using their private keys in the distributed ledger, and the transaction is added to the block. And all the entities in blockchain should verify the transaction before adding another transaction so that no one can deny or tamper this transaction in future. Now the only objective of the distributor is to distribute the drugs according to the requirement of hospitals and pharmacy stores through transport agencies (Table 2).

Once the distributor receives the drugs from the manufacturer, all the parties in the blockchain will know that distributor has received the drugs. The transport agency will now take the drugs from the distributor to pharmacy stores and hospitals. The hand-off between distributor and the transport agencies will happen after they both digitally sign in the shared ledger using their private keys. It is crucial period during the transport because during the transportation, the probability of counterfeiting is maximum. Here, we can take the help of IoT and wireless sensor devices to track and check. A GPS system can be attached to the drug package and that can be monitored constantly before it reaches its next destination. A wireless sensor can also be placed inside package which will constantly monitor temperature humidity and other parameters because some drugs are sensitive to temperature, and higher temperature will spoil the medicine (Table 3).

Table 2 Drug details

Name	Aspirin
DOM	08/09/2017
DOE	07/09/2020

Table 3 Transaction details

Owner	Date	Status	Sign
Manufacturer	02/02/2019	Verified	Xg78$9op
Distributor	04/03/2019	Verified	IK#02Gj!
Transport	07/03/2019	Verified	mGr9#kfR
Pharmacy	15/03/2019	Verified	cEj4#98i
Hospital	06/04/2019	Verified	uYt#$03Kw

In a similar fashion, the hand-off of drugs between transport agency and hospital or transport agency and pharmacy is to be carried out. Now, the drug is available for its final usage. The drugs can be now administrated to the patients, or the consumers can buy from the pharmacy stores.

6 Conclusion

With the recent advancement in network and Internet technologies, there is an urgent need of improvement in medical health care services. In this paper, various issues related to drug supply chain management have been discussed. This work also discusses how blockchain can prevent the issues in a transparent and secure manner. Blockchain can be used to add traceability and visibility to the drug supply chain to overcome the issues of drug counterfeiting. A model of decentralized blockchain architecture has also been presented with a shared ledger system, which will not only prevent drug counterfeiting but also will make the drug supply more robust, transparent and trustworthy. Using this proposed model loopholes in our current drug supply chain can be blocked. Moreover, the proposed model of blockchain framework is not only limited to countering the counterfeiting of drugs but also be useful in real-time tracking such as scheduling delivery of products.

References

1. Salman, T., M. Zolanvari, A. Erbad, R. Jain, and M. Samaka. 2018. Security services using blockchains: A state of the art survey. *IEEE Communications Surveys and Tutorials* 21 (1): 858–880.
2. Nakamoto, S. 2008. *Bitcoin: A peer-to-peer electronic cash system.*
3. Zheng, Z., S. Xie, H.-N. Dai, and H. Wang. 2016. Blockchain challenges and opportunities: A survey. *International Journal of Web and Grid Services* 1: 1–25.
4. Roehrs, A., C.A. da Costa, and R. da Rosa Righi. 2017. Omniphr: A distributed architecture model to integrate personal health records. *Journal of Biomedical Informatics* 71: 70–81.
5. Pharmaceutical industry in India. https://en.wikipedia.org/wiki/Pharmaceutical.
6. Mettler, M. 2016. Blockchain technology in healthcare: The revolution startshere. In *2016 IEEE 18th International conference on e-Health networking, applications and services (Healthcom)*, 1–3. IEEE.

7. Azaria, A., A. Ekblaw, T. Vieira, and A. Lippman. 2016. Medrec: Using blockchain for medical data access and permission management. In *2016 2nd International conference on open and big data (OBD)*, 25–30. IEEE.
8. Zhao, H., Y. Zhang, Y. Peng, and R. Xu. 2017. Lightweight backup and efficient recovery scheme for health blockchain keys. In *2017 IEEE 13th international symposium on autonomous decentralized system (ISADS)*, 229–234. IEEE.
9. Liang, X., J. Zhao, S. Shetty, J. Liu, and D. Li. 2017. Integrating blockchain for data sharing and collaboration in mobile healthcare applications. In *2017 IEEE 28th annual international symposium on personal, indoor, and mobile radio communications (PIMRC)*, 1–5. IEEE.
10. Dubovitskaya, A., Z. Xu, S. Ryu, M. Schumacher, and F. Wang. 2017. How Blockchain could empower eHealth: An application for radiation oncology. In *Data management and analytics for medicine and healthcare. DMAH 2017*, ed. E. Begoli, F. Wang, G. Luo, Lecture Notes in Computer Science, vol. 10494. Cham: Springer.

A Comparative Analysis of Predictive Modeling Techniques: A Case Study of Device Failure

Jayasri Santhappan and Pragadeesh Chokkalingam

Abstract In the modern lifestyle, a person's life is full of gadgets and appliances. At some period of time, every device will have downtime. During the downtime of a device, its performance keeps on degrading up to any level. Preventive maintenance in every regular interval can only save a machine from performance degradation or sometimes any kind of accidental hazard. Hence, if the device downtime can be detected automatically, then some kind of preventive measure can be taken to recover it from its degradation state. Here, a machine learning-based approach is proposed to detect the state of any device. The Kaggle dataset is used here for experimentation. The whole system is divided into three different phases. In the first phase, the data is collected and cleaned using various statistical methods. In the next phase, the model is trained using the training dataset. Finally, the trained model is evaluated using the testing dataset. In this work, various machine learning algorithms are used to find the best algorithm showing the best performance against the given dataset.

Keywords Random forest · K-nearest neighbor · Support-vector machine

1 Introduction

At the present time in modern lifestyle, a person's life is surrounded by different gadgets and machines. In different industries, maintenance management became a very big issue. Maintenance normally carries out whenever any kind of failure occurs in any machinery. Sometimes, it happens in a regular interval of time. In some other situations, this activity needs a prediction [1].

With the use of various smart systems built using IoT technology, a number of sensors are used for predicting the downtime of any machine. During the downtime of any machine, its performance keeps on decreasing up to any extent. If this downtime

J. Santhappan (✉)
NorthBay Solutions, Houston, TX, USA
e-mail: jstudent921@gmail.com

P. Chokkalingam
CFA Level 3—CFA Institute, Houston, TX, USA

© Springer Nature Singapore Pte Ltd. 2020
D. Swain et al. (eds.), *Machine Learning and Information Processing*,
Advances in Intelligent Systems and Computing 1101,
https://doi.org/10.1007/978-981-15-1884-3_21

state of any device can be detected, then it becomes very convenient to recover it through maintenance. Through predictive maintenance, the accidental failure can be avoided and the productivity gets increased [2].

Presently, banking organizations are having a lot of concern about providing qualitative service to the customers. For that reason, they are trying to maintain good quality in their ATMs, online transaction platforms, and other web-related applications. While improving their services, sometimes older versions need enhancement in terms of its functionalities. During enhancement in service, some older components need to interact with some new components. Due to this, any service may encounter some error at any particular point of time due to incompatibility between different components. As a solution to this problem, nowadays technical support team keeps track of the technical maintenance activity of different components [3].

In some cases, for some particular device, the past data related to its performance at a different instance of time is collected. This data should be brief in nature. This data can give information about the characteristic of the device explaining its healthy and degraded performance condition. This data can be useful for doing some prediction about the downtime of the device and also suggest the time when maintenance is required for the device. But while collecting the data, its quality needs to be checked because sometimes exceptional data also gets introduced during data collection. Exceptional data creates a special case study which creates some error during predictive maintenance. Hence, precaution needs to be taken during data collection [4].

Due to failure, there is always some intervention created in different applications. Due to this, a company generally faces a lot of loss. In industries, various transmitters are used for different purposes like temperature, pressure, and level. These are the different crucial parameters for any production-based industry. Sometimes, intelligent agents like intelligent field devices are used for the detection of possible failure before it occurs. This detection helps in avoiding an emergency trip [5].

Predictive maintenance plays a major role in medical industries also. For the detection of the diseases, different methods are used such as MRI, CT scan, and X-rays. At some period of time, the devices used in these methods undergo a downtime. During this period, its performance gets affected. Because of this, it delays the diagnosis process, and sometimes, errors also observed in the reports generated by the devices. In such kind of applications, sometimes sensor-based systems are used for fault detection of any part of the device. In some other places device, log data is used for fault detection [6].

In telecommunication systems, different devices like AD converter, DA converter, and signal conditioning systems are used. Due to the lesser capacity of the capacitors used in these systems generally, the performance of the system gets affected. Sometimes, the lives of the equipment are affected by temperature, humidity, and other environmental factors. For such applications, equipment monitoring programs are used to get equipment data at a different instance of time. Hence, if any faulty status data is recorded, then immediately maintenance activity is carried out to avoid any kind of failure [7].

For modern industrial applications, proper maintenance activity is carried out to minimize the cost of defects occurs in different components. The heavy cost that an industry has to bear because of the failure of any device can be avoided by the use of adequate predictive maintenance activity. It is generally done by collecting the statistical data and then applying a machine learning model to do prediction about the failure of any device [8].

With the intention to develop an appropriate predictive maintenance system, here a machine learning-based system is proposed. Here, the different device dataset available in Kaggle repository is used for the model training and testing. Machine learning is an approach to create intelligence in any model by using a set of training examples. After the training is over, some unseen testing data are given to the model for prediction. In the remaining section of the paper, a number of literatures are discussed, data set description is done, and finally, the result of different machine learning models is shown.

2 Literature

Susto et al. [1] proposed an effective predictive maintenance method using support-vector machine (SVM) and K-nearest neighbor (KNN) algorithm. Here, first the predictive maintenance module receives physical variables information of different processes. Through this data, the faulty status of different processes is known. After that, the data is given to the classifiers to train the model. Then the inputs about different process are given to the classifier to predict about the future status of the processes. The SVM classifier creates a hyperplane to differentiate between the different classes. If the hyperplane is not able to linearly separate the instances, then the additional kernel is added for proper separation of the instances. In the KNN algorithm, the class of k-nearest neighbors of any test instance is determined first. The class having the majority is assigned to the test record.

Jose et al. [2] proposed an IoT-based predictive maintenance system. In this work, the wrong alignment of bearings and machinery operational fail due to metal fatigue are analyzed. Here, in the proposed system, different sensors are used to monitor the system. The system is used for vibration analysis, oil level analysis, and electrical parameter analysis. After that, the captured data is given to the machine learning system to do predictive maintenance.

Mishra et al. [3] developed a useful predictive maintenance method for financial organizations. The system is divided into three layouts. Component-level failure probability is determined at a regular time interval using machine learning method. In the data flow layout, the parameters related to failure are collected. In the data preparation approach, the error event is taken as input and prediction of each device level failure is made. For every device, the component-level errors are determined. Here, the observation, transition, and prediction window methods are used to prepare the features of failure. The final architecture layout talks about the high-level architecture of the whole system. Various services are associated with a different layer of the main architecture. Hence, all the parameters related to failure are analyzed and then given to the machine learning model for prediction of failure.

Chen et al. [4] have used a virtual metrology-based predictive maintenance scheme for the minimization of accidental failure. Through BPM scheme, the faulty system prediction can be done. The quality of prediction depends upon the past failure data. The data sample consists of concise and good data collected before and after maintenance is done. If some exceptional data gets introduced in the data set, then it affects the prediction process of BPM scheme. Hence, while the data selection processes, an ABSS scheme was used that selects healthy and proper data by removing the exceptional data samples.

Fornaro et al. [5] have developed a device diagnosis algorithm to predict the future device failure and future abnormal behavior to be shown by the different transmitter. The failure and mal-operational data of different pressure transmitters and echo transmitters are first collected. Then using this data rule base for the prediction is developed.

Patil et al. [6] implemented a machine learning-based method for the failure forecasting of different imaging device used in the medical field. The behavior of these devices is observed by the original equipment manufacturer remotely to collect different features like downtime, accidental failures. After that, the data is used to train the machine learning model and predict the failure of any device. Here, the author has done a case study on PHILIPS iXR system.

Sisman et al. [7] have developed a predictive maintenance system for different electronic devices used in the telecommunication system. Statistical analysis is carried on failure data to identify the critical factors responsible for the malfunction of the devices that cause the system to stop working. The initial analysis identified the capacity of the capacitors as a crucial factor for the failure of the devices. Then after using fuzzy graphs also, the same capacitor capacity is again identified as the crucial parameter. Hence, predictive maintenance is suggested for the capacitors to avoid any kind of failure.

Susto et al. [8] have proposed an effective maintenance technique for the fabrication of semiconductors used in electronics industries. In the predictive maintenance technique, the historical data is used for predicting the capability of the device. In the predictive maintenance technique, support-vector machine algorithm is used to forecast the life of the filament used in semiconductor fabrication.

3 Dataset

The data set is collected from Kaggle repository. The dataset consists of a total of 575 records and nine features. The dataset consists of a total of 575 records and 11 attributes. The attribute1 and attribute2 are the device id and data of failure. These two features are not having a correlation with the target which is identified through Karl Pearson method of correlation. Hence, these two features are removed from the dataset. Out of the nine attributes, eight are independent and one is dependent. Now, out of the remaining, eight independent attributes 3, 7, 8 are categorical, whereas rest all attributes are continuous. Categorical attributes are those whose values are distributed among a few specific classes. But continuous values are those where the value of the feature ranges between 2 values. The dependent attribute is the status of failure. It is a binary attribute. It has value 0 and 1. 1 indicates failure, whereas 0 indicates non-failure.

4 Proposed System

The proposed architecture is shown in Fig. 1.

Fig. 1 Proposed system

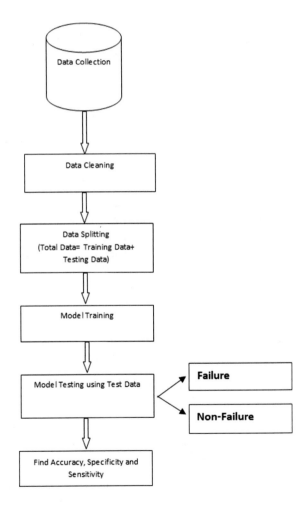

4.1 Proposed Algorithm

Data Collection

In this step, the data is collected from the device failure data set available in Kaggle.

Data Cleaning

The dataset consists of a total of 575 records and nine features. Out of the nine features, eight are independent features, whereas one is a dependent feature. Feature1, feature2, and feature3 contain total of 90, 75, and 62 missing values. The number of null value counts is determined by using isnull.conut() function. The null values in the feature1 are imputed with the mean, whereas missing values in feature2 and feature3 are imputed with their mode. So, many missing values are found in 45 records; hence, those records are removed from the dataset. Hence, finally the total cleaned records found in the dataset are 530. For the important feature selection, here forward feature selection method is used. During this, it is found that all nine features are really important. Hence, none of the features is dropped from the dataset.

Data Splitting

The whole dataset with 530 records is divided into two parts. 90% (477) of the records are used for training, whereas 10% (53) of the records are used for testing.

Model Training

In this step, the model training is carried out using the training data. During the training of the model, it finds the best values of the parameters related to the model. This process is known as hyperparameter tuning. The actual knowledge base of any model is created during its training. The model uses its knowledge base to do accurate predictions during its testing. The issue of overfitting and underfitting is handled properly during the training process.

Model Testing

In this step, the trained model is evaluated using the test data. Generally, the test data is unseen for the model. Here, the model categorizes the test data into two classes: (1) failure class and (2) non-failure or healthy class.

In this step, the accuracy, specificity, and sensitivity are calculated. Through these scores, the effectiveness of the model can be known.

4.2 Model Details

In the proposed system, the different classification models used are decision tree, random forest, K-Naïve Bayes.

Decision Tree. A decision tree is a supervised machine learning model which does the prediction of any data by forming a tree-like structure. In the tree at each node, some decision is made upon the value of some feature present in the test data. In the tree, each leaf represents some target class label. At every node, the possible values of every feature are tested. The selection of the root node is based on the information gain factor. A node is selected as root for which information gain is more and entropy is less. The calculation of entropy and information gain is shown in Eqs. (1) and (2).

$$\text{Entropy } E = -\sum_{j=1}^{n} q_j \log q_j \tag{1}$$

where

$E =$ The group for which entropy is calculated.
$q_j =$ Probability of the group to be there in jth state.

$$\text{Information Gain } I = E - \sum \frac{N_i}{N} E_i \tag{2}$$

$E =$ Entropy of all attributes at root node.
$N_i =$ Number of elements at ith level in group.
$N =$ Total number of elements in the group.
$E_i =$ Entropy at ith level.

Random Forest. A random forest is a supervised machine learning model formed by the ensemble of decision trees. It forms a number of distinct decision trees by taking different features subset from the original feature set. Finally, it combines the predictions made by different decision trees and uses the voting algorithm to select the class having the majority for the given test data.

Naïve Bayes Classifier. Naïve Bayes classifier is a simple probability-based classifier that uses Bayes theorem during its prediction process. It takes the assumption that all the features are independent of each other. It finds the most likely occurrence of the test data for all the output classes and assigns the class having the highest score. The calculation of probability for a class is shown in Eqs. (3) and (4).

$$P(c|X) = \frac{P(x|c)P(c)}{P(x)} \tag{3}$$

$$P(c|x) = P(x1|c) \times P(x2|c) \times P(x3|c) \times \ldots \times P(xn|c) \tag{4}$$

(Probability calculation for Naïve Bayes classifier)
where

$P(c|x) =$ Probability of output class 'c' when given attribute 'x.'

$P(c)$ = Observed probability of class 'c' for all the observations.
$P(x|c)$ = Probability of 'x' likely to occur in class 'c.'
$P(x)$ = Predictor prior probability.

5   Results

The efficiency of any machine learning algorithm is evaluated using different performance metrics are used. In the proposed work, the different metrics considered
are accuracy, recall, and precision.

Accuracy is the ratio of the total number of correct predictions to the total number
of predictions done [9]. Recall is the ratio between true positive to the total number
of positive cases. Precision is the ratio between true positive and the sum of true
positive and false positive [10]. The device which is detected to have failure referred
to as positive cases. The devices which are detected as non-failure cases referred
to as negative cases. The following Table 1 and Fig. 2 show the details of different
performance-related factors of all the above-discussed methods.

In the above table TP, FP, TN, FN, accuracy, precision, and recall score are given.
The highest accuracy and recall are obtained in the case of random forest classifier,

Table 1 Accuracy, precision, recall comparison table

Classifier	TP	FP	TN	FN	Accuracy (%)	Precision (%)	Recall (%)
Naïve Bayes	17	0	24	12	77.53	100	58.62
Decision tree	21	3	23	6	83.01	87.05	77.77
Random forest	22	2	24	5	86.07	91.66	81.48

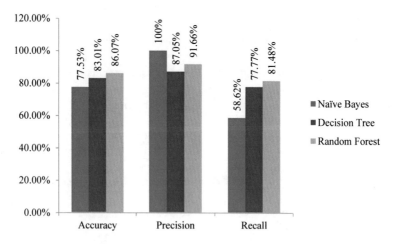

Fig. 2 Performance analysis graph

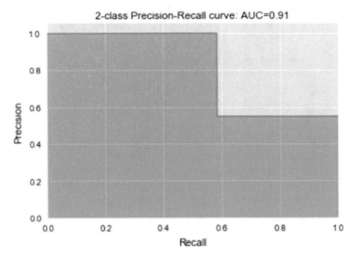

Fig. 3 AUC of Naïve Bayes

whereas precision score is highest in the case of Naïve Bayes classifier. For all of the models, the false positive value is less. False positive is the count that is predicted as non-failure device but actually, the device will get a failure. This creates a problem because as it is predicted not to have failure; hence, no preventive maintenance activity will be carried out. By considering all these above factors, random forest can be considered as the most efficient algorithm. In random forest algorithm, the overfitting and the underfitting issues are properly handled by the use of model-related parameters like the number of splits, the number of subtrees used.

ROC Curve. It is a curve created by drawing a graph between precision and recall value of the classifier. The area present under the graph is called AUC. This value generally ranges between 0.5 and 1. If it is closer to 1, then it indicates that the classifier is doing better prediction. In the following, the AUC curve of different classifiers is shown. For the random forest and decision tree, the AUC value is obtained as 0.89, whereas, in the case of Naïve Bayes, the AUC value obtained as 0.91. This AUC is demonstrated in Figs. 3, 4 and 5.

6 Conclusion

In this study, a number of failure detection methods are discussed. In the proposed system, a machine learning-based predictive model is applied. Here, three different models are taken like Naïve Bayes, decision tree, and random forest. Out of all the models, random forest method has shown a better result in terms of accuracy and recall. Hence, it is suggested here to use a machine learning-based algorithm for the detection of failure. Once the failure device is detected, then some predictive

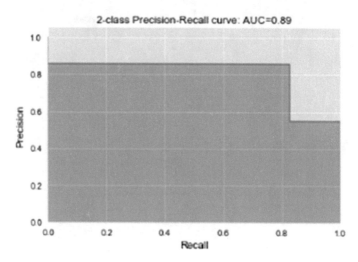

Fig. 4 AUC of random forest

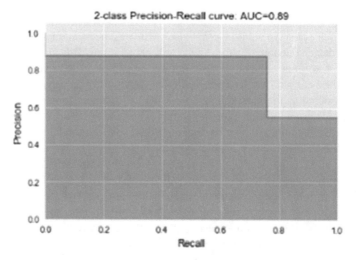

Fig. 5 AUC of decision tree

maintenance can be planned and it can be saved from any failure. Through machine learning algorithms, failure can be detected within less amount of time with high accuracy. Finally, it is concluded to use random forest algorithm for the prediction of real-time applications where failure causes a huge loss.

References

1. Susto, Gian Antonio, Andrea Schirru, Simone Pampuri, and Se´an McLoone. 2013. Machine learning for predictive maintenance: A multiple classifier approach. *IEEE Transactions on Industrial Informatics*. https://doi.org/10.1109/tii.2014.2349359.
2. Jose, Tinku Malayil, and Roshan Zameer. 2018. A novel sensor based approach to predictive maintenance of machines by leveraging heterogeneous computing. *IEEE Sensors 2018*. https://doi.org/10.1109/ICSENS.2018.8589620.
3. Mishra, Kamala Kanta, and Sachin Kumar Manjhi. 2018. Failure prediction model for predictive maintenance. In: *2018 IEEE International Conference on Cloud Computing in Emerging Markets (CCEM)*.
4. Chen, Chun-Fang, and Yao-Sheng Hsieh. 2013. Automatic baseline-sample-selection scheme for baseline predictive maintenance. In: *2013 IEEE International Conference on Automation Science and Engineering (CASE)*.
5. Fornaro, Oscar, Micaela Cascrza Magrot, and Paolo Pinceti. 2004. Diagnostics for measure transmitters. In: *IMTC 2004 Instrumentation and Measurement Technology Conference Como, Italy, 18–20 May 2004.*
6. Patil, Ravindra B., Meru A Patil, Vidya Ravi, and Sarif Naik. 2017. Predictive modeling for corrective maintenance of imaging devices from machine logs. In: *2017 39th Annual International Conference of the IEEE Engineering in Medicine and Biology Society (EMBC)*. https://doi.org/10.1109/embc.2017.8037163.
7. Sisman, George Roberti, and Oproescu Mihai. 2017. Monitoring the parameters of the electronics devices to assure the predictive maintenance of equipment. In: *10th International Symposium on Advanced Topics in Electrical Engineering*, Mar 23–25, Bucharest, Romania.
8. Susto, Gian Antonio, Andrea Schirru, Simone Pampuri, Daniele Pagano, Se´an McLoone, and Alessandro Beghi. 2013. A predictive maintenance system for integral type faults based on support vector machines: An application to Ion implantation. In: *2013 IEEE International Conference on Automation Science and Engineering (CASE)*.
9. Swain, Debabrata, Santosh Pani, and Debabala Swain. 2019. An efficient system for the prediction of coronary artery disease using dense neural network with hyper parameter tuning. *International Journal of Innovative Technology and Exploring Engineering (IJITEE)*, 8: (6S). ISSN: 2278-3075.
10. Swain, Debabrata, Santosh Pani, and Debabala Swain. 2019. Diagnosis of coronary artery disease using 1-D convolutional neural network. *International Journal of Recent Technology and Engineering (IJRTE)* 8: (2). ISSN: 2277-3878.

Python and OpenCV in Automation of Live Surveillance

Nilesh Navghare, Pratik Kedar, Prasad Sangamkar and Manas Mahajan

Abstract There have been many uses of image recognition and feature detection in recent days. With the growth in the popularity of Python and simplification of automation allows us to bring the live surveillance in the domain. Nowadays, the situation, with the sectors like banking, needs high amount of security due to their increasing importance. Especially on remote location where services like ATM are provided, the security factor becomes main concern. By automation of surveillance, there will be an efficient way to reduce stress on security.

Keywords Image and video processing · Haar Cascades [1] · Haar-like feature [1] · Integral images · OpenCV–Python

1 Introduction

Nowadays, automation has become very important thing in day-to-day life. It becomes very necessary in case of surveillance as any human intervention is error going to be error prone. Python and OpenCV together provide very powerful tools for video and image processing. This paper reflects the importance of automation in surveillance system with the help of Python and OpenCV. It aims on the detection of persons wearing helmet by developing XML classifier called Harr cascade file which is quite simple and powerful tool if trained and tested correctly.

N. Navghare (✉) · P. Kedar · P. Sangamkar · M. Mahajan
Computer Engineering Department, Vishwakarma Institute of Technology, 666, Upper Indra Nagar, Bibwewadi, Pune 411037, India
e-mail: nileshnavghare93@gmail.com

P. Kedar
e-mail: pratikkedar555@gmail.com

P. Sangamkar
e-mail: prasadsangamkar@gmail.com

M. Mahajan
e-mail: manas.mahajan99@gmail.com

© Springer Nature Singapore Pte Ltd. 2020
D. Swain et al. (eds.), *Machine Learning and Information Processing*,
Advances in Intelligent Systems and Computing 1101,
https://doi.org/10.1007/978-981-15-1884-3_22

235

The main objective of our work is to detect suspicious activities in public places. Currently, we are working on the object detection process by creating Haar cascade file for specific object like the helmet. The Harr cascade file which was created is XML classifier which is used in various detection processes like detection of helmet wearing persons in front of ATM's, in traffic monitoring, etc.

This requires the collection of datasets of multiple images of the objects for the detection. It basically uses the images as the reference to detect alike objects. It requires different datasets such as dataset of positive (to be detected object) images and negative (not to be detected) images, etc. To achieve high accuracy, dataset must be large enough so that it can be trained at maximum level.

The main motive is to create efficient Harr cascade file for detection of person wearing helmet. The surveillance system will be further enhanced to increase the accuracy and efficiency by training the object dataset model to further stages.

2 Object Detection

Today, pictures and video are available in plenty. The field of vision research has been overwhelmed by AI, measurements, and statistics. Making use of pictures to recognize and characterize and track objects or events so as to discern a genuine scene. We consider picture as the scene comprising of objects of interest and the foundation background represented by everything else in the image. Object identification decides the presence of the object or its extension and its areas in the picture.

Object recognition field is commonly completed via looking through each section of a picture to limit parts, whose various properties like geometric, photometric, etc. coordinate those of the objective article in the preparation database. This can be cultivated by examining an object layout over a picture at various areas, scales, and turns, and location is proclaimed if the likeness between format and picture is sufficiently high.

3 Image Processing

Image processing is a strategy to change an image into computerized structure and perform certain activities on it, so as to get an improved image or to extract some helpful data from it [1]. It is kind of sign regulation in which info is image, similar to photo and output might be image or qualities or characteristics related with that image. Image preparing fundamentally incorporates the accompanying three stages:

- Importing the image with optical scanner or by computerized photography.
- Analyzing and controlling the image which incorporates information pressure and image upgrade and spotting designs.

Fig. 1 Block diagram for object detection

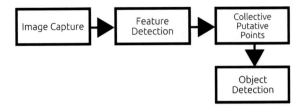

- Output is the last stage where result can be modified image or report that depends on image examination (Fig. 1).

4 Haar Cascade [2]

4.1 Haar-Like Feature and Viola–Jones

The use of Haar-like feature for detection is proposed first time by Paul Viola and Michael Jones in 2001. The thought behind the Haar detection cascade is to discard negative images with very negligible computation [3, 4]. A series of classifiers are processed to each subsection of the image. If that subsection fails to pass through any classifier, then that image is discarded, and further calculations are done. Once the subsection passes the stages, amount of computations required will be more. If the image passes the classifier, it is passed to the next subsection. Classifiers' accuracy can be increased by more training.

Haar-like characteristics are an over complete arrangement of two-dimensional capacities, which can be used to encode local appearance of object. They consist of at least two rectangular areas encased in layout.

Haar-classifier is a binary classifier; it simply detects whether the object is present in given a image or not (Fig. 2).

4.2 Mathematical Model

As shown in Fig. 3, the feature is calculated as the difference of the sum of the pixel intensities of black and white. The ideal Δ is 1. While as in figure for real image, we have the $\Delta = 0.74 - 0.18$ which is equal to 0.56. So closer the value to 1, the more likely we have found a Haar-feature.

$$\Delta = \frac{1}{n} \sum_{\text{DARK}}^{n} I(x) - \frac{1}{n} \sum_{\text{WHITE}}^{n} I(x)$$

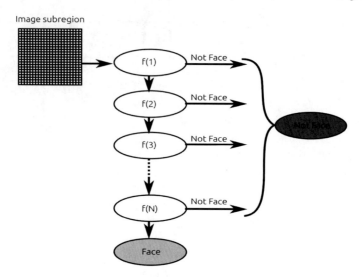

Fig. 2 Haar classifier diagram for face recognition

Fig. 3 Haar-feature pixel intensities

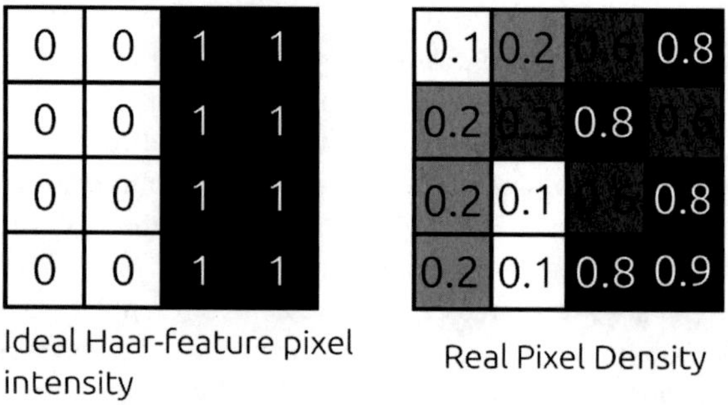

0	0	1	1
0	0	1	1
0	0	1	1
0	0	1	1

Ideal Haar-feature pixel intensity

0.1	0.2		0.8
0.2		0.8	
0.2	0.1		0.8
0.2	0.1	0.8	0.9

Real Pixel Density

We have to calculate the Δ, but we need to do it several times; as to use Haar features with all possible sizes and location, there are approximately 200 k features to calculate and time complexity of these operations is O(N2). So, to reduce the time complexity, we have to use a quadratic algorithm.

In Fig. 4 as you can see, the pixel intensities have been added to the [5] integral value, and in that way to detect the yellow block, we just need to subtract 3.7 from the remaining red boxes to calculate the Δ. By using [5] integral images, we reduce the time complexity to O(1) and will make the calculations simple.

Fig. 4 Original to integral conversion

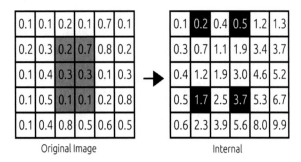

Original Image   Internal

5 Literature Review

- In 1984, Franklin C. Crow proposed summed-area tables for texture mapping [6]. This paper describes the following features. Given a method to map the antialiasing texture of egg-shaped surfaces.
- The design and use of steerable filters in 1991. This paper describes following features: analysis of orientation and phase, angularity adaptive filtering, edge detection and shape from shading.
- Coarse to Fine Face Detection 1999 describes following features. Detect and roughly locate the generic objects class. The detection is rapid but high rate of false alarms is raised.
- In 2001, Viola, P. and Jones, M. found rapid object detection technique using a boosted cascade of simple features [1]. The paper describes the new concept of integral images which results less computation time. Adaboost algorithm reduces number of features from larger one to give efficient classifier.

6 Methodology

6.1 Collection of Images

We were supposed to collect the positive images of the object which was to be detected (Here in this case, images of helmet) [7]. Having resolution between 300 * 300 and 400 * 400, we kept resolution of positive images lesser than negative images because, at time of creation of sample, we are going to supper impose positive image on negative image [8].

We have used 30 positive images and stored them in raw data directory and nearly 400 negative images and stored them in …\negative directory. There is also one descriptor file for negative images named as bg.txt which can be created by running create_list.bat file.

We have taken care that the object which we want to detect its images (positive images) is covered in all direction and in all possible cases. If we consider for gun detection, we have collected images like gun placed on the table, in pocket, in the hand, etc [9].

Similarly, we have taken all way possible images of helmet (positive images), i.e., helmet placed at different angle, wore by different people, at different angles, etc [10].

6.2 Cropping and Marking Positives

Before training, we need one vector file that contains path to positives and location of object in each image [11]. To do this run objectmarker.exe file which is present at.../training/positives. Draw rectangle over required object press "SPACE" to add object and press "ENTER" to save that object.

6.3 For Creation of Vector File

Run samples_creation.bat file present at.../training which will create vector of all positives [12]. The content of this file is as follows.

createsamples.exe-info positive/info.txt-vec vector/facevector.vec-num 20-w 24-h 24.

6.4 Training

The training command has the parameters like the location of info file and vector file, number of positive and negative images, the memory used the minimum height and width of window for detect and number of stages unto which we are supposed to train the object [13]. More the stages better the detection. The stages also depend upon the size of database as more the stages more the feature to extract, thus more example images required. Run the haartraining.bat file present at.../training [4].

haartraining.exe -data cascades -vec vector/vector.vec -bg negative/bg.txt -npos 30 -nneg 400 -nstages 9 -mem 1024 -mode ALL -w 24- h 24–nonsym.

7 Experimental Setup

S. no.	CPU specifications
1	Intel Core i5-7th gen
2	Intel Core i5-8th gen

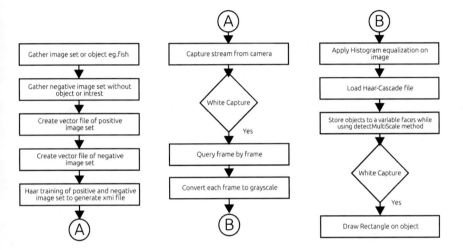

Dataset	Algorithm
Images (positive, negative)	Rapid object detection using boosted cascade of simple features

8 Results and Conclusion

Level no	Background processing time (ms)	Precalculation time (ms)	Stage training time (ms)
0	0.01	7.43	2.96
1	0.03	7.03	2.83

(continued)

(continued)

Level no	Background processing time (ms)	Precalculation time (ms)	Stage training time (ms)
2	0.07	6.65	2.72
3	0.10	6.61	3.09
4	0.43	6.99	3.27
5	0.83	6.92	3.69
6	1.48	6.74	4.73
7	7.84	6.93	5.72
8	27.05	6.44	8.83

Above table indicating training of helmet detection Harr cascade XML classifier. The experimental readings are mentioned above. In this experiment, we have successfully crated Harr cascade XML classifier up to ninth stage [1]. From the above reading, we can clearly see that as the level increases in training of Harr cascade file, background processing time gets increased exponentially but precalculation time remains constant.

Along with this stage, training time gets increased gradually.

Here, precalculation time varies between 6.5 and 7.5 ms. Which means, it is nearly constant in that range.

NOTE: For above reading, we have used Intel i5-8th gen processor.

Below graphs can show that how background processing time and stage training time increase exponentially.

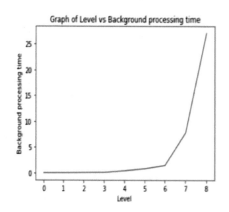

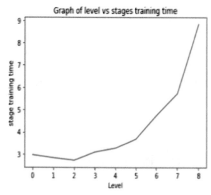

9 Future Scope

Image processing has wide range of application in current developing industry and day-to-day life also [1]. In order to do image processing, we need to have XML classifier. In this experiment, we have created Harr cascade file for helmet detection and this XML classifier can be used for surveillance in banking industries for detection of helmet wearing person in premises of ATM machines.

And through this process, we can create different Harr cascade XML classifier as per the need of industries. For example, in product line, for detection of any other object except the product, we can use this approach in order to solve this issue and many more.

References

1. Viola, Paul, and Michael Jones. 2001. Rapid object detection using a boosted cascade of simple features. *CVPR* (1): 511–518.
2. Amit, Yali, Donald German, and Kenneth Wilder. 1997. Joint induction of shape features and tree classifiers. *IEEE Transactions on Pattern Analysis and Machine Intelligence* 11: 1300–1305.
3. Rezaei, Mahdi. 2013. Creating a cascade of haar-like classifiers: Step by step. *Aplikasi Pendeteksian Ras kucing dengan mendeteksi wajah kucing dengan metode viola jones.*
4. Mahdi, Rezaei. Creating a cascade of Harr like classifiers. Department of computer science, Auckland.
5. Sung, Kah K., and Tomaso Poggio. 1994. *Example based learning for view-based human face detection.* Massachusetts Institute of Technology Cambridge Artificial Intelligence Lab, No. AI-M-1521.
6. ElSaid, W.K. A System for managing attendance of academic staff members in university development programs using face recognition.
7. Greenspan, Hayit et al. 1994. Overcomplete steerable pyramid filters and rotation invariance, 222–228.
8. Schneiderman, Henry, Takeo Kanade. 2000. *A statistical approach to 3D object detection applied to faces and cars.* Carnegie Mellon University, the Robotics Institute.
9. Tsotsos, John K., et al. 1995. Modeling visual attention via selective tuning. *Artificial Intelligence* 78 (1-2): 507–545.
10. Itti, Laurent, Christof Koch, and Ernst Niebur. 1998. A model of saliency-based visual attention for rapid scene analysis. *IEEE Transactions on Pattern Analysis and Machine Intelligence* 11: 1254–1259.
11. Fleuret, Francois, and Donald Geman. 2001. Coarse-to-fine face detection. *International Journal of computer vision.* 41 (1–2): 85–107.
12. Schapire, Robert E., et al. 1998. Boosting the margin: a new explanation for the effectiveness of voting methods. *The Annals of Statistics* 26 (5): 1651–1686.
13. Osuna, Edgar, Robert Freund, and Federico Girosi. 1997. Training support vector machines: an application to face detection. *cvpr.* 97: 130–136.

Privacy Preservation in ROI of Medical Images Using LSB Manipulation

Madhusmita Das, Debabala Swain and Bijay Ku. Paikaray

Abstract In the current era of digitization, the medical images are getting transmitted over the communication channel for diagnosis. During this process, it can be modified accidentally or intentionally. This occurs because of the insecure Internet network, which ultimately affects the sent image information. Thus, it results in wrong diagnostic decisions. Here, the basic purpose is to secure image transmission, without changing the sensitive areas of images that carry medical information. One of the possible watermarking methods to avoid erroneous diagnosis and checking the reliability of the received medical image is region-based analysis using ROI. The ROI of medical image is irregularly figured which contains important information. This paper tries to analyze these issues and attempts to overcome such problems by recovering secure medical images.

Keywords Medical image · Region-based analysis · ROI · LSB manipulation · Data hiding · Zero distortion

1 Introduction

Nowadays, immense progress is being done in a health care environment using distributed networking and cloud facilities but significant patient safety and security are major challenges [1]. Medical image, which can be stored for different purpose like diagnosis as well as long-time storage and research, plays a vital role in proper diagnosis inpatient treatment. In the mode of transmission, most of the cases, medical images are modified intentionally or accidentally by some attack or some noise, which makes a wrong diagnosis conclusion and that affects the patients [2]. Here,

M. Das · D. Swain (✉)
Department of Computer Science, Rama Devi Women's University, Bhubaneswar, Odisha, India
e-mail: debabala.swain@gmail.com

B. Ku. Paikaray
Department of CSE, Centurion University of Technology and Management, Bhubaneswar, Odisha, India
e-mail: bijaypaikaray87@gmail.com

© Springer Nature Singapore Pte Ltd. 2020   245
D. Swain et al. (eds.), *Machine Learning and Information Processing*,
Advances in Intelligent Systems and Computing 1101,
https://doi.org/10.1007/978-981-15-1884-3_23

a new method is proposed to get the exact original image when it needs without affecting the affected area. The sensitive as well as important areas in a medical image are named region of interest (ROI), which is meant for diagnostic decision making and on the other hand, remaining part is referred to as region of non-interest (RONI) [2]. Now, diagnostic of the specialist depends upon the ROI contents of the different medical image, so it has to be tamper-free or should not be modified [2]. Excluding the rest part of the paper is structured as follows: Sect. 2 covers the image security techniques, different ROI techniques in Sect. 3, and in Sect. 4, the proposed ROI hiding technique is discussed and the conclusion with possible future work is described in Sect. 5.

2 Image Security Techniques

2.1 Watermarking

Watermarking is an application which works widely with image to enhance image security, especially the information that has not been amended or destroyed unauthorized. Watermarking is an efficient way that is used for providing copyright protection into a secret or private communication. Every multiple copies should have different watermarking processes, which are found to be successful and hopeful mechanisms. Different watermarking processes like embedding and extracted have different mechanisms where slight modification of digital data is done after embedding the watermarking data and the extracted process recovers the original data. The watermarking process is found to be an efficient and effective mechanism in our context.

2.2 Steganography

On the other hand, the steganography method is widely accepted which is meant to be a proposed technique of both covered and secret writing. It is a technique of hiding messages from the others where the only intended recipient can only know that a message has been sent, where the cover messages are encoded. This encoding message is called a secret message after that it is covered into steganography messages. In this case, image steganography lossy and lossless compressions are used for saving the storage space but have performed different effects on any decompressed hidden data in their image. As per the name suggests, lossy provides high compression but not properly maintains the original image integrity, where a lossless compression performed the opposite principle of the loss compression. The best part of this method is to provide better security for sharing data faster with a large number of software which is difficult to detect except the receiver.

2.3 Reversible Data Hidings

The main propose behind the hiding information in an image is to protect original image pixels from a vast distortion on a permanent basis after extracting that information. This proposals is commonly called a reversible data hiding technique. It extracts the original state of the image at the receiver end from the cover image, as well as it helps the embedding of messages in a host image without any losses of the host content. So this technique is not only used for data hiding but also the entire recovery of the original image from the encrypted image. This technique is most desirable in those applications where degraded restoration is not allowed, like, medical images, forensic images, military maps, etc.

2.4 Region-Based Analysis

ROI is commonly known as region of interest. That means it defines certain areas or region which is a very important area of an image. The regions can be polygons, circles, etc. Selecting ROI of an image means to select that area; this is not intended for any kind of loss in pixels or any kinds of distortions. RONI is commonly known as region of non-interest. Due to the consideration of the important and non-important areas of an image, an image is divided into two parts. Though ROI is considered an important part of an image, on the other hand, RONI is an area that is left after selecting the ROI part. In the case of RONI, the pixels of that area may get compressed or data that may lose.

3 Data Hiding Techniques Using ROI Analysis

3.1 ROI Lossless Colored Medical Image Watermarking Scheme with Secure Embedding of Patient Data

In 2016, Deepa S. et al. proposed an ROI lossless colored medical image watermarking scheme with secure embedding of patient data [3]. This proposed system consists of two types of watermarks BP and BP+ at the two different LSB planes for creating and embedding the first watermark in the bit plane BP+1, patient health record, and color medical image is provided as input [3].

As a result of which color medical image with watermark1 is come out. Secondly, for generating watermark in the bit plane BP, a color medical image with watermark1 is provided as input and on the other hand, as a result of its output, it is ready for transmission. On the process, the receiver side is the same occurred in a reverse manner. A similar process is done in the case of a noisy environment except that there is an addition of noise at the transmitter side and it will be removed by the filter

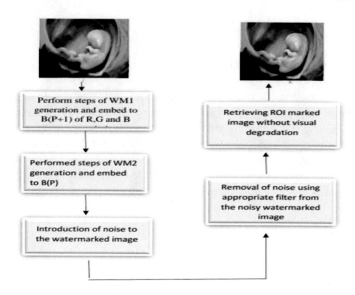

Fig. 1 Block diagram for a colored image with noise addition [3]

at the received side. A colored image environment with noise addition and removal complete block diagram is shown Fig. 1.

As a result of it the accurate health record of the patient at the receiver side with the marked ROI can be accurately recovered for diagnosis [3].

3.2 Watermarking and Encryption in Medical Image Through ROI Lossless Compression

Nithya S. et al. proposed a method, watermarking and encryption in medical image through ROI—Lossless compression. Here, without affecting the RONI part, ROI of an image is gone through the hash of ROI value where SHA-256 is used by users to verify the medical image integrity and computing hash of the health image. As a result, this encrypted output is combining with a hash value of ROI and produced a compressed image and fed into the ROI image. Finally, a combination of both watermarked ROI and RONI part gives a watermarked output image and the secret key is applied with AES and produced final encrypted image [4].

In this process, the watermark should be extracted from the output. In this process, encrypted image goes through the decryption mode, where one of the secret keys (key2) is applied. As a result of which watermarked image is produced. The watermarked image is divided into two parts, such as ROI and RONI. After that, ROI extraction process is to be done by decompressing it. At the stage of decryption, a secret key (key1) is added and it produces patient information, hash ROI, and ROI of

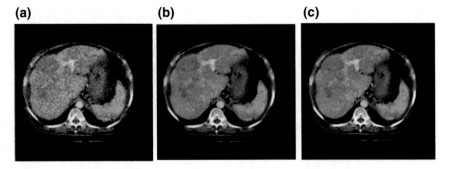

Fig. 2 **a** Embedded watermarked image, **b** extracted image, **c** output image [4]

an image. The ROI is calculated by the hash of ROI if decrypted image. On the other hand, the received ROI is calculated by hash of ROI. Calculated value is compared with the extracted value. If both of the values are the same, then it will be the original retrieved image otherwise the image is discarded. As a result of which

$$MSE = \frac{1}{mn} \sum_{i=0}^{m-1} \sum_{j=0}^{n-1} (I(i, j) - I_w(i, j))^2 \tag{1}$$

MSE defined the decrypted original image. In this case, PSNR is used to show the loss and lossless compression of the image. The PSNR ratio is maximum, easily clear by the MSE. The PSNR can be measured from the MSE as follows (Fig. 2):

$$PSNR(\text{in dB}) = 10 \log_{10} \left(\frac{255^2}{MSE} \right) \tag{2}$$

3.3 A Novel Medical Image Watermarking Technique for Detecting Tampers Inside ROI and Recovering Original ROI

In 2014, Eawaraiah R. et al. proposed a novel medical image watermarking technique for detecting tampers inside ROI and recovering original ROI. In this technique to recover the ROI data of an image, each pixel is split into 3×3 blocks [2]. Center blocks with its eight neighboring blocks represent either positive (0) or negative (1) value. As a result of which 56 bits of recovery data are generated, where 8 bits are meant for center block and others have 6 bits, respectively. In the case, if 3×3 blocks of ROI image in binary form have to be recovered first, consider the first 8 bits as center pixels and the other 6 bits are for eight neighboring pixels. The decimal value of each segment will be subtracted from the center and added with the center depending whether first bit of the segment is zero or one simultaneously.

After improving data of ROI, it embeds within RONI. Then the medical image of the RONI area is divided into block of 8 × 8 size each and these division steps are continued for each blocks B in RONI. After all, the bits of recovery data are embedded in the ROI, CDF (2, 2) integer wavelet transfer up to second level on B is applied. There are two bits of recovery data in ROI is embed in every coefficient of the middle frequency sub-bands LH_1, HL_1, LH_2, and HL_2 of B and finally, inverse integer wavelet transform is applied on B [2]. In case of to avoid overflow/underflow problem, histogram shifting process is used by

$$B(x, y) = \begin{cases} B(x, y) + k \text{ if } B(x, y) \leq k \\ B(x, y) - k \text{ if } B(x, y) \geq k \end{cases} \tag{3}$$

where k is considered as a threshold value and for block B, $B(x, y)$ is the pixel value. To avoid overflow and underflow, always the value which is set in place of k will be greater than 15 or similar to it.

As a result of which after conducting experiments on medical image like MRI scan, ultrasound, CT scan, etc., as shown in the below Fig. 3, this method properly recognizes tampered areas within ROI and get back to its original ROI as per the necessity. No embedding distraction will be there after generating the watermarking medical image. Also, find out the tampers within ROI and without any losses the original ROI recovers during the time of tempered, as well as provides robustness.

4 Proposed ROI Hiding Algorithm

4.1 ROI Selection and Data Hiding

Here, the original grayscale image is $I_{M \times N}$ whose pixel values range from 0 to 255.

Step-1: Let the ROI image be $I_{ROI} = (I)_{n \times n}$, where the sides of the square are **n**.
Step-2: Convert each pixel of I_{ROI} format to 8-bit ASCII format; let it be I_{ROI-8}.
Step-3: Let us extract the 2 LSB of each pixel of I_{ROI-8} and store in a matrix I_{LSB-2}, circular row shift them and store in I_{LSB-2R} of order $[n^2 \times 2]$.
Step-4: Let us replace the matrix I_{LSB-2} with I_{LSB-2R} in I_{ROI}.
Step-5: Now, convert the I_{ROI-8} matrix to I_{ROI} by 8-bit ASCII to decimal conversion.
Step-6: Finally, select and replace the I_{ROI} in $I_{M \times N}$ and let it be $I'_{M \times N}$.

After hiding data into the ROI segment, now the image will be sent to the receiver. At the receiver end, the ROI region will be again selected and recovered.

Actual image   Watermarked image   Reconstructed Image

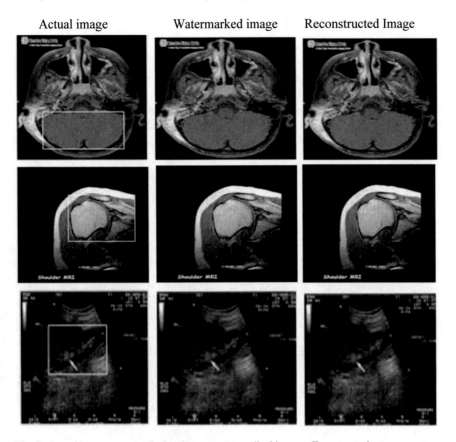

Fig. 3 Actual image watermarked and reconstruct medical images. From top to bottom sequence: CT scan, MRI scan, and ultrasound images [2]

4.2 ROI Selection and Data Recovery

Here, the received grayscale image is $I'_{M \times N}$ whose pixel values range from 0 to 255.

Step-1: Now, the ROI image selected from $I'_{M \times N}$ be $I'_{ROI} = (I'_{M \times N})_{n \times n}$, where the sides of the square are n.

Step-2: Convert each pixel of I'_{ROI} format to 8-bit ASCII format; let it be I'_{ROI-8}.

Step-3: Let us extract the 2 LSBs of each pixel of I'_{ROI-8} and store them in a matrix I'_{LSB-2}, now reverse circular shift the rows and store in I'_{LSB-2L} of order $[n^2 \times 2]$.

Step-4: Let us replace the matrix I'_{LSB-2L} with I'_{LSB-2} in I'_{ROI}.

Step-5: Now, replace the I'_{ROI} matrix in $I'_{M \times N}$.

5 Experimental Results

The proposed method demonstrates the prospective experiments were carried out on standard 512×512 test X-ray images. Figure 4a depicts an original X-ray image, Fig. 4b shows the extracted ROI of an X-ray image. Then, the extracted ROI image is marked using the embedding procedure to generate the embedded image, shown in Fig. 4c. Figure 5a depicts the embedded X-ray image which is visually similar to the original image. Finally, the recovered image is generated by reversing the marked ROI in Fig. 5c.

The complete experimental data of our proposed scheme is given below in Table 1 showing the PSNR, SSIM, and MSE of recovered image with reference to an input image.

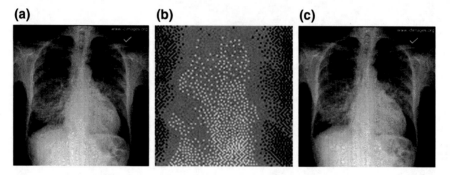

Fig. 4 **a** Original Image, **b** extracted ROI image, and **c** embedded image, respectively, at sender end

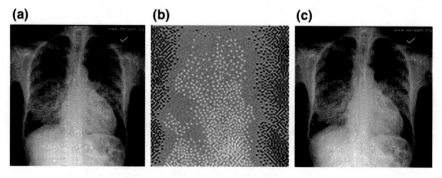

Fig. 5 **a** Embedded image, **b** extracted ROI image, and **c** recovered Image, respectively, at receiver end

Table 1 MSE, PSNR, and SSIM of recovered image with reference to input image

MSE	0
PSNR	Inf dB
SSIM	1.0

6 Conclusion and Future Works

The proposed technique is completely reversible, which can preserve the ROI in medical images, and can be recovered as in the original image. A simple data hiding capability and creating enough embedding capacity are the two most important areas that have scope for improvement. In the existing methods, the embedding process follows a uniform circular shift technique of the LSB bits of ROI of the input image, this largely improves the data hiding capacity. In the RONI segment, no such data hiding is done. It is a concern of the future scope to increase the data preservation on sensitive images. Color images can be taken as test images. The test images can be further expanded into JPEG, DICOM, and higher than 512×512 grayscale images. Other evaluation parameters can be effectively measured.

References

1. Rocek, Ales, M. Javornik, K. Slavicek, and O. Dostal. 2017. Reversible watermarking in medical imaging with zero distortion in ROI. In: *2017 24th IEEE International Conference on Electronics Circuits and Systems (ICECS)*, 356–359. https://doi.org/10.1109/icecs.2017.8292071.
2. Rayachoti, E. and S.R. Edara. 2014. A novel medical image watermarking technique for detecting tampers inside ROI and recovering original ROI. In: *IEEE International Symposium on Signal Processing and Information Technique (ISSPIT)*, 321–326. https://doi.org/10.1109/isspit.2014.7300608.
3. Deepa, S., and A. Sandeep. 2016. ROI lossless colored medical image watermarking scheme with secure embedding of patient data. In: *2016 International Conference on Communication Systems and Network (ComNet)*, 21–23 July 2016 Trivandrum, 103–106.
4. Avcibas, I., B. Sankur, and K. Sayood. 2002. Statistical evaluation of image qualitymeasures. *Journal of Electronic Imaging* 11 (2): 206–224. https://doi.org/10.1117/1.1455011.

Novel Approach for Measuring Nutrition Values Using Smartphone

Sashikala Mishra, Prashant Gadakh, Chinmay Gosavi, Shivam Bhoskar, Shrinit Padwal and Vishal Kasar

Abstract Food is the source of energy, and it plays a vital role in human existence. The quality of food is suffering day by day such as adulteration and heavy use of various pesticides. The traditional approach to analyze food nutritional values involves the use of various sensors and laboratory procedures to detect the quality, but such sensors and methods really take lots of time. There is a need of a system which we can use to quickly evaluate the quality of food by methods which are ubiquitous. The number of handheld devices and their processing capabilities has increased manifolds over the last few years. In this paper, the novel methodology has been proposed which uses the smartphones to take the image, and instantly, it provides the nutrition value. The proposed model helps detect the nutritional quality of the food by utilizing the various sensors which are present in smartphones such as cameras and microphone. The model uses the classifiers to detect the type of food and process all the algorithms in cloud. Four datasets are used with multi-class level. Machine is trained with various algorithms such as CNN and RNN, and we have used transfer learning. The whole system is implemented successfully, and the accuracy of 82% has been achieved.

Keywords Statistics · Data mining · NLP · Object detection · Machine learning · DietCam

1 Introduction

As per the economist in 2016, more than 1.9 billion adults aged 18 years and older were overweight. Of these, over 650 million adults were obese [1]. Similarly, about 13% of the world's adult population (11% of men and 15% of women) were obese in 2016 [2]. Obesity increases the risk of various diseases and health conditions such

S. Mishra · P. Gadakh (✉)
International Institute of Information Technology, Hinjawadi, Pune, India
e-mail: prashantgadakh31@gmail.com

C. Gosavi · S. Bhoskar · S. Padwal · V. Kasar
International Institute of Information Technology, Hinjawadi, Pune, India

© Springer Nature Singapore Pte Ltd. 2020
D. Swain et al. (eds.), *Machine Learning and Information Processing*,
Advances in Intelligent Systems and Computing 1101,
https://doi.org/10.1007/978-981-15-1884-3_24

as heart disease, stroke, high blood pressure, diabetes and some cancers. Similarly, more than one-third of the world's malnourished children live in India. Among these, half of the children under three years old are underweight and one-third of wealthiest children are over-nutriented [3]. Malnutrition also adverse effect on just children but adult as well, some them include stunted growth, poor immune system, micronutrient deficiency, weak immunity, inactivity of muscles, apathy, depression, kidney function impairment. Thus, there is a need for a method which can quickly determine the nutrition value of food without using complicated laboratory procedures. Such methods can help users quickly determine nutrition information and thus make better diet choices. While laboratory procedures are most accurate and reliable, they are often time consuming and cannot be carried out in all situations. Smartphones are becoming ubiquitous and more powerful with better cameras and processors. The aim of this paper is to leverage the technologies like object detection and ML to provide a method of detection of nutritional values on smartphones by utilizing the cameras to identify the food and subsequently provide these values to the user. Using object detection, now we can identify the food items easily. While annotating the images, we create bounding boxes and label each food item in the photograph. So, using this, we can detect more than two food items from a single image. Transfer learning is also used here, for finding accuracy on trained models like MobileNet and Inceptionv3. ML is used to improve the accuracy of the system based on the previous interactions.

1.1 Novel Approach

We aim to create a system (android application) which will be able to detect the nutritional values of the food. This will be done by utilizing the cameras which are presented in smartphones. We aim to leverage machine learning and object detection to help detect the nutritional quality of the food. The traditional approach involves the use of various sensors and laboratory procedures to detect the quality, but such sensors and methods are not ubiquitous. The model created will utilize object detection to recognize instances of food items and help in better classification, while machine learning will be utilized to make sure that the system improves based on the previous interactions in order to increase the accuracy over time.

2 Approach (TensorFlow)

We are using TensorFlow open-source software for model training and classification. TensorFlow provides us with a trained model, which we then port into our Android application for food detection, classification and then to obtain the nutritional values of the given food items.

Table 1 Dataset information

Name	Dimension	Information
UPMC_Food20 [4]	612 × 432	20 categories of food with 100 images per category. Accuracy is 72.4%
UBCF100 [5]	448 × 336	100 kinds of food photographs, accuracy of 54.2%
Food 11 [6]	512 × 512	11 categories of food with images in range of 800–1500 pet category. MobileNet 1.0 224 gives accuracy 79.8%

Steps:

- Create database of food images.
- Annotate or label the database.
- Create xml files using image labeling softwares, e.g., LabelImg.
- Create a csv file and then convert it into TFRecord file.
- Create a pipeline, i.e.,. build a.config file.
- Use cloud services to train the classifier using the images, tfrecord file and some configuration steps.
- Model is created and then ports it into Android application, i.e., copy the.pb files, labels.txt file and the assets into the android project.
- Create activities in Android to exploit the model and obtain nutritional value results.

Dataset Information
See Table 1.

3 Graphs for MobileNets and Inception

MobileNet provides enough accuracy as compared to Inception considering the low computation power and the time it takes. Inception provides more accuracy but is not much suitable for mobile devices.

- This architecture uses depth-wise separable convolutions which significantly reduces the number of parameters when compared to the network with normal convolutions with the same depth in the networks. This results in lightweight deep neural networks.
- The normal convolution is replaced by depth-wise convolution followed by point-wise convolution which is called as depth-wise separable convolution (Figs. 1 and 2).

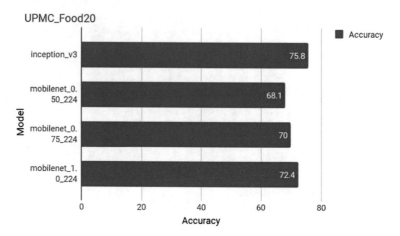

Fig. 1 Analysis of UPMC_Food20

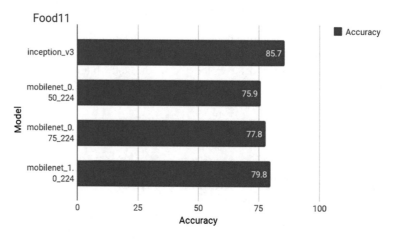

Fig. 2 Analysis of food11

So, what we did here is train datasets on various MobileNet models and Inception v3 and did analysis based on that. MobileNets are suitable for low-compute power devices, whereas Inception requires more computation power (Fig. 3).

Analysis

MobileNet provides enough accuracy as compared to Inception considering the low computation power and the time it takes. Inception provides more accuracy but is not much suitable for mobile devices.

(1) **UPMC_Food20**

This dataset has 20 categories of food with 100 images per category. MobileNet 1.0 gives the maximum accuracy here when we consider dimension = 224.

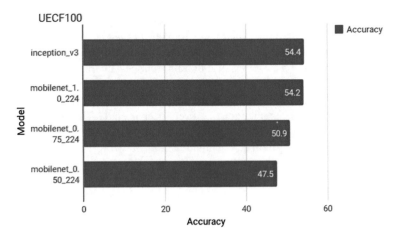

Fig. 3 Analysis of UECF100

Accuracy is 72.4% compared to Inception's 75.8% which requires more computation and time. The thing to be considered here is the trade-off between speed and accuracy. As the speed increases, the accuracy will decrease.

(2) **UECF100**

The dataset contains 100 kinds of food photograph. Each food photograph has a bounding box indicating the location of the food item in the photograph. This dataset gave less accuracy as compared to others, because there is an increase in the number of classes. This tells us that there is also a trade-off between the number of classes and the number of images in it. As the number of images per category is only 100, the accuracy reduces. Even the highly powerful Inception gives accuracy as less as 54.4%.

(3) **Food11**

The dataset contains 11 categories of food with lots of images in the range of 800–1500 per category. Due to the increase in the number of images per category, accuracy increases. Also the number of classes is less, thus increasing accuracy even more. MobileNet 1.0 224 gives accuracy 79.8% the highest among the other datasets. Hence, we can conclude that as the number of images per category increases, accuracy of classification of the images into the correct class increases.

4 Implementation

We have implemented the aforementioned system by means of an Android application. The user can select various meals of the day and categories of food. The application gives the suggested calorie intake along with the actual calorie count.

The application uses the smartphone camera to get the input which is provided to the object detection algorithm which helps to identify the respective food item. To improve the accuracy over time, we have used machine learning algorithms. The user has various options, and the user can use the smartphone camera, choose the desired food item from the list of options provided in—app and choose from the recently searched item or frequently searched items. Everyday intake can also be monitored using the application. The application has a very user-friendly GUI and can function online as well as offline.

The nutritional value system can be divided into following components:

(1) Object detection
(2) Machine learning
(3) Barcode scanning
(4) Nutritional information
(5) User diary (daily log).

• *Object Detection*

Object detection is a computer technology related to computer vision and image processing that deals with detecting instances of semantic objects of a certain class (such as humans, buildings or cars) in digital images and videos. We are using it for detection of food items. The main advantage of using object detection is that in one food image, more than one category of food can be identified, e.g., object detection using MobileNet SSD (Fig. 4).

• *Machine Learning*

Machine learning is used to make sure that the system improves based on the previous interactions in order to increase the accuracy overtime.

• *Barcode Scanning*

We have integrated a barcode scanning option for the user, if there is a barcode available on the food product. If the food product is not getting detected or classified due to some reason, the user can avail the barcode scanning option. Just point the camera toward the barcode on the food product, and the application will display the detailed information about the product.

• *Nutritional Information*

After all the detection and classification is over, the application displays the nutritional information about the classified food item. Depending upon the classified category and quantity of the food item, it will display the calories present in the food and also suggested calories. Then, the food is being divided into macros, i.e., display of protein, fats and carbohydrates. Each food item will show different, and it also depends on the quantity. The nutritional info here is shown as per 100 g or 500 g and is then scaled according to the user consumption.

Fig. 4 Object detection

- *User Diary*

A daily diary or logbook can be maintained by the user, where he can see the breakfast, lunch, snacks and dinner intake. What foods has he/she eaten, how much calories consumed, what all food items are present. The user can see up to one week's previous entries and can plan up to two days ahead of time. Using this diary, user can find out trends and the level of calorie consumption or his day-to-day food habits (Fig. 5).

5 Conclusion

The problem of increasing food adulteration is posing a serious harm to the society. While laboratory procedures are the ultimate test for checking quality, it is not feasible to use them every time. The final system, if properly implemented, will be a huge benefit in day-to-day life as we will be able to quickly decide whether the food is fit for consumption and what its nutritional value is. The main advantage of such a system being that no extra investment is required for this procedure as smartphones are widely available.

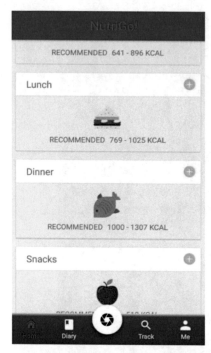

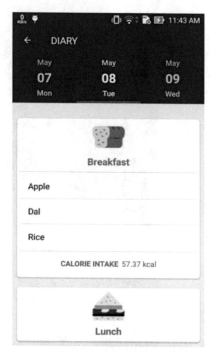

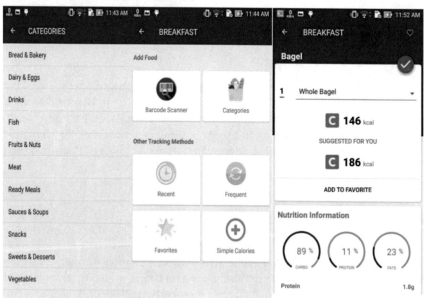

Fig. 5 Object detection and analysis

References

1. World Health Organization. 2018. Obesity and overweight. Available at: http://www.who.int/news-room/fact-sheets/detail/obesity-and-overweight. Accessed May 29, 2018.
2. The Economist. 2018. Putting the smallest first. Available at: https://www.economist.com/node/17090948 Accessed May 29, 2018.
3. The Economist. 2018. The Indian exception. Available at: https://www.economist.com/node/18485871. Accessed May 29, 2018.
4. Singh, D.P., P.J. Gadakh, P.M. Dhanrao, S. Mohanty, D. Swain, and D. Swain. 2017. An application of NGBM for forecasting indian electricity power generation. *Advances in Intelligent Systems and Computing* 556: 203–214.
5. Pathan, A., R. Kokate, A. Mutha, P. Pingale, and P. Gadakh. 2016. Digital India: IoT based intelligent interactive super market framework for shopping mall. *Engineering Science* 1(1): 1–5.
6. Singh, V., and S. Dwivedi. Ultrasonic detection of adulteration in fluid foods. In: *Proceedings of the First Regional Conference, IEEE Engineering in Medicine and Biology Society and 14th Conference of the Biomedical Engineering Society of India. An International Meet.*
7. Pouladzadeh, P., S. Shirmohammadi, and R. Al-Maghrabi. 2014. Measuring calorie and nutrition from food image. *IEEE Transactions on Instrumentation and Measurement* 63 (8): 1947–1956.
8. He, H., F. Kong, and J. Tan. 2016. DietCam: Multiview food recognition using a multikernel SVM. *IEEE Journal of Biomedical and Health Informatics* 20 (3): 848–855.
9. Villalobos, G., R. Almaghrabi, P. Pouladzadeh, and S. Shirmohammadi. 2012. An image processing approach for calorie intake measurement. In *Proceedings of IEEE Symposium on Medical Measurement Applications*, 1–5, Budapest, Hungary.
10. Villalobos, G., R. Almaghrabi, B. Hariri, and S. Shirmohammadi. 2011. A personal assistive system for nutrient intake monitoring. In: *Proceedings of ACM Workshop Ubiquitous Meta User Inter,* 17–22, Scottsdale, AZ, USA.
11. Ginesu, G., D. Giusto, V. Margner, and P. Meinlschmidt. 2004. Detection of foreign bodies in food by thermal image processing. *IEEE Transactions on Industrial Electronics* 51 (2): 480–490.
12. Anthimopoulos, M., L. Gianola, L. Scarnato, P. Diem, and S. Mougiakakou. 2014. A food recognition system for diabetic patients based on an optimized bag-of-features model. *IEEE Journal of Biomedical and Health Informatics* 18 (4): 1261–1271.
13. Karuppuswami, S., A. Kaur, H. Arangali, and P. Chahal. 2017. A hybrid magnetoelastic wireless sensor for detection of food adulteration. *IEEE Sensors Journal* 17 (6): 1706–1714.
14. Dave, A., D. Banwari, S. Srivastava, and S. Sadistap. 2016. Optical sensing system for detecting water adulteration in milk. In: *2016 IEEE Global Humanitarian Technology Conference (GHTC).*
15. Nandi, C., B. Tudu, and C. Koley. 2016. A Machine vision technique for grading of harvested mangoes based on maturity and quality. *IEEE Sensors Journal* 16 (16): 6387–6396.
16. Kong, F. and J. Tan. 2011. Dietcam: Regular shape food recognition with a camera phone. In: *2011 International Conference on Body Sensor Networks (BSN)* 127–132.
17. Casasent D., M.A. Sipe, T.F. Schatzki, P.M. Keagy, and L.L. Lee. 1998. Neural net classification of X-ray pistachio nut data. *Lebensmittel Wissenschaft and Technologie* 31: 122–128.
18. Davidson, V., J. Ryks, and T. Chu. 2001. Fuzzy models to predict consumer ratings for biscuits based on digital image features. *IEEE Transactions on Fuzzy Systems* 9 (1): 62–67.
19. Neelamegam, P., S. Abirami, K. Vishnu Priya, and S. Valantina. 2013. Analysis of rice granules using image processing and neural network. In: *2013 IEEE Conference on Information and Communication Technologies.*
20. Anami, B.S., V. Burkpalli, S.A. Angadi, and N.M. Patil. 2003. Neural network approach for grain classification and gradation. In: *Proceedings of the Second National Conference on Document Analysis and Recognition* 394408.

21. O'Farrell, M., C. Sheridan, E. Lewis, C. Flanagan, J. Kerry, and N. Jackman. 2007. Online optical fiber sensor for detecting premature browning in ground beef using pattern recognition techniques and reflection spectroscopy. *IEEE Sensors Journal* 7 (12): 1685–1692.
22. Khosa, I. and E. Pasero. 2014. Artificial neural network classifier for quality inspection of nuts. In: *2014 International Conference on Robotics and Emerging Allied Technologies in Engineering (iCREATE)*.
23. Sasano, S., X. Han, and Y. Chen. 2016. Food recognition by combined bags of color features and texture features. In: *2016 9th International Congress on Image and Signal Processing, Bio-Medical Engineering and Informatics (CISP-BMEI)*.
24. Mueen, A., M.S. Baba, and R. Zainuddin. 2007. Multilevel feature extraction and X-ray image classification. *Applied Sciences*, 7: (8) 1224–1229.
25. Lee, K., Q. Li, and W. Daley. 2007. Effects of classification methods on color-based feature detection with food processing applications. *IEEE Transactions on Automation Science and Engineering* 4 (1): 40–51.
26. Singh, V., and S. Dwivedi. 1995. Ultrasonic detection of adulteration in fluid foods. In: *Proceedings of the First Regional Conference, IEEE Engineering in Medicine and Biology Society and 14th Conference of the Biomedical Engineering Society of India. An International Meet*.
27. Howard, Andrew G., Menglong Zhu, Bo Chen, Dmitry Kalenichenko, Weijun Wang, Tobias Weyand, Marco Andreetto, and Hartwig Adam. 2017. MobileNets: Efficient convolutional neural networks for mobile vision applications. CoRR, abs/1704.04861.

Detection of Hate Speech in Hinglish Language

Rahul S. Varade and Vikas B. Pathak

Abstract As mobile phones and Internet become more and more popular, the number of social media users in India continues to go up. Majority of Indian social media users use Hinglish as their medium of communication. The Hinglish language is a mixture of Hindi words (typed in English) and English words. However, with increasing numbers, there is also an increase in the amount of hate-filled messages, posts, and comments put up on social media platforms. Hate speech is usually done to target an individual or group of individuals on the basis of caste, community, ethnicity, religion, gender, or any other discriminating factor. It can have negative impacts on the individuals facing it and consequently on the society as well. As the amount in which such kind of content is generated is huge, it becomes necessary to automatically detect hate speech so that preventive measures can be taken to control it. Although there has been quite a lot of research on hate speech detection in English texts, not much work can be found on hate speech detection in Hinglish language. This paper presents an approach of detecting hate speech in Hinglish texts using long short-term memory (LSTM), which works on word embeddings generated by gensim's word2vec model.

Keywords Long short term memory · Hinglish · Transliteration · Word embedding · Hate speech detection

1 Introduction

The online space is rapidly growing by leaps and bounds year after year. The Indian social media space is a highly diverse community as people hailing from different ethnicities and religions are a part of it. As of 2018, there were 326.1 million social media users in India. It is estimated that it will cross 400 million by 2021 [1]. Lowering

R. S. Varade · V. B. Pathak (✉)
Vishwakarma Institute of Technology, Pune, India
e-mail: vjpathak59@gmail.com

R. S. Varade
e-mail: rahulvarade96@gmail.com

© Springer Nature Singapore Pte Ltd. 2020
D. Swain et al. (eds.), *Machine Learning and Information Processing*,
Advances in Intelligent Systems and Computing 1101,
https://doi.org/10.1007/978-981-15-1884-3_25

of mobile data rates resulting in cheaper data plans has largely contributed to this increase in numbers. Facebook, Twitter, and Instagram are few of the most widely used social media applications in India. When people from such a wide variety of backgrounds come together, discussions and disagreements are inevitable. However, many a times it is observed that these disagreements transform into heated debates where people start using harsh and abusive words. Such situations become the source of hate speech.

In Indian social media, hate speech can be mostly observed in forums or threads related to politics and religion. People who are politically or communally polarized often resort to using hate speech. Statements like 'ise pakistan bhej dena chahie' or 'aise deshdrohiyo ko bharat mein rehne ka koi haq nahi hai' are very common. Such statements can create social unrest as they can easily offend someone and can even lead to violent incidents like riots. In recent past, there was a rise in number of mob lynching and communal violence incidents in India fueled by propaganda-driven hate messages, which would usually get viral through WhatsApp groups [2]. Social scientists have found a correlation between online hate speeches and hate crimes [3]. Thus, the problem of hate speech is worrying and needs to be kept in check.

Facebook, Instagram, and Twitter have taken measures from their own end by letting users report offensive posts, comments, or tweets. A team of experts then reviews such reports and takes appropriate actions. However, all these measures are reactionary in nature and not proactive. Besides, almost all sentiment analysis tools are for English texts, whereas very few works of sentiment analysis for Hinglish texts can be found on the Internet. A Hinglish word is a Hindi word spelled in Roman script. Thus, the same Hindi word is bound to suffer from spelling variations. For example, for the Hindi word "प्यार" there could be different spellings like pyaar, pyar, pyara, etc. Thus, a lot is dependent on phonetics. Due to all these reasons, there arises a need for making an effort in the direction of automatic hate speech detection in Hinglish language.

In this paper, a variant of recurrent neural network called long short-term memory (LSTM) is used, which takes word embedding values as input for training and then classifies Hinglish documents into two categories—(1) hate and (2) not hate. LSTM is a deep neural network which captures long-term dependencies between words of the same text document. In the following sections, the paper describes the related work, the process of collecting and constructing the dataset, the proposed methodology which is followed by performance analysis. The paper concludes by mentioning the scope of future works that can be pursued for the same problem statement.

The main goals of this work are as follows:

• Classification of Hinglish tweets as hate or not hate and
• Compilation of a dictionary of over 11k Hindi words and phrases with their corresponding English translations.

2 Related Work

Mathur et al. [4] used multi-input multi-channel transfer learning-based model for detecting offensive Hinglish tweets. They used a hybrid CNN-LSTM architecture which was pre-trained on English tweets and then used for transfer learning. Ravi and Ravi [5] concluded that a combination of TF-IDF features with gain ratio-based feature selection when passed to radial basis function neural network gives best results for sentiment classification in Hinglish text. Bohra et al. [6] did supervised classification on Hinglish tweets using support vector machine and random forests on character-level, word-level, and lexicon-based features. Each word of every tweet was labeled with its language, i.e., English or Hindi, and all the tweets were classified into two classes—hate speech and normal speech.

Kunchukuttan et al. [7] developed Brahmi-Net, a publicly available transliteration tool on the Internet for interconversion of words in scripts of different Indian languages as well as English. The services of Brahmi-Net were made available via a REST API.

Pan et al. [8] suggested that how transfer learning can be used to solve various text classification problems. Bali et al. [9] did analysis on various Facebook posts, generated by Hinglish multilingual operators. They concluded that many posts contain significant amount of code-mixing.

3 Dataset Collection

Table 1 illustrates the detailed dataset distribution. There are a total of 6431 text documents out of which 2211 are categorized as hate speech and remaining 4431 are categorized as non-hate speech. Ratio of hate text documents to non-hate text documents is approximately 1:2, since in real life as well the proportion of usage of non-offensive language is more as compared to that of offensive langauge.

Table 2 illustrates how multiple documents from various sources are combined to create the dataset. Hate tweets are gathered from HOT dataset [4] and Aggression dataset [10]. HOT dataset consists of manually annotated Hinglish tweets (Hindi tweets written in English script), primarily classified into two categories: '1' for tweets which do not express any kind of hate or offense and '0' for tweets which express hate or offense. Similarly, aggression dataset consists of three distinct categories of tweets which are listed as follows: covertly aggressive (CAG), overtly

Table 1 Tweets distribution

Label	Tweets
Hate	2211
Non-hate	4431
Total	6642

Table 2 Tweets collection description

Class	Dataset used	Selected labels	Extracted number of records	Total
Hate	HOT dataset	0	303	2211
	Aggression dataset	CAG/OAG	1908	
Non-hate	HOT dataset	1	1121	4431
	Hindi–English code mixed dataset	Neutral/positive	3310	

aggressive (OAG), and non-aggressive (NAG). In the present paper, tweets which are having label '0' in HOT dataset and tweets having labels 'CAG' or 'OAG' are considered as hate tweets.

4   Proposed Methodology

This section focuses on brief description about system workflow overview, LSTM architecture, and detailed description system workflow.

4.1   System Workflow Overview

Figure 1 gives an overview of the workflow of the present system. A raw Hinglish dataset annotated with '1' (hate) and '0' (non-hate) labels is taken and made to go through some cleaning steps like removal of URLs, usernames, special symbols, etc.

Then, two separate dictionaries—profanity list and Hindi–English dictionary—are used to replace all the Hindi words in the dataset with their appropriate English translations. Both the dictionaries are self-constructed and will be discussed later in detail in Sect. 4.3.1. This is followed by stop words removal, lemmatization, and generation of word embeddings using gensim's word2vec model. These word embeddings are used later for creating an embedding layer for the LSTM network. The word2vec model also generates a vocabulary list from the dataset, which is used to replace each word in the dataset by its corresponding integer index in the vocabulary list. A 75:25 train–test split is performed on the resultant dataset. The training data along with the word embedding layer is used to create and train the LSTM network, followed by making predictions using testing data. All these steps are discussed in detail in Sect. 4.3.

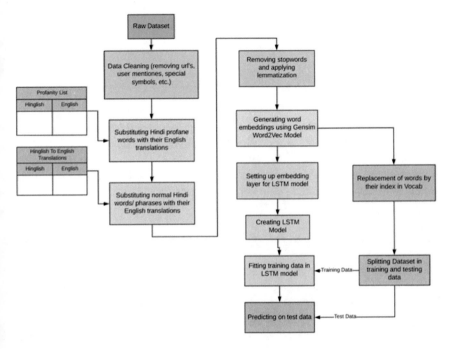

Fig. 1 System workflow diagram

4.2 LSTM Architecture

Long short-term memory (LSTM) is a special kind of RNN proficient in handling long-term dependencies as it has the ability to store information for long periods of time. LSTMs have chainlike structure containing a repeating module—the LSTM cell. Every repeating module consists of four different neural network layers. Out of these four, three layers have sigmoid activation function and the fourth one has tanh activation function. Figure 2 describes the detailed chainlike architecture of the LSTM repeating module. The cell state is a key element in LSTM. It is a vector containing relevant information throughout the multiple time steps of LSTM. LSTM has the capability to remove and add new data/information in cell state by using

Fig. 2 Detailed architecture
of LSTM repeating module

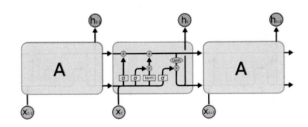

Fig. 3 Removing Irrelevant
Information from LSTM
Cell State

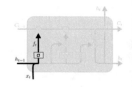

gates. Unlike RNNs, LSTM does not fail to retain information from many time steps
ago. Rather, LSTM is made to forget irrelevant old information.

Steps of LSTM

In every time step, new data (x_t) and output from previous time step (h_{t-1}) are
combined to form the input data for current time step.

4.2.1 Removing Irrelevant Information from LSTM Cell State

To decide which information is irrelevant and should be thrown out, forgetting gate
layer is used. The forgetting gate layer uses the sigmoid activation function which
takes input of a combination of new data (x_t) occurring at the new time step and the
output of the hidden layer from the previous time step (h_{t-1}). The sigmoid activation
function in the forgetting gate layer outputs a bunch of values ranging from 0 to 1,
where each value corresponds to a value stored in the cell state vector c_{t-1}. Here,
0 suggests completely discard the data, while 1 represents completely keeping the
data in the cell state. Only that data for which f_t . C_{t-1} is nonzero will be kept in the
cell state. It is demonstrated in Fig. 3.

$$f_t = \text{sigmoid}\,(W_f.[h_{t-1}, X_1] + b_f)$$

4.2.2 Updating the LSTM Cell State

For deciding and adding new information in the cell state, LSTM uses two different
neural network layers—one with sigmoid activation and the other one with tanh
activation. The first input gate layer (with sigmoid activation function) decides which
values should get updated. Next layer with tanh activation function generates a vector
of newly generated values that could be added to the cell state. To add new information
in cell state, $(I_t * C \sim t)$ is added in current cell state as shown in Fig. 4.

Fig. 4 Updating the LSTM
Cell State

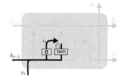

Fig. 5 Output of Current
Time Step

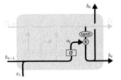

$$I_t = \text{sigmoid}\,(W_i \,.\, [h_{t-1}, x_t] + b_i)$$
$$C \sim t = \tanh\,(W_c \,.\, [h_{t-1}, x_t] + b_c)$$

4.2.3 Deciding the Output of Current Time Step

The current cell state is filtered through tanh, and the resulting vector is multiplied point by point with the result obtained by applying sigmoid layer on combination of x_t and h_{t-1}. Tanh outputs values between -1 and 1. The resultant product gives the output of current time step (h_t) which is passed as h_{t-1} for the next time step. It is demonstrated in Fig. 5.

4.3 System Workflow (In Detail)

4.3.1 Preprocessing

A dataset of 6642 tweets is passed through the following steps:

Removing URLs, removing username mentions, converting 'n't' abbreviation, removing numbers, removing 'RT' abbreviation, removing Devanagari literals, multiple occurrences of same characters are replaced with a single character (for example, 'pyaaaaaaar' is converted to 'pyar').

The dataset contains English words, normal Hindi words, as well as profane Hindi words. In order to deal with normal and profane Hindi words, two dictionaries are constructed. Firstly, a dictionary of profane Hindi words along with their English translations is created by exploring various online sources and selecting the most common profane words in Indian social media. As of now, it contains 224 unique words. Secondly, an 11662-length long dictionary storing English translations of Hindi words or phrases from the dataset is constructed. The procedure to construct dictionary is as follows:

- Extract all non-English words in corpus.
- Using Selenium Automation tool, all those extracted words were sent for translation on the Google translate page where source language as Hindi and destination language as English were programmatically selected and their translations were stored in a file.

After the above two dictionaries are ready, the entire corpus is scanned document by document, which is in turn scanned word by word. If any of the words is found in either of the two dictionaries, it is replaced by its corresponding English translation.

4.3.2 Generating Word Embeddings

The word2vec is trained on the preprocessed dataset to generate a vocabulary of 3498 words occurring at least twice in the corpus. The word2vec outputs a 300-dimensional vector for each word in the vocabulary. Using these vectors, an embedding matrix of dimensions 3498 × 300 is created. Each word in corpus is replaced with its index in the vocabulary generated by word2vec. Thus, each document becomes a sequence of numeric literals. If vectors of two words are close to each other, then it implies that those words mostly occur in the same context in the corpus. Thus, vector representation of words in word2vec is based on their relative similarity of contexts. For example, in a corpus, the word *salman* occurs with the word *bajrangi* mostly in the same context. Using the embedding matrix, a Keras Embedding layer is created which stores mapping of each integer (representing a word in a corpus) and the corresponding embedding vector of the word that the integer represents.

4.3.3 Training LSTM Model and Making Predictions

A Keras Sequential model is created by adding the embedding layer as the first layer. Then, it is followed by adding an LSTM layer of seven units, which implies that each of the hidden layers (input gate layer, forget gate layer, etc., in the LSTM cell) will have 7 activation units. The number of LSTM units was determined with the following formula:

$$N_h = N_s/(\alpha * (N_i + N_o))$$

where N_h is count of LSTM units, N_i is count of activation units in the input layer, N_o is count of activation units in the output layer, N_s is number of samples in training dataset, and α is an arbitrary number between 2 and 10.

Here, values taken are $N_i = 300$, $N_o = 1$, $N_s = 4823$, $\alpha = 2.3$.

Therefore, the number of LSTM units is **N_h= 7** (approx.)

Then, a dropout layer with dropout rate as 0.4 is added which means that 40% of the input layer units will be set to 0 at each update during training, to prevent overfitting. Finally, a dense layer with 1 activation unit and sigmoid activation is added to the network. Then, it is compiled using 'binary cross entropy' loss function, 'adam' optimizer algorithm, and 'accuracy metrics' as the evaluation metrics.

'binary cross entropy' loss function is grouping of 'sigmoid activation' function along with 'cross entropy loss' function. It is basically used for binary classification as it isolates decisions from multiple classes. 'adam' optimizer function is enhanced version of stochastic gradient optimizer. It is taken with its default parameter values, so that it can optimize parameters such that it can reduce loss. The 'accuracy metrics' is used to judge the performance of the model when it is compiled.

A train–test split of 75/25 is performed. The training data is used for training the above designed deep neural network in a single epoch. The test data is used to evaluate the trained model using the accuracy metrics.

5 Performance Analysis

Multiple experiments were carried out by tuning different hyperparameter values to obtain best accuracy on test data. The following hyperparameters were tuned during the experiments—number of LSTM units (number of activations in hidden layers of one LSTM cell), dropout rate, and number of epochs. Since the training data contains only 4982 samples, the model started to overfit as the number of epochs was increased. Hence, it is concluded that the number of training epochs should be 1. Then, different values for number of LSTM units were tried while keeping dropout rate at 0.2 and 0.4. Table 3 demonstrates the training and testing accuracies recorded for dropout rate at 0.2, and Table 4 illustrates the training and testing accuracies recorded for dropout rate at 0.4. The number of training epochs was kept at 1 in both the cases.

Table 3 Training and testing accuracies with dropout rate = 0.2

No. of LSTM units	Dropout rate	Training accuracy	Testing accuracy
5	0.2	94.09	84.70
7		94.78	84.14
9		94.38	84.51
11		95.23	84.27
13		95.23	84.20
15		94.22	84.45

Table 4 Training and testing accuracies with dropout rate = 0.4

No. of LSTM units	Dropout rate	Training accuracy	Testing accuracy
5	**0.4**	93.26	84.83
7		95.53	84.89
9		**91.83**	**85.07**
11		93.74	84.64
13		93.33	84.20
15		92.29	84.64

Bold shows highest accuracies and the corresponding hyper parameter values

Table 5 Precision and recall values with dropout rate = 0.2

No. of LSTM units	Dropout rate	Precision	Recall
5	0.2	0.7593	0.6673
7		0.7409	0.7342
9		0.7381	0.6980
11		0.7464	0.7396
13		0.7673	0.7034
15		**0.7912**	0.6854

Bold value shows highest precision and recall values respectively

Table 6 Precision and recall values with dropout rate = 0.4

No. of LSTM units	Dropout rate	Precision	Recall
5	0.4	0.7737	0.6491
7		0.7155	**0.7504**
9		0.7604	0.6600
11		0.7339	0.7432
13		0.7642	0.7034
15		0.7794	0.6835

Bold value shows highest precision and recall values respectively

From above tables, it can be seen that maximum testing accuracy of 85.07% was achieved for the following combination of hyperparameter values:

- Number of LSTM units—9 and
- Dropout rate—0.4.

Besides, other important evaluation metrics like precision and recall were also computed for dropout rates at 0.2 and 0.4. Table 5 demonstrates precision and recall values for dropout rate at 0.2, and Table 6 demonstrates precision and recall values for dropout rate at 0.4. The number of training epochs was kept at 1 in both the cases. Maximum precision of 0.7912 was achieved for 15 LSTM units and 0.2 dropout rate. Maximum recall of 0.7504 was achieved for 7 LSTM units and 0.4 dropout rate.

It can be seen that best results for accuracy, precision, and recall were obtained for different combinations of hyperparameter values. Precision represents the number of correct positive predictions out of total positive predictions. Here, positive prediction corresponds to the class with label 1, i.e., hate class. Out of total predictions for hate class, 79.12% were actually from hate class. Likewise, recall represents the fraction of actual positives that were correctly identified. Out of all the actual hate class samples, 75.04% samples were identified to be from the hate class. Misclassifying non-hate samples as hate samples is always less risky than misclassifying hate samples as non-hate. Hence, correctly identifying all the hate samples is more important. Thus, out of all evaluation metrics, recall value is chosen for setting the best combination of hyperparameter values.

6 Conclusion and Future Scope

In this paper, an approach to detect hate speech in Hinglish texts using LSTM was proposed. The dataset used comprised of hate and non-hate Hinglish texts, collected from various sources. The LSTM network was trained on word embeddings generated by gensim's word2vec model. A maximum recall value of 0.7504 was achieved for a specific set of hyperparameters of the LSTM network. Besides, two dictionaries namely—profanity list dictionary and Hindi-to-English dictionary—were also constructed. The profanity list dictionary contains English translations of 224 profane Hindi words, and the Hindi-to-English dictionary contains English translations of 11662 Hindi words or phrases. These dictionaries were used in the system to translate the Hinglish documents in the dataset into English, word by word.

It can be noted that this paper presents an approach in which the Hinglish sequence of words in the dataset was translated word by word, resulting in loss of proper sentence structure and thereby making the sentence meaningless. To overcome the problem of losing structure and sense from the sentence, a proper language translation model needs to be developed that can translate a Hinglish sentence into meaningful English sentence. Once the translation is done, various sentiment analysis and NLP tasks can be conveniently conducted on such data. Apart from using only word embeddings as input features, other semantic and sentimental features can also be included in the feature set. Those semantic features may include number of different types of punctuations, different types of emoticons among others. Sentimental features like sentiment score of text document and score of profane words in the text document can be included. Thus, by creating a diverse set of features, one can develop an even more robust hate speech detection model.

References

1. https://www.statista.com/statistics/278407/number-of-social-network-users-in-india/.
2. https://www.washingtonpost.com/graphics/2018/world/reports-of-hate-crime-cases-have-spiked-in-india/?noredirect=on&utm_term=.cfbf7bbb0432.
3. https://www.cfr.org/backgrounder/hate-speech-social-media-global-comparisons.
4. Mathur, Puneet, Ramit Sawhney, Meghna Ayyar, and Rajiv Shah. 2018. Did you offend me? Classification of offensive tweets in Hinglish language. In: *Proceedings of the 2nd Workshop on Abusive Language Online (ALW2)*.
5. Ravi, Kumar, and Vadlamani Ravi. 2016. Sentiment classification of Hinglish text. In: *3rd International Conference on Recent Advances in Information Technology (RAIT)*.
6. Bohra, Aditya, Deepanshu Vijay, Vinay Singh, Syed S. Akhtar, and Manish Shrivastava. 2018. A dataset of hindi-english code-mixed social media text for hate speech detection. In: *Proceedings of the Second Workshop on Computational Modeling of People's Opinions, Personality, and Emotions in Social Media*.
7. Kunchukuttan, Anoop, Ratish Puduppully, and Pushpak Bhattacharyya. 2015. Brahmi-Net: A transliteration and script conversion system for languages of the Indian subcontinent. In: *Conference of the North American Chapter of the Association for Computational Linguistics—Human Language Technologies: System Demonstrations*.

8. Pan, Weike, Erheng Zhong, and Qiang Yang. 2012. Transfer learning for text mining. *Mining text data*. Boston: Springer.
9. Vyas, Yogarshi, Spandana Gella, Jatin Sharma, Kalika Bali, and Monojit Choudhary. 2014. Pos tagging of english-hindi code-mixed social media content. In: *Conference on Emperical Methods in Natural Language Processing (EMNLP)*, 974–979.
10. Kumar, Ritesh, Aishwarya N. Reganti, Akshit Bhatia, and Tushar Maheshwari. 2018. Aggression-annotated corpus of hindi-english code-mixed. In: *Proceedings of the 11th Language Resources and Evaluation Conference (LREC)*.

Enhancing Multi-level Cache Performance Using Dynamic R-F Characteristics

Akshay Motwani, Debabrata Swain, Nikhil Motwani, Vinit Vijan, Akshar Awari and Banchhanidhi Dash

Abstract Cache memory or CPU memory is a high-speed static random access memory that a computer microprocessor can access more quickly than it can access regular random access memory. Hence, the high-performance cache memory is used to bridge the performance gap between the processor and main memory. Multi-level caches refer to a type of memory hierarchy which uses memory stores with different access speed to cache data. Our proposed work uses combination of different tuning algorithms considering the R-F characteristics of page to provide an efficient solution for cache replacement in multi-level cache hierarchy which has an easy implementation and a better performance compared to traditional cache replacement policies like clock with adaptive replacement (CAR), least recently used (LRU), and first in, first out (FIFO) on a cache of equivalent size.

Keywords Clock with adaptive replacement (CAR) · Least recently used (LRU) · First in, first out (FIFO) · Classical weight ranking policy (CWRP)

1 Introduction

The pessimal utilizations of memory in computers are causing a significant problem these days. To revamp the utilization of memory, caching is a method that can be applied. Caching is considered to be an elementary method used in today's computers. Access time and hit rate are the two factors that affect the utilization in caching [1]. Caching is applied to enhance the speed by decreasing the access time. Disk cache, processor cache, and Web cache are few of the exemplars of it. The pace of the main memory is significantly lacking than that of cache.

A. Motwani · N. Motwani · V. Vijan · A. Awari
Department of Computer Engineering, Vishwakarma Institute of Technology, Pune, Odisha, India

D. Swain (✉)
Information Technology Department, Vishwakarma Institute of Technology, Pune, India
e-mail: debabrata.swain@vit.edu

B. Dash
School of Computer Engineering, K.I.I.T. University, Bhubaneswar, India

© Springer Nature Singapore Pte Ltd. 2020
D. Swain et al. (eds.), *Machine Learning and Information Processing*,
Advances in Intelligent Systems and Computing 1101,
https://doi.org/10.1007/978-981-15-1884-3_26

An easiest way to enhance cache performance is to increase the size of it, but it is not an inexpensive solution. Therefore, this effect is seen on the size of cache which is very small than that of the main memory. The working of cache is as follows, if accessed data is present in the cache, that means the data is found in cache and is called as 'cache hit.' Otherwise, if the data is not present 'cache miss' takes place [1]. The percentage of overall cache hit along overall access is called as 'hit ratio' or 'hit rate.' The dominant variables that determine the cache utilization are access time and hit rate [1].

Various substitution techniques are applied to revamp cache potency. The cache does not contain any pages or blocks at the beginning so the inceptive access is a miss and this is constant until the cache is filled with pages or blocks, where pages or blocks are consistently sized articles in the cache memory. After this gradually when a page or block is to be put on to cache which does not exist in cache, if a cache miss takes place then this particular page or block is to be restored with latest page or block that is there in the cache.

The execution of the restoration of a page can be done by substitution techniques where to choose a page with less probability of being looked for ahead in the cache and to restore the particular page with a latest page is its vital part.

There are three categories of mapping methods which correspond with the cache: (1) Blocks within the main memory are mapped to a lone block of cache is entitled as direct mapping. (2) Rather than mapping to a distinctive block of cache, fully associative gives the permission to save the data in any block. (3) The method where the overall cache is divided into sets is called as set associative. Set associative is a combination of fully associative mapping and direct mapping in which sets are directly mapped to the particular number of blocks in main memory and fully associative mapping is used within each set [2–8].

An optimal substitution technique should be easy to enact and should lessen the overhead of the central processing unit. The use of multi-level cache has become pervasive in computing systems. Increasing the cache hit ratio would greatly enhance the performance of the system. Our observation of cache memory tells us that selecting the victim page for page replacement plays an important role in determining the cache hit ratio. The recency and frequency of the page play a critical value in predicting whether that particular page would be referenced in the ensuing future [9]. We have taken this into consideration while formulating a method that would enhance the performance of multi-level cache. We have considered a cache having three levels, namely L1, L2, and L3. L1 cache is near to the processor and having less access time, while L3 is far from the processor and is having more access time compared to L1 and L2. L2 level cache is in between L1 and L3 having moderate access time. L2 and L3 level caches are exclusive to L1 cache. The pages evicted from L2 are placed into L3 cache rather than going to main memory.

Our method uses a combination of cache replacement algorithms to increase the cache hit ratio of the overall multi-level cache. In case of L1 cache, frequency, recency, and freshness of the page would be considered and each page in L1 would be given a R-F score accordingly. The one having least R-F score would be selected for replacement in case of L1 cache. For L2 cache, least frequently used page would be selected as a victim for replacement, and in case of L3 cache, least recently used page would be selected for replacement.

The main thing to note about our proposed method is that blocks from the main memory are brought into L1 cache rather than placing them in L3 cache. The method formulated has a better performance than that of FIFO, CAR, LRU, and CWRP [9].

2 Background

The major applications of caching in a computer system can be found in operating system and database management systems. For application of page replacement in modern systems, LRU has been the first choice worldwide. LRU algorithm works on the assumption that if a page has not been referred to in a while, it is not imminent to be referenced to. In other words, the pages that have been used the most are in fact the ones referenced the most.

The LRU algorithm suggests that the pages that will be referenced in the future are actually the ones that have been majorly used. Thus, implying that, the pages that have been used the most are in fact the ones referenced the most. Hence, in the situation of a page fault being occurred, the course of action is to replace the page least referred, by the one being queried [10].

Drawbacks of the 'Least Recently Used' algorithm [12]:

1. Given a sequence of pages being referred only once, it might result them in leading to total depreciation in its evaluation.
2. As LRU does not focus on frequency of the pages being referred but only the pastness of it, performance issues are caused in the event of certain pages being referred repeatedly but with a wide time span, such as greater than the size of cache [11].
3. When it is a hit, the page is moved to most recently used (MRU) location. Now, as a computer system has many tasks and processes being carried out parallelly, there is a feature of lock on MRU to preserve the page and make certain of accuracy and regularity. But as the hits are serialized after the lock, it leads to great amount of conflicts, which is undesirable when it comes to its application in high turnout systems for storage such as file systems, VMs, and database systems [12].
4. In the case of page hit being occurred in a virtual memory, moving the page to MRU location is relatively more expensive [1].

Another cache replacement method which is also straightforward to execute is first in, first out (FIFO). FIFO, unsurprisingly, suggests to choose the earliest page for replacement. On the event of a page fault, old page is popped from the front of the queue and new page is pushed at the end of it. This algorithm results in lesser loads to the setup. FIFO is not a good fit when it comes to vast physical memory. Neither does it follow program model [13].

Least frequently used (LFU) algorithm works on the assumption that if a page has been used least frequently, it might not be referred to in the ensuing future and hence suggests on removal of that page. As miscellaneous fragments of the system memory have varying time variants, LFU is inappropriate.

Downsides of 'least frequently used' (LFU) algorithm [2–8] are as follows:

1. This method is not fruitful in the case where the page calling style varies as the cache is still populated with the pages with most frequency, which no longer remain useful.
2. For calculating the cache size, the method requires some application using logarithms.
3. The method ignores recency.

Adaptive replacement cache (ARC) uses both periodicity and recentness characteristics of the stack, balanced both effectively and continuously, and hence is less expensive and self-adjusting. This method also avoids continuous monitoring, so when a particular pattern of page calls takes place, cache memory remains unshaken [14].

Following is how ARC is implemented: Let 'x' be the size of the cache memory. Then, according to ARC, a cache directory of two times 'x' pages, i.e., '2x', is supported, namely 'x' for the pages as record to store history and remaining 'x' for the pages in cache. For dealing with the 'x' pages pertaining to the cache, ARC maintains two lists, that is, L1 and L2. L1 stores the pages that have been called very recently. L2 stores those pages which have been called with high frequency. Those pages are defined as having high frequency that have been called multiple times in a short interval of time. Hence, pages in both L1 and L2 when summed up give the total pages in the cache. The number of pages in the individual lists L1 and L2 varies and is determined by the other 'x' of the record pages stored in the cache directory of ARC. The 'x' pages in the directory that store the history are classified into two lists, namely B1 and B2. B1 stores those pages that were removed from L1, whereas B2 stores those pages that were removed from L2. In the case where any of the two lists L1 and L2 are full, least recently used (LRU) algorithm of page replacement is implemented to move the removed page into B1 or B2, respectively. If a page being called is present in L1, it is popped and is pushed into L2. When the page being called is present in B1 (also called as Phantom hit), then size of L1 is incremented by 1 and L2's size is reduced by 1, but if L2 was full before the operation, a page is then moved from L2 to B2, as per LRU. Contrarily, when a page being called is present in B2, then size of L2 is incremented by 1 and L1's size is reduced by 1, but if L1 was full before the operation of reduction, a page is then moved from L1 to B1, as per LRU. Hence, allowing the cache to tune as per high periodicity or the recentness characteristic of the stack [14].

Adaptive replacement cache (ARC) has a disadvantage such that the most recently used (MRU) location is heavily loaded as all the four lists L1, L2, B1, and B2 use LRU method of page replacement when either of them becomes full.

Another cache replacement policy known as clock with adaptive replacement (CAR) is also widely used because of it not only being identical to ARC, but also assuming all the benefits of it. It performs great on workloads that iterate over a working set larger than the size of the cache and is also less expensive in implementation.

Clock with adaptive replacement (CAR) maintains a cache directory with a count of '2x' pages, where first half 'x' denotes pages as record to store history and remaining 'x' for the pages in cache, just like the adaptive replacement cache (ARC) does. The 'x' pages in the directory that store the history are classified into two LRU lists, B1 and B2, similar to ARC. For dealing with the 'x' pages pertaining to the cache, CAR maintains two circular buffers or CLOCKs that are L1 and L2, with every page having a reference bit. Just like ARC, L1 stores the pages that have been called very recently and L2 stores the pages which have been called with high frequency. B1 stores those pages that were removed from L1, whereas B2 stores those pages that were removed from L2. When the page being called is present in B1, L1's size is increased by 1 and L2's size is reduced by 1, so as to maintain the cache's size upon their addition. Contrarily, when a page being called is present in B2, L2's size is increased by 1 and L1's size is reduced by 1. In such a manner, CLOCKs manage as per the high requests. This policy also prevents the addition of CLOCK and history list to overshoot the size of the cache. Insertion of latest pages is done behind clock hands for both the CLOCKs and their respective reference bit being set to 0. The reference bit of a page is changed to 1 in the event of a cache hit, whereas in the event of a page fault, the clock hand searches in clockwise direction, and when it finds a page with reference bit 0, it replaces it with the one referenced page and moves the one replaced in the MRU location of either B1 or B2, whichever applies. In the event of cache miss, when the clock hand of L1 encounters a page with reference bit 1, it moves it to L2 and modifies its value to 0, and thereon continues in clockwise direction as long as it locates a page with reference bit 0. The same is done with L2, when there is a cache miss; when the clock hand of L2 encounters a page with reference bit 1, it resets its reference bit to 0. L2 and keeps on searching in clockwise direction as long as it locates a page with reference bit 0 [12].

Both clock with adaptive replacement (CAR) and adaptive replacement cache (ARC) fail to perform in the case of two consecutive hits, as they both move the page from the recency list to high periodicity list, but quick and continuous hits are recognized as 'correlated references' and do not tend to be as frequent in the near future, hence decreasing the performance of the cache and making it more expensive for this case [12].

3 Proposed Work

We have considered a three-level cache hierarchy model in which L1 cache is closest toward the processor. L3 is farther away from the processor and nearer to the main memory. Figure 1 illustrates the three-level cache memory hierarchy model considered by us. Before proceeding further to the algorithm that is being employed to

Fig. 1 Three-level cache
hierarchy model

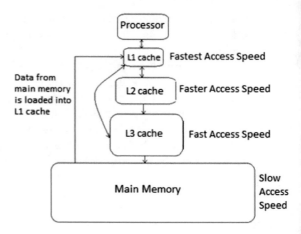

make this multi-level cache functional, following are the assumptions that need to be considered for our proposed algorithm to work successfully.

1. The size of L3 is triple as compared to size of L1.
2. The size of L2 is double as compared to that of L1.
3. All blocks are of same size.
4. When a memory block is needed by the system or the application, the processor first looks it into the cache.
5. Finite amount of memory is available in the system having specific number of blocks.

We have considered the recency and frequency values commonly known as R-F score of the pages [9]. Based on our analysis, we have used different tuning algorithms for different levels of cache that would increase the hit ratio of the overall multi-level cache and thus would increase the overall system performance.

3.1 Working

Whenever a block (say b) is needed by the system, the processor will first look into the L1 cache.

1. If it is not present in L1 level cache, then the block b is searched in the L2 level cache.

As L2 being exclusive L1, if the block b is present in L2 cache, then block is moved from L2 to L1. This movement of block from L2 to L1 may cause eviction from L1 level cache. The replacement block for L1 level cache is selected based on the dynamic weight ranking policy in which weights are assigned to the blocks based on their R-F scores. Following formula is used to assign weights or R-F score to the blocks.

$$Wi = fi/(Rh - Ri + 1) \qquad (1)$$

where is R_h latest recorded recency value, R_i is recency value for last access request made for ith block, and is the frequency value of ith block. Initially, W_i, f_i and R_i are set to 0.

The block having minimum weight value from the L1 cache is chosen for replacement with block b. The evicted block from L1 is then placed into L2 cache.

2. If block is not present in L2 cache, then L3 cache is searched. If the block is present in L3 cache, similar action is taken and the block b in L3 cache is swapped with the block having minimum R-F score from L1 level cache.
3. If the block b is not found in either of the caches, then it is a cache miss. The block b, then, has to be fetched from the main memory and it is placed in L1 level cache. This may cause removal of existing block from L1 level cache, given that it is full. The block having minimum R-F score is removed, and block b is placed in its position. The removed block from L1 is then placed into L2 cache.

If L2 cache is full. The replacement block in L2 cache is selected based on the LFU cache replacement policy. The evicted block from L1 cache is then replaced with the least frequently used block from L2 cache. The removed block from L2 cache is then positioned into L3 cache.

If L3 cache is full. The evicted block from L2 cache is replaced with least recently used block from L3 cache. The expelled block from L3 level cache is brought back to the main memory.

4. If it is found in either of the caches, then there is a cache hit. The frequency of the block b would be incremented by one.

The recency of that block would hold the latest clock access value.

Our proposed algorithm takes into account the combination of recency and frequency [9], only frequency and only recency of the blocks present in different levels of cache. For L1 level cache, both recency and frequency are analyzed and R-F scores of the block are then calculated accordingly. L2 level cache takes care of frequency feature of blocks, whereas L3 cache takes into account of recency feature of blocks. In this way, our multi-cache model takes the holistic view of capturing the recency and frequency features of the blocks and also considering them individually, as they play a decisive role in determining the cache hit ratio.

4 Performance Analysis

We have simulated our proposed algorithm with other traditional cache replacement algorithms such as CAR, FIFO, and LRU. The simulation results were calculated based on hit ratio, which takes into consideration the size of cache, locality of reference, and replacement algorithm being used during cache miss. Our algorithm gave better results than other algorithms.

4.1   Input Traces

For analyzing the performance, the address trace (dataset) used had one thousand memory addresses formed by the actual execution of the program in CPU. Usually, an address trace has data addresses (store and load) and instruction fetch addresses, but as simulation is performed on data cache only, traces having only data addresses are taken into consideration. Mapping scheme considered for simulation is set associative mapping. Based on the traces being used, simulation was executed with seven different sizes of cache and results obtained from the simulation were analyzed. As our proposed algorithm works for multi-level cache, the cache size was divided in a ratio of 1:2:3 among L1, L2, and L3 level caches, respectively.

4.2   Simulation Results

Results were then compared with CAR, FIFO, LRU, and CWRP [1, 9]. Figure 2 provides the comparison summary of all results obtained by using different cache replacement algorithms. It is clearly observable from Fig. 2 that our proposed algorithm performs better than CAR, FIFO, LRU, and CWRP. The average gain compared to CAR is 5.3%, whereas the average gain compared to LRU, FIFO, and CWRP is 10.33%, 8.47%, and 4.35% respectively [9]. Table 1 provides the hit ratio of all the block accesses implemented by using different cache replacement algorithms.

The maximum hit ratio obtained by our proposed algorithm is 76.12% for a cache size of 210 blocks, and the minimum hit ratio is 41.23% for a cache size of 30 blocks.

In comparison with CAR, minimum gain obtained by our proposed work is 0.93% for a cache size of 210 blocks and maximum gain is 11.38% for a cache size of 60

Fig. 2 Performance evaluation of CAR, FIFO, LRU, CWRP, and our proposed work with different cache sizes

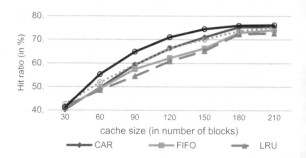

Table 1 An assessment between hit ratio of CAR, LRU, FIFO, CWRP, and our proposed work

Cache size (in number of blocks)	CAR	LRU	FIFO	CWRP	Proposed work
30	40.24	41.6	40.93	42.51	41.23
60	49.65	48.6	49.26	51.83	55.30
90	59.27	54.5	57.48	59.36	64.82
120	66.20	60.81	62.14	66.40	70.96
150	70.96	65.21	66.30	70.06	74.43
180	75.22	72.30	72.84	73.63	75.92
210	75.42	72.70	74.03	75.12	76.12

blocks. Minimum gain obtained from FIFO is 0.73% for a cache size of 30 blocks, and maximum gain is 14.19% for cache size of 120 blocks. Minimum gain obtained with respect to LRU is 4.70% for a cache size of 210 blocks, and maximum gain is 18.94% for a cache size of 90 blocks. From CWRP, minimum gain obtained is 1.33% for cache size of 210 blocks and maximum gain is 9.2% for cache containing 90 blocks. There is one exceptional case for cache size of 30 blocks in which LRU performs 0.9% better than our proposed method and CWRP performs 3.1% better than our proposed work.

We can observe from Fig. 2 that as the cache size is increasing, the performance of our algorithm is also increasing significantly as compared to other cache replacement algorithms. The reason for such behavior to occur is that we are considering combinations of recency and frequency which helps in accurately deciding the page that would not be used in the near future, and this increases our overall hit ratio as there would be less occurrences of cache misses.

5 Summary

This paper introduces a new algorithm which enhances the performance of a multi-level cache. It gave better results than CAR, FIFO, LRU, and CWRP [9]. We observed that reference rate of a block depends on the freshness of the block present in the cache along with recency and frequency features of the block. As multi-level cache has become an inevitable part of modern computing systems, we have devised a method that uses a combination of algorithms that takes into account the R-F score of the block along with the level in which the cache is present. Our proposed method would work very well in case of multi-level cache. We have used L1 cache to capture the R-F score of the block along with its freshness. L2 cache is used for capturing only frequency, and L3 cache is used for capturing only recency. Our assumption was that we should not only take the R-F score and freshness of the block but also consider taking, recency and frequency of a block, individually, would increase the cache performance and multi-level cache design model would be a best fit for our assumption to hold true. Simulation results validated our assumptions, and we got better results than CAR, FIFO, LRU, and CWRP [9]. Our proposed method can be used for different address traces, and results can be used for comparison with other cache replacement algorithms. It is worth mentioning that if reference rate depends on other features of the cache, then these features can be added in a multi-dimensional cache and can be tuned according to the application domain. Our proposed work would perform even much better in such application-specific domain.

References

1. Tanenbaum, A., and A. Woodhull. 1997. Operating systems: Design and implementation. *Prentice Hall* 3: 373–410.
2. Yang, Q., H. Zhang, and H. Zhang. 2001. Taylor series prediction: A cache replacement policy based on second-order trend analysis. In: *Proceedings of 34th Hawaii Conference on System Science* 5: 5023.
3. Belady, A. 1966. A study of replacement algorithms for a virtual storage computer. *IBM systems Journal* 5 (2): 78–101.
4. O'Neil, E., P. O'Neil, and G. Weikum. 1999. An optimality proof of the lru-k page replacement algorithm. *Journal of the ACM* 46 (1): 92–112.
5. Jihang, S., and X. Zhang. 2002. LIRS: An efficient low inter reference recency set replacement policy to improve buffer cache performance. In: *Proceedings of ACM Sigmetrics Conference on ACM Presentation* 2: 31–42.
6. Hosseini-khayat, S. 2000. On optimal replacement of nonuniform cache objects. *IEEE Transactions on Computers* 49: (8).
7. Glass, G., and P. Cao. 2003. Adaptive page replacement based on memory reference behavior. In: *Proceedings of ACM SIGMETRICS Conference on Overhead Replacement Cache. Proceedings of Usenix Conference on File and Storage Technologies (FAST 2003)*, Usenix, 115–130.
8. Irani, S. 1997. Page replacement with multi-size pages and applications to WebCaching. In: *Proceedings of 29th Annual ACM Symposium of Theory of Computing* 50: 701–710.
9. Swain. D., S. Marar, N. Motwani, V. Hiwarkar, and N. Valakunde. 2017. CWRP: An efficient and classical weight ranking ploicy for enhancing cache performance. In: *IEEE Fourth International Conference on Image Information Processing*.
10. Swain, Debabala, Bijay Paikaray, and Swain Debabrata. 2011. AWRP: Adaptive weight ranking policy for improving cache performance. *Journal of Computing* 3: (2).
11. Dash, B., D. Swain, BK. Paikaray. (2017) *International Journal of Computational Systems Engineering (IJCSYSE)* 3: (1/2).
12. Bansal, Sorav, and Dharmendra Modha. 2004. CAR: Clock with adaptive replacement. In: *USENIX File and Storage Technologies (FAST)*, Mar 31–Apr 2, San Francisco, CA.
13. Swain, Debabrata, Bancha Nidhi Dash, Debendra O Shamkuwar, Debabala Swain. 2012. Analysis and predictability of page replacement techniques towards optimized performance. *International Journal of Computer Application Proceedings on ICRTITCS-2011* 12–16.
14. Megiddo, N., and D. Modha. 2003. ARC: A self-tunning, low overhead replacement cache. In: *Proceedings of Usenix Conference on File and Storage Technologies (FAST 2003)*. Usenix, 2: 115–130.

Denoising Documents Using Image Processing for Digital Restoration

Mohit Kulkarni, Shivam Kakad, Rahul Mehra and Bhavya Mehta

Abstract This paper develops an algorithm that will help decipher the text in decrepit documents, which if put in simpler terms aims at converting stained, blotted, creased, and faded documents into a cleaner and legible format. Handwritten or printed records are carelessly stacked away without undertaking measures for preserving them. They are subjected to degradation because of mishandling and improper storage conditions. This ultimately results in the loss of important documentation owing to the inability of reproduction or recovery of original data. Digital image preprocessing techniques are used to convert a color (RGB) image into a grayscale image for further processing. Image denoising is one of the most sought areas after research in image processing, and in this paper, we use image segmentation and median filter to achieve this. In this paper, we attempted to come up with an approach to remove noise from the image by applying image segmentation and thresholding, histogram, and median filter.

Keywords OCR · Denoising documents · Text predictor · Histogram · Thresholding · Laplacian · Median · Regression · OpenCV

M. Kulkarni (✉) · S. Kakad · R. Mehra · B. Mehta
Electronics and Telecommunication Department,
Vishwakarma Institute of Technology, Pune, India
e-mail: mohitvkulkarni@gmail.com

S. Kakad
e-mail: shivamkakad05@gmail.com

R. Mehra
e-mail: mehrar12@gmail.com

B. Mehta
e-mail: bhavya.y.mehta@gmail.com

© Springer Nature Singapore Pte Ltd. 2020   287
D. Swain et al. (eds.), *Machine Learning and Information Processing*,
Advances in Intelligent Systems and Computing 1101,
https://doi.org/10.1007/978-981-15-1884-3_27

1   Introduction

Visual distortion is usually described as the 'noise' in the image. It is similar to the grains we see in photographs when the image is captured in low lights or when there is some arbitrary variation in the color or brightness in the image [1]. Improving the image quality majorly concerns itself with image restoration by eliminating unwanted pixels. Noise filtering has numerous approaches, the most popular being—replacing the noisy pixel by a befitting one often dependent on the values of the neighborhood pixels. This paper proposes a method which focuses on restoring the image by detecting noise. This is done using median filters, averaging, Gaussian, or bilateral techniques.

Averaging is a technique where the center pixel is replaced by the average value of all the pixels surrounding it. Convolution of the image with a box filter can help us achieve this. Gaussian operator uses the weighted mean; pixels closer to the subject pixel contribute more weight. Here, the kernel size needs to be specified and must necessarily be an odd positive integer. Median filtering replaces the central pixel with the median value of the surrounding pixels, and bilateral technique uses two Gaussian distributions and is a function of space [2]. The texture is not preserved, but the edges are.

The median filter is a nonlinear digital image filtering technique which is very effective in removing the salt-and-pepper noise as well as the impulse noise without harming the details of the image. The edges are preserved as this filter only removes the outlier points. The only drawback of median filtering is its computational complexity.

2   Literature Review

2.1   *Image Inpainting*

Image inpainting is a method by which we can recreate the spoiled part of the image by recognizing the patterns and curvatures present in the other parts of the image. Image inpainting can be used to denoise the stained part of documents by analyzing the patterns present between the letters and white spaces in the text [3].

After the image inpainting, the image can be refurbished to the real and accord with the human ocular sight. Image inpainting has a broad application scope, such as protection of valuable authentic artifacts, past photographs, and error in images.

In our case, images are considered to have coffee or tea patches across the document. We used image inpainting because the portion, which was filled by these specks of dirt, required to be restored by the original pure data.

2.2 Optical Character Recognition (OCR)

Optical character recognition identifies, generates, and interprets the individual character in a scanned copy of the document. This method is very complicated and requires that the OCR program maps each letter present in the image to a computer-generated transcription that resembles it [4].

We tried using OCR to extract text from our test document. As most of the texts on the document was covered with coffee stains, OCR misattributed the characteristics of individual letters due to which it was giving unreliable results.

3 Proposed Methodology

3.1 Median Filter

The median filter is a nonlinear robust filter, which is used to remove noise. This filter is usually used as a preprocessing step which helps the results of later steps [5]. This method does not harm the edges which makes this filter to be used widely. The main concept of the median filter in image processing is to run through the image matrix, pixel by pixel, and to replace the entries with the median of neighboring pixels. Median is nothing but the center value of entries when the entries are sorted numerically. If there are even entries, there can be more than one median [6, 7]. In this case, cv2.medianBlur function from OpenCV was used to blur the image. This removed the text from the image leaving behind only the noise [8].

An approximate threshold constant value is subtracted from the output of the median filter for differentiating between the text and noise. The later output is then compared with the original image to recover the original text in the output. If the pixel value of the original image is less than the subtracted value, then we restore the pixels from the original image in the output. Remaining everything is made equal to the pixel value of white color, i.e., 255 [9].

Application of the median filter is the first step in this technique to overcome noise in the image. The median filter is used to detect noisy background. The filter is applied twice on decrepit images. This smoothens the image background while keeping the edges near the letters sharp. The step is repeated to get pure noise (Fig. 1).

Output of median filter iterated twice is shown in Fig. 2.

3.2 Noise Subtraction

In noise subtraction, the noise detected in the first step is subtracted from the original image to get a grayscale foreground image.

There exist several methods to design forms with fields to
fields may be surrounded by bounding boxes, by light rectan
These methods specify where to write and, therefore, minimi
overlapping with other parts of the form. These guides can
sheet of paper that is located below the form or they can b
form. The use of guides on a separate sheet is much better fro
quality of the scanned image, but requires giving more instructio
restricts its use to tasks where this type of acquisition is used.
the form are more commonly used for this reason. Light re
more easily with filters than dark lines whenever the handwritt
Nevertheless, other practical issues must be taken into accoun

Fig. 1 Original image

Fig. 2 Output of median filter

The noise detected in the first step is considered as the background, which contains
noise such as coffee stains. By performing matrix subtraction, the noisy background
is subtracted from the original image pixel by pixel to get a foreground image which
contains only the text (Fig. 3).

3.3 Histogram Approach

A histogram is a graph or plot, which gives an overall idea about the intensity distribution of an image. It is a plot with pixel values on the X-axis and the corresponding
number of pixels in the image on the Y-axis [10]. Histogram analysis is just another
way of understanding the image graphically. The histogram gives us an overall idea
of all the pixel values present in the image. Information about contrast, brightness,
and intensity distribution can be obtained from the histogram of any image [11].

In this step, histogram of the original image was plotted which looked as shown
in Fig. 4.

There exist several methods to design forms with fields to
fields may be surrounded by bounding boxes, by light rectan
These methods specify where to write and, therefore, minimi
overlapping with other parts of the form. These guides can
sheet of paper that is located below the form or they can b
form. The use of guides on a separate sheet is much better fro
quality of the scanned image, but requires giving more instructio
restricts its use to tasks where this type of acquisition is used.
the form are more commonly used for this reason. Light re
more easily with filters than dark lines whenever the handwritt
Nevertheless, other practical issues must be taken into accoun

Fig. 3 Output of noise subtraction

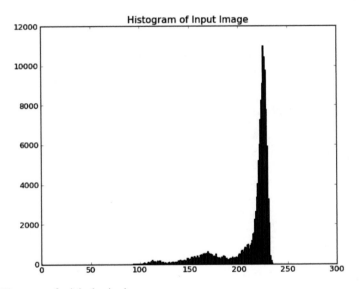

Fig. 4 Histogram of original noisy image

By analyzing the histogram, it can be seen that all the noise lies between pixel
values 50–225. Because our image is grayscale, we have any pixel as text either
in black color (pixel value lying in between 0 and 40) or white background (pixel
value lying in between 225 and 255). We can remove the noisy pixels with intensity
varying from 50 to 225 and get a clean image.

After removing the corresponding noisy pixels with image segmentation (next
step), the output image and its histogram are as shown in Figs. 5 and 6, respectively.

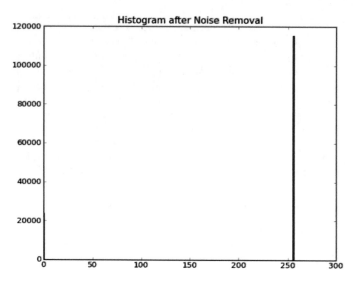

Fig. 5 Histogram of output image

3.4 Image Segmentation

Image segmentation involves partitioning of an image into multiple segments which make the image representation to be more analytical. Objects and boundaries in the images are typically detected using image segmentation [12]. This involves assigning a label to each pixel in the image such that pixels with the same labels share common visual characteristics [12]. Image segmentation is useful where one wants to retain the pixels of the same characteristics in an image and remove the rest of them.

Considering a grayscale image, the values of pixels of gray levels of the background of the image differ significantly than the values of pixels in the foreground. Adaptive thresholding is an effective tool which can be used to separate the foreground from the background of the image. The pixel's value is compared to the predecided threshold value, and if the value is greater or lesser than the threshold, respective values (white or black) are assigned to the pixel [11].

Adaptive thresholding calculates the threshold for small regions of the input image. Better results can be obtained for an image with varying illumination for the different regions of the same image. The function used for this is cv2.adaptiveThreshold [8]. The sample output after successful thresholding is shown in Fig. 6. After thresholding, it was observed that only pixels, where we see the text were kept and rest all pixels of coffee stains were removed. The peak in the histogram, i.e., in Fig. 5, represents the text.

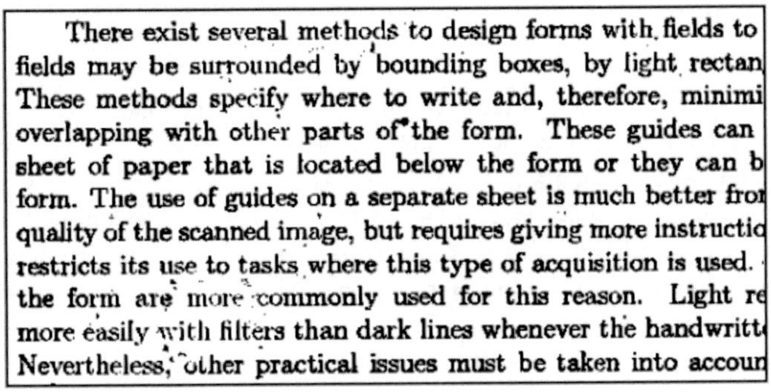

There exist several methods to design forms with. fields to
fields may be surrounded by bounding boxes, by light rectan
These methods specify where to write and, therefore, minimi
overlapping with other parts of the form. These guides can
sheet of paper that is located below the form or they can b
form. The use of guides on a separate sheet is much better from
quality of the scanned image, but requires giving more instructio
restricts its use to tasks where this type of acquisition is used.
the form are more commonly used for this reason. Light re
more easily with filters than dark lines whenever the handwritt
Nevertheless, other practical issues must be taken into accoun

Fig. 6 Output image after histogram analysis and image segmentation

3.5 Salt-and-Pepper Noise Reduction

In the technique defined in this paper, salt-and-pepper noise is removed considering
the color intensity of pixels surrounding a pixel. A 3 × 3 kernel is used for this
calculation.

$$\text{kernel} = \begin{bmatrix} 1 & 1 & 1 \\ 1 & 0 & 1 \\ 1 & 1 & 1 \end{bmatrix}$$

This kernel considers the pixel as black and measures the number of black pixels
surrounding the pixel. If a pixel in the text is black, we assume that six or more
pixels surrounding it are black. If the pixel is a part of the salt-and-pepper noise, the
number of pixels in the surrounding must be less than six. The salt-and-pepper noise
is reduced considering this assumption. It can be seen that significant salt-and-pepper
noise has been reduced as shown in Fig. 7.

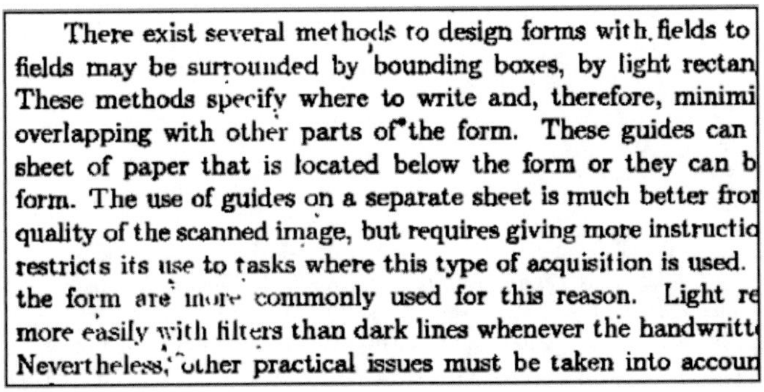

There exist several methods to design forms with. fields to
fields may be surrounded by bounding boxes, by light rectan
These methods specify where to write and, therefore, minimi
overlapping with other parts of the form. These guides can
sheet of paper that is located below the form or they can b
form. The use of guides on a separate sheet is much better from
quality of the scanned image, but requires giving more instructio
restricts its use to tasks where this type of acquisition is used.
the form are more commonly used for this reason. Light re
more easily with filters than dark lines whenever the handwritt
Nevertheless, other practical issues must be taken into accoun

Fig. 7 Output after reducing salt-and-pepper noise

4 Conclusion

A new and easy method to suppress noise from distorted images has been presented. In this work, four different techniques, i.e., median filter, noise subtraction, histogram analysis, and image segmentation, were applied on coffee-stained, noisy, as well as dog-eared document which reduced the noise. It can be concluded that the median filter gives better results for compound images after the performance analysis of different filters. The median filter requires less computational power when compared to image inpainting as image inpainting uses deep learning in the backend. Furthermore, to get better results, curvelet transform [13] and sparse modeling [14] techniques can be used.

5 Future Scope

This project can be improvised using many machine learning algorithms as well as the correlation. Correlation is an amazing technique by which we can extract characters on the digital document. Training and testing data can be created and used to train our machine learning model which can be used to denoise the document quite efficiently [15]. Sparse modeling [14] and curvelet transform [13] are also some of the methods by which background subtraction is possible. Recent techniques in machine learning like convolutional neural networks can also be used to train a model for a particular use case [16]. The model can automatically detect noise based on the learnings and denoise the required documents for us. Furthermore, traditional machine learning techniques can also be used.

Acknowledgements The authors feel grateful and wish their profound indebtedness to their guide **Prof. Milind Kamble**, Department of Electronics and Telecommunication, Vishwakarma Institute of Technology, Pune. The authors also express their gratitude to **Prof. Dr. R. M. Jalnekar**, Director, and **Prof. Dr. Shripad Bhatlawande**, Head, Department of Electronics and Telecommunication, for their help in completion of the project. The authors also thank all the anonymous reviewers of this paper whose comments helped to improve the paper.

References

1. Reka, Durai, and V. Thiagarasu. 2014. A study and analysis on image processing techniques for historical document preservation. *International Journal of Innovative Research in Computer and Communication* 2 (7): 5195–5200.
2. Afrose, Zinat. 2012. A comparative study on noise removal of compound images using different types of filters. *International Journal of Computer Applications* (0975–888) 47 (14): 45–47.
3. Mallick, Satya. 2019. Image Inpainting with OpenCV. Available via https://www.learnopencv.com/image-inpainting-with-opencv-c-python/. Accessed 13 Apr 2019.

4. Vamvakas, G., B. Gatos, N. Stamatopoulos, S.J. Perantonis. 2008. A complete optical character recognition methodology for historical documents. Document Analysis Systems, IAPR International Workshop, 525–532. doi:10.1109/DAS.2008.73
5. Rajasekaran, Angalaparameswari, and P. Senthilkumar. 2014. Image denoising using median filter with edge detection using canny operator. *International Journal of Science and Research* 3 (2): 30–33.
6. Malothu, Nagu, and Shanker N. Vijay. 2014. Image de-noising by using median filter and weiner filter. *International Journal of Innovative Research in Computer and Communication Engineering* 2 (9): 5641–5645.
7. Sandeep, Kumar, Kumar Munish, and Agrawal Neha Rashid. 2017. A comparative analysis on image denoising using different median filter methods. *International Journal for Research in Applied Science & Engineering Technology* 5 (7): 231–238.
8. OpenCV Documentation, Available via http://docs.opencv.org/. Accessed 7 March 2017.
9. Suman, Shrestha. 2014. Image denoising using new adaptive based median filter. *Signal & Image Processing: An International Journal (SIPIJ)* 5 (4): 1–12.
10. Govindaraj, V., and G. Sengottaiyan. 2013. Survey of image denoising using different filters. *International Journal of Science Engineering and Technology Research (IJSETR)* 2 (2): 344–350.
11. Senthilkumaran, N., and S. Vaithegi. 2016. Image segmentation by using thresholding techniques for medical images. *Computer Science & Engineering* 6 (1): 1–6.
12. Sujata, Saini, and Arora Komal. 2014. A study analysis on the different image segmentation techniques. *International Journal of Information & Computation Technology* 4 (14): 1445–1452.
13. Jean-Luc, Starck, J. Candès Emmanuel, and Donoho David. 2002. The curvelet transform for image denoising. *IEEE Transactions on Image Processing* 11 (6): 670–684.
14. Julien, Mairal, Bach, Francis, Ponce, Jean. 2014. Sparse modeling for image and vision processing. arXiv 1411.3230: 76–97.
15. Priest, Colin. 2015. Denoising Dirty Documents. Available via https://colinpriest.com/2015/08/01/denoising-dirty-documents-part-1/. Accessed 27 Jan 2017.
16. Kaggle. 2015. Denoising Dirty Documents. Available via https://www.kaggle.com/c/denoising-dirty-documents. Accessed 13 April 2017.

Humming-Based Song Recognition

Shreerag Marar, Faisal Sheikh, Debabrata Swain and Pushkar Joglekar

Abstract Today, the Internet is the only medium through which songs are accessed by the majority, resulting in myriad of songs and artists on the Internet. The only effective way to search music on the Internet still requires at least some of the details about a song or its artist. But, what if we don't know any such detail about the music? What if we heard it somewhere and know only how to sing or hum? Humming-based song recognition (HBSR) addresses this conundrum. It is a system which takes the audio file or audio recording as input and predicts the song present in the database based on the humming sound analysed. The fundamental ideas of HBSR, its implementation and some techniques to improve its performance are introduced in this paper.

Keywords Data set · Spectrogram · Librosa · CNN · Shazam

1 Introduction

Humming-based song recognition (HBSR) is a system that takes hum audio as input and analyses the audio to predict the song that rhymes with the input audio. Music retrieval techniques have been developed in recent years since signals have been digitized. Searching by song name or singer's name is often considered as an easy task to find the song. However, many times it happens that we cannot recall the wordings of the song and all we can remember is the tune of the song. We can use HBSR in this case. We can easily create a hum sound for the song we want to search for. HBSR will analyse the hum sound and output the song name.

S. Marar (✉) · F. Sheikh
Department of Computer Engineering, Vishwakarma Institute of Technology, Pune, India
e-mail: shreerag.marar15@vit.edu

D. Swain
Information Technology Department, Vishwakarma Institute of Technology, Pune, India
e-mail: debabrata.swain@vit.edu

P. Joglekar
Computer Engineering Department, Vishwakarma Institute of Technology, Pune, India

© Springer Nature Singapore Pte Ltd. 2020   297
D. Swain et al. (eds.), *Machine Learning and Information Processing*,
Advances in Intelligent Systems and Computing 1101,
https://doi.org/10.1007/978-981-15-1884-3_28

Various approaches are made in the field of song recognition by tune. Shazam [1] is a well-known system that searches for a song by listening to the music. Shazam [1] uses fingerprinting algorithm to generate a unique hash for each song in its database usually known as fingerprint. This fingerprint is unique in terms of songs similar to unique human fingerprint in terms of people. Whenever the input is fed to the system, it generates the fingerprint of that music using the same algorithm used to generate fingerprints of the songs in the database. This fingerprint is then matched with the stored fingerprints, and the most similar song is given as output.

The HBSR approach is different as compared to the fingerprinting. HBSR uses neural network [2] approach to achieve the results. HBSR system has multiple stored images of many songs in its database. These images are the spectrograms of the audio samples generated using the humming audio samples. These images are fed to the CNN [3] model, and it is trained on these images. A new hum audio is then fed to HBSR system which first creates a spectrogram of the audio which is then fed to the model which predicts the song which is similar to the humming sound. The detail working of the HBSR is discussed in this paper.

2 Background

A hum is a sound arising forcefully from the nose with the mouth either opened or closed. The hum by a human most probably is based on some already existing song or music, and hence, it is melodious. It is common to hum when it is hard to recall lyrics of a particular song. Considering the fact that hum of a particular song is as per its melody, a song can be identified based on its hum. Based on this finding, there have been numerous approaches proposed in the past for identification of songs based on hum input, each with their pros and cons.

Over the past decade, the Internet has expanded exponentially, resulting in colossal amount of data being generated every day. As a result, some of the recent successful approaches for identification of songs by hum heavily rely on song data and machine learning [5] algorithms. Machine learning [5], as the name suggests, allows a machine to learn. It is a field of study involving statistics and algorithms which enables a machine to carry out a particular task without explicit instructions, relying on patterns in data. Larger the amount of data with variations in records, better the accuracy (in most cases) of the machine learning [5] algorithms while predicting output. Prediction by machine learning [5] algorithms can either be a discrete value or continuous value. Song identification by hum is better viewed as a classification problem as songs can be viewed as discrete values to be predicted.

In the field of machine learning [5], classification problem is a problem of identifying set of categories to which a new observation belongs to. It is used for predicting discrete responses. While viewing song identification problem as a classification problem, each song is viewed as a category and then prediction involves classifying

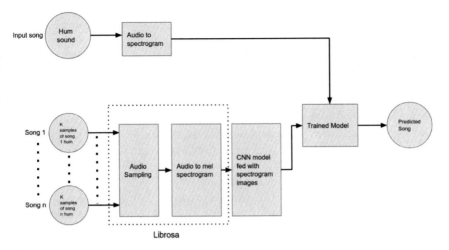

Fig. 1 Overview of the HBSR system

the hum input to one of the categories. Over the years, solutions to classification problems have evolved, and some of the recent approaches perform much better than before.

3 System Overview

Figure 1 depicts the overview of the HBSR system and the flow of the system. The working of the system and how the input song is processed till the final prediction state are shown in Fig. 1. The usage of external libraries like Librosa [4] and CNN [3] model is shown. K-samples of audio data belonging to each song are collected and converted to spectrogram images using Librosa [4]. Further, these images are trained using CNN [3]. The trained model can be used to predict the new spectrogram images. The entire system was implemented and tested against six songs with 42 instances (hum audio) belonging to each song.

4 Implementation Phases

There are three major phases involved in the implementation of HBSR:

1. Data collection phase—collection of data for model training
2. Data conversion phase—audio samples are converted to images
3. Model training phase—CNN [3] model is trained on the generated images.

4.1 Data Collection Phase

To train a CNN [3] model for HBSR, a large amount of data is required to cover as much songs as possible. Hence to collect samples, audio samples from different people were collected. The age group ranged from 9 to 60 where majority of voices belonged to the group of 20–24 including both male and female voices. The variation in voice of different people helped to generate variety in audio files. Although the voice data collected was not enough to train the model, we used Audacity [5] to tune the pitch of existing audio files and generated multiple samples per audio file. These changes in pitch ensured that the audio samples are different from the existing audio samples and also helped to increase the number of samples. A total of 252 data belonging to six different songs (42 each) were collected for experimentation.

4.2 Data Conversion Phase

The first phase through which the sound input goes through in the HBSR system is the data conversion phase. The data conversion phase, as the name suggests, carries out the data conversion task, converting sound input data to images. As CNN [3] does not work on audio, the audio data must first be converted to some suitable form. The conversion of audio inputs to images boils down the sound identification problem to image identification problem. The conversion phase comprises two steps—audio sampling and audio to mel spectrogram images.

The uniqueness of a music note lies within its frequency. Hence, the audio files are first passed through sampling. Audio sampling involves mapping of audio file into set of consecutive numbers. The size of set depends on the duration of audio in audio file. The generated set is stored in the form of an array, which is then converted into an image. Audio to mel spectrogram involves converting sampled data to mel spectrogram images which will further be used as an input to the CNN [3] model. Frequency data of each audio sample is plotted in the form of a spectrogram to generate images.

Both processes, audio sampling and audio to mel spectrogram conversion, are carried out with the help of an external Python library for music and audio analysis called Librosa [4]. Librosa [4] is a musical information retrieval library that provides all the necessary building blocks required for data sampling and plotting the sampled data in amplitude time domain or frequency domain or other forms of graphical representation.

The generated spectrogram images are then used to create a data set which will further be used to train the CNN [3] model. Input audio is also converted to spectrogram and fed to the model for the prediction of a similar song. As discussed in Sect. 1, Shazam's [1] fingerprinting algorithm used spectrogram data to predict the song. This was the motivation behind the approach to train the model on spectrogram images. Figure 2 shows the spectrogram of humming audio files belonging to four songs.

4.3 Model Training Phase

Model training phase, as the name suggests, involves the training of a machine learning [5] model. The sole purpose of training a model is to make the model capable of predicting songs based on the user's audio input. HBSR uses CNN [3] algorithm to train the model. The spectrogram images of the audio files (generated during the conversion phase) are used to train the CNN [3] model. The trained CNN [3] model is a classification model which treats each song as a separate class, i.e. each class represents one song.

The image data set is divided into two parts, testing data and training data. 20% of the data is used as testing data, and 80% of the data is used as training data. The CNN [3] model comprises the following layers:

- Two convolutional layers for finding patterns in the images. The convolutional layer consisted of 32 3 * 3 filters more specifically known as feature detectors used against the input shape of 64 * 64 with RGB colour code. Activation function used for convolutional layer is 'relu'.
- One max pooling layer of size 2 * 2 for improving the performance by progressively reducing the spatial size of the representation, resulting in reduction in the amount of parameters and computations in the network.
- One flatten layer for converting the output of the convolutional network part of the CNN [3] into a 1D feature vector to be used by the ANN part of it.
- Two dense layers for full connection in which each input node is connected to each output node, so all the neurons in one layer are connected to all the neurons in the next layer. A total of 128 nodes were used in the hidden layer with 'relu' activation function. The second dense layer had an output dimension of 6. 'Softmax' activation function was used as it was more suitable for categorical classification rather than 'sigmoid' which is suitable for binary classification.
- One dropout layer in which randomly selected neurons are ignored during training. It results in their contribution to the downstream neurons to be removed temporarily on the forward pass, and any weight updates are not applied to the neurons on the backward pass. Dropout percentage used was 20%.
- One fully connected layer in which each neuron receives input from every element of the previous layer.

Once the CNN [3] model has finished training, the HBSR system is then ready for taking hum audio input from the user and predicting the song. The user's audio input is fed to the HBSR system. It then goes through data conversion phase, which takes the user's audio input and converts it into respective spectrogram image. The

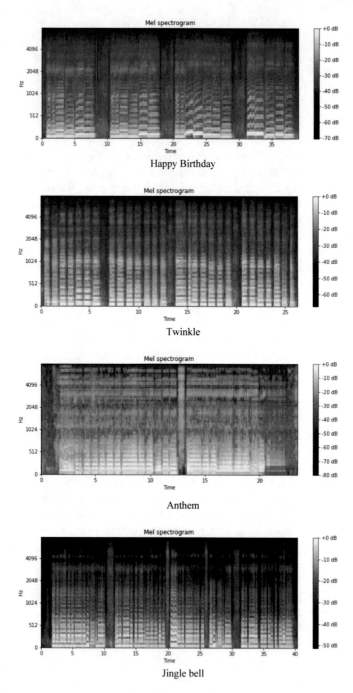

Fig. 2 Spectrogram representation of four different audio files

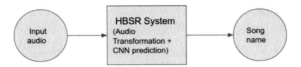

Fig. 3 Overview of song prediction

```
Epoch 42/45
186/186 [==============================] - 38s 205ms/step - loss: 0.0041 - acc: 0.9987 - val_loss: 0.0627 - val_acc: 0.9583
Epoch 43/45
186/186 [==============================] - 38s 204ms/step - loss: 0.0033 - acc: 0.9993 - val_loss: 0.1654 - val_acc: 0.9792
Epoch 44/45
186/186 [==============================] - 38s 205ms/step - loss: 0.0095 - acc: 0.9973 - val_loss: 0.0293 - val_acc: 0.9792
Epoch 45/45
186/186 [==============================] - 38s 207ms/step - loss: 0.0048 - acc: 0.9993 - val_loss: 0.0401 - val_acc: 0.9792
Out[52]: <keras.callbacks.History at 0x7f0ea9b3bf98>
```

Fig. 4 CNN model accuracy

spectrogram image is then passed to the trained CNN [3] model, which then outputs its prediction. Figure 3 depicts the overview of the song prediction.

5 Performance Analysis

CNN [3] was trained against 234 spectrogram images belonging to six songs. Training set consisted of 186 spectrograms, and test set consisted of 48 spectrograms. HBSR system achieved accuracy of 97.92% in predicting the songs correctly on a data set consisting of six songs. This implies that out of 48 test set images, only one spectrogram image was predicted incorrect (Fig. 4).

6 Experimental Results

A total of 186 humming data collected by us were used for training the CNN [3] model. The trained model was then saved to memory so that it can be used later. For the saving of model, an external Python library called joblib [6] was used. By saving the model on the disc, we eliminated redundant training of model on the same data set.

A total of 18 new humming data were then collected by us and were provided as input to the trained model for its evaluation. This new humming data was not a part of train or test set. Out of the 18 new humming data inputs which belonged to six songs, the model was successful in predicting 17 of them correctly, i.e. the model predicted the correct song names for 17 out of 18 given inputs.

7 Future Scope

HBSR currently works accurately if a similar humming of particular part of a song is already present in the database. However, HBSR system can be extended in future to predict a song by using humming of any part of the song. As the number of songs increases, HBSR will require tons of audio samples so as to build a robust model. Also, the system has the potential for real-time usage on portable devices. That is, the hum can be recorded on a mobile device which will then intimate that song to the user.

8 Conclusion

This approach of using machine learning [5] and neural networks [2] to predict songs can be used in real time. Mel spectrogram representation of audio uniquely differentiates one song from another. The system thus has the potential to predict the songs based on humming input. It predicted correctly for six songs in the database. It can be expanded to predict more songs by adding data for more songs and then training the model on this data. The system can become more robust by training it against tons of audio samples representing a huge category of songs.

References

1. Wang, Avery Li-Chun. 2003. An industrial-strength audio search algorithm. In *Proceedings 4th International Conference on Music Information Retrieval*, pp. 7–13, 2003-Oct.
2. Bishop, J.M., R.J. Mitchell. 1991, January 25–25. Neural networks-an introduction. IEE Colloquium on Neural Networks for Systems: Principles and Applications.
3. Al-Zawi, S., T. Mohammed, S. Albawi. 2017. Understanding of a convolutional neural network. In *Proceedings of ICET*, Aug 2017.
4. McVicar, Matt, P.W. Ellis Daniel, Dawen, Liang, Colin, Raffel, Brian, McFee, Eric, Battenberg, Oriol, Nieto. 2015. librosa: audio and music signal analysis in python. In *Proceedings of the 14th Python in Science Conference*.
5. Mitchell, Thomas M. 1997. *Machine Learning*. New York, NY: McGraw-Hill Inc.
6. Joblib: For running Python functions as pipeline jobs, https://joblib.readthedocs.io/en/latest/.

A New Data Structure for Representation of Relational Databases for Application in the Normalization Process

Ranjana Jadhav, Priyadarshan Dhabe, Sadanand Gandewar, Param Mirani and Rahul Chugwani

Abstract In this paper, a new data structure named relational tree is proposed for the representation of the relational database in computer memory. Relational database schema represented using relational tree(s) appears to be more promising for semi-automating the process of normalization in a very efficient manner, which is prime motive of this paper. This paper provides all the fundamental concepts required for the understanding of the representation of relational schema using relational tree so that efficient algorithms of various normal forms can be designed by using this representation. This automation will considerably reduce manual efforts and errors in the process of normalization in software industries. Space requirements also improved, compared to previously proposed approaches. It is expected that application of various normalization algorithms on this way of representation is very efficient since the relational tree can be easily manipulated.

Keywords Relational schema · Relational tree · Normal forms · Representation of relational databases

R. Jadhav · P. Dhabe · S. Gandewar · P. Mirani (✉) · R. Chugwani
Department of Information Technology, Vishwakarma Institute of Technology, Pune, Maharashtra, India
e-mail: param.mirani17@vit.edu

R. Jadhav
e-mail: ranjana.jadhav@vit.edu

P. Dhabe
e-mail: priyadarshan.dhabe@vit.edu

S. Gandewar
e-mail: sadanand.gandewar17@vit.edu

R. Chugwani
e-mail: rahul.chugwani17@vit.edu

© Springer Nature Singapore Pte Ltd. 2020
D. Swain et al. (eds.), *Machine Learning and Information Processing*,
Advances in Intelligent Systems and Computing 1101,
https://doi.org/10.1007/978-981-15-1884-3_29

305

1   Introduction

Increment in the profit of any commercial organization can be seen by increasing the productivity and quality of the product. We have to maintain the quality of the product while increasing its productivity. To achieve this kind of production, automation of the tasks becomes necessary. Organizations can involve automation in their design and development sectors to gain the required amount of products with maintained quality criteria.

Normalization plays a vital role in database management system, but before normalizing any relational schema, its proper representation is very essential. Appropriate representation of attributes and functional dependencies provides easier way to apply the normalization algorithms on the given relational schema. Representation of the relational schema provides better understanding of the database and thereby simplifies the process of normalization.

In this proposed approach of representing a database given the attributes and relations, these attributes share a new data structure is suggested and named as "relational tree". This new data structure is an alternative approach for the tool proposed in [1].

Using tree as data structure is a better option compared to linked list as we know we do not have to traverse to whole tree to get a particular node. Application of the normalization algorithms on relational schema represented by the proposed tool would be quite efficient and simpler to understand as compared to the tool proposed in [1].

We have known that widely used relational databases used in recent times were proposed by Dr. Codd [2]. The different applications of major organizations are storage and manipulation of data for decision making by using the bulk of data. They are capable of managing an enterprise in an efficient manner which is also simple and trusted. Their capability in these areas has really increased their scope in software industries, which are trying to develop relational databases according to the requirements of their clients.

Design of relational schema is one of the important factors on which success of the relational database model depends. "Normalization" is a key step in the design of relational database. This process takes a relation as input and decomposes the bigger relation into smaller ones which are free from redundant data along with anomalies like deletion and insertion [2]. Each step of normalization has a name, first, second and third normal forms which are represented as 1NF, 2NF, 3NF, respectively.

Normalization is not limited till the third normal form, and there are further normal forms of normalization, 3.5NF also known as Boyce–Codd Normal form (BCNF) and 4NF. 3.5NF, as the name suggests, it is stricter than 3NF but not as strict as 4NF. 4NF has its own rules, and it is even stricter than 3.5NF.

The paper has been organized in the following way. Section 2 describes the work that has been previously done in this domain. We also found some useful resources which would help students and researchers understand normalization along with basic concepts required for relational trees. Section 3 gives detailed explanation of node structure which has been used in this approach for representation of relational databases in computer memory. Section 4 explicitly explains advantage of using regular trees as a basic data structure in comparison with previous approaches which used linked list. Section 5 explains creation of relational tree along with a real-world example. Section 6 compares ideas and methods in previous approach and current approach along with a comparison between the node structures. Conclusions drawn are mentioned in Sect. 7. Future scope is mentioned in Sect. 8, and references are cited at the end.

2 Related Work

Software industries mostly carry out normalization manually, which requires employees with proficiency in normalization. In order to satisfy today's business enterprise's demands, large number of relations and functional dependencies are required.

So, the process of normalization requires more than one person when it is carried out manually. The drawbacks of carrying out normalization manually are as follows.

1. It makes the process slower and thus less productive: many relations containing many attributes are required to model an enterprise which makes it slower manually.
2. It could have many errors: reasons are declared in 1.
3. Requirement of skilled people: requires skilled people but since we can automate the process there is no need.

Several researchers have tried to eliminate these drawbacks by automating the normalization process by proposing new tools/methods. There is a patent [3], which proposes a database normalizing system. It normalizes the given database by observing record source input for the tool is records which are already present in table. Another tool named *Micro* which was proposed by Du and Wery [4] in which the representation of a relational schema is done using two linked list. These linked lists are used to represent attributes and functional dependencies in a relation. Jmath Norm is another tool which was made using Jlink library's inbuilt functions given by Mathematica. Jmath Norm was proposed by Yazici et al. [5]. Dependency graph and dependency matrix were created by an automatic database normalization system, and this was proposed by Bahmani et al. [6]. One can define normalization algorithms on them now. Generation of primary keys and relational tables is also done.

It is difficult to motivate students and researchers to study normalization because the process is difficult, theoretical. Maier [7] also claimed that normalization can be hard for even average designers. LBDN [8] is another tool we came to know which is abbreviated for Learn Database Normalization and which is also web-based. The tool is very interactive since tool is client–server model. This tool provides a platform to students to understand normalization process by providing them lectures and hands-on practice. This tool also provides them some assignments to solve. In this tool, the representation is done in the form of sets and the keys, functional dependencies and attributes get stored as array of strings. We found another similar web-based tool which is mentioned in [9]. This tool is found useful in data management courses and also in system design and analysis. "This tool has a positive effect on students" claim the authors of the tool.

Representation of a relational schema is done using only a singly linked list in the tool RDBNorma [10]. Since the requirement for representation is only one single linked list, it is really a good effort which does the job using less time and space in comparison with Micro. This tool operates at schema level.

RDBNorma uses a single linked list, and thus, it requires less memory and time to process. It is an honest attempt for the development of a new way to represent relational schema. Since RDBNorma works on schema level, it really helps to eliminate the manual process of normalization, in a very efficient and speedy way. The work behind RDBNorma really increases the automation of normalization and helps to eliminate the drawbacks of manual process of normalization.

In this paper, we are proposing an alternate method to implement RDBNorma. We are well aware that RDBNorma uses single linked list as data structure, and we are proposing that if we use trees as data structure instead of linked list, the results will be better. The reason behind that is the complexity of trees is more as compared to linked, but searching a node is simpler and less time-consuming compared to linked list. The trees which will store the functional dependencies are called as relational trees.

3 Node Structure for Relational Tree

The node structure of the relational tree is as shown in Table 1.

Following is the detailed information about every field present in the node structure:

1. *AttributeName*: Attribute name is stored in this field. It can hold up to 20 characters. We assume unique attribute names in a single relational tree.

Table 1 Node structure for relational tree

AttributeName
AttributeType
Determiner
ChildofNode [0]
ChildofNode [1]
ChildofNode [2]
ChildofNode [3]
KeyAttribute
ptr_to_next

2. *AttributeType*: This is where we store type of attribute. It is set to store "*" for multi-valued attributes and "1" when attributes will be of atomic type.
3. *Determiner*: This field is used to hold the binary value, and it will store 1, if the given attribute is determiner in functional dependencies (FDs), and it will store 0, if it is not a determiner.

   For example, consider the following FDs

$$B \to A$$
$$B \to C$$
$$E \to D$$

   In these FDs, B and E are the determiners, while A and D are not determiners so value 1 will be stored in determiner field of attribute B and E and 0 for A and D nodes.
4. *childofnode [0]*, *childofnode [1]*, *childofnode [2]* and *childofnode [3]*: These fields will hold the pointer to all the children of the attribute. This assumes that there can be at most four children of any given attribute.

   For example,

$$A \to D$$
$$B \to E$$
$$A \to BC$$

   For node A, nodes D, B and C will be children which will be pointed by pointers in child field, and for node B, child pointer points to node E.
5. *KeyAttribute*: This field is used to know if the node is part of primary key or not. It stores 1 if it is part of the primary key else it stores 0.
6. *ptr_to_next*: This is a pointer variable which is used as a link to travel to the next node in linked list.

4 Advantages of Trees Over Linked List

1. The time complexity of searching a node in regular trees is comparatively less than that of linked list, and hence, the searching and accessing time required will be reduced.
2. Since we are reducing the access time, the handling will be easy and less time-consuming.
3. In linked list, every node needs to be travelled at least once, and since tree is a nonlinear data structure, there is no such need for traversal in tree.
4. Because of this traversing restriction, linked list becomes less beneficial than trees.

5 Working Principle Along with and Example

Following are the steps of creating the relational tree.

5.1 Temporary Storage of Attributes and Functional Dependencies of a Relational Schema

Attributes names and functional dependencies will be taken as input, and it will get temporarily stored in different arrays separately. We have used two different 2-D arrays, one for attributes names and other for functional dependencies.

5.2 Automatic Key Detection

By using the given FDs and the attributes, automatic key detection will be done for each relation schema. The key which will get detected will act as a primary key for that relation schema. The program will go through all the attributes and FDs, and the determiners which will determine all the attributes present in relational schema will be selected as key.

5.3 Linked List Formation for Key

The key detected in automatic detection can be of the composite form or non-composite form. Either way we have to store the key in linked list to create the relational tree. If the key is not composite in that case, only head of the linked list will store the key for the given relational schema. In case of composite key, each attribute in the key will get stored in linked list as a different node.

This step is the first step towards getting minimal cover of FDs, and by doing automatic key detection, we are going to calculate minimal cover of the given relation schema.

5.4 Relational Tree Formation

The key which gets detected automatically will get stored in the linked list, and by using given FDs, the child of key node in the linked list will get detected. The child pointer of a particular node in linked list will point to the child of that node. There will be maximum four child node pointers for each node.

After allocating child to the node, program will traverse the FDs for the child considering it as a parent, and if there is child present for that node, then child pointer of that node will point to the child, and if child is not present, then child pointer will point to NULL. Recursively, program will search the FDs for each node by traversing linked list and FDs.

Step 1: In this first step, algorithm will traverse the linked list of keys, and after taking of each node, it will search that node in the linked list.

Step 2: After getting that node, it will traverse that FD and will look for the child of that node, i.e. RHS side of the FD. Algorithm will detect the right side of FD, and it can be of composite form or non-composite. If the right side is of composite form, then the algorithm will detect each attribute from composite form and make each different node as a child from composite. If it is not composite, then it will directly make a new node for that attribute and will make that node as a child of left side of FD.

Step 3: Then, algorithm will recursively check for the child node as a parent, it will traverse the FDs for that node and detect its child, and its child pointer will point to that child.

Step 4: Before the Step 2 and Step 3, it will check that all child node pointers are full or not. It will point new child pointer node to any new node if any child pointer is null before that.

Following section explains an example of representing a real-word relation and its FDs using relational trees which would improve understanding of the algorithm discussed above.

Consider a relational schema

EMPLOYEE = (*supp_id, supp_name, street, supp_phone, produc_id, produc_name, produc_price, produc_quantity*)

Functional Dependencies

supp_id→supp_name, street, supp_phone
produc_id→produc_name, produc_price
supp_id, produc_id→produc_quantity

Process of Formation of Relational Tree

Step 1: In the above given relational schema and FDs, all the attribute names and FDs will get temporarily stored in the array. Here, for "node [] []", one 2D-array will get created which will store the attribute names, and for "fd [] []", another 2D-array will get created which will store the given FDs temporarily.

Detailed information of the attributes is shown in Table 2.

Step 2: By using the above two arrays, the algorithm will detect the primary key of the relational schema. In the above example, the primary key will be composite key consisting *supp_id* and *produc_id*. Through primary key, we can access all the attributes nodes of relational schema.

Table 2 Detailed information

Attribute name	KeyAttribute	NodeType
supp_id	1	Atomic
supp_name	0	Atomic
Street	0	Composite
Supp_phone	0	Multi-valued
Product_id	1	Atomic
Product_name	0	Atomic
Product_price	0	Atomic
Product_quantity	0	Atomic

Fig. 1 Linked list of relational tree

Step 3: Primary key will get stored in the linked list of relational tree.
In the FD: *supp_id, produc_id→produc_quantity*, both (*supp_id, produc_id*) will get stored in the one node of linked list for the representation.

The other two keys will get stored after that in the linked list, and *supp_id* will be the head of linked list (Fig. 1).

Step 4: Now, while traversing the linked list, first node will be *supp_id*, and by traversing the FD's array, program will get the child nodes of the attribute *supp_id*. There are three child nodes for *supp_id* which are *supp_name*, *supp_phone* and *street*. Child field's pointer of *supp_id* will point to these three child nodes.

Therefore, relational tree will be as shown in Fig. 2.

Step 5: Now, program will recursively search for all child nodes for each node in the linked list, i.e. primary key. Step by step, child nodes of each primary node will get attached to the its parent node in the linked list (Fig. 3).

Final Relational Tree
See Fig. 4.
Memory Representation for *EMPLOYEE* Schema.
Memory representation for *EMPLOYEE* schema is shown in Fig. 5.

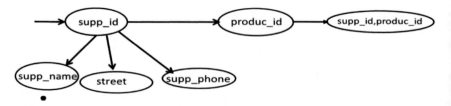

Fig. 2 Stage 1

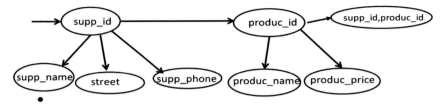

Fig. 3 Stage 2

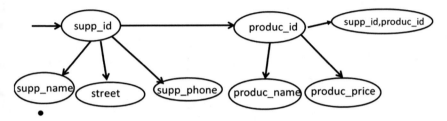

Fig. 4 Relational tree

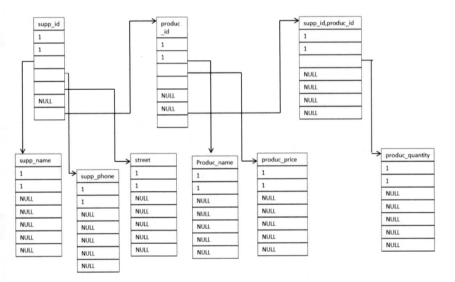

Fig. 5 Memory representation for *EMPLOYEE* schema

6 Comparison of Relational Tree with Previous Tools

Basic difference between relational trees and previous tool lies in the basic idea behind using trees and linked list data structure, respectively. Technique proposed in [2] uses single linked list to store attributes and FDs. *Micro* uses two linked lists,

Table 3 Relational tree node structure

AttributeName
AttributeType
Determiner
ChildofNode [0]
ChildofNode [1]
ChildofNode [2]
ChildofNode [3]
KeyAttribute
ptr_to_next

one of which is used to store attributes, and the other is used to store FDs. However, when we use single linked list to store relational schema and FDs, accessing the required node becomes way faster than expected. Using a singly linked list probably uses a list to store all the presently created nodes along with a pointer pointing to the address of the node, or it travels the whole linked list to find a particular node. But the basic problem with this technique lies with the structure of the node. A node contains pointers to determiner of that node. However, this is not how logical flow takes place. Basic flow should be from determiner to its children.

In relational trees, linking is from determiner to its children. Any given node of relational tree stores pointers to its children. Also, this allows us to create a structure which can be simply understood as a linked list of roots of trees, where all the roots together act as primary key.

Along with this, relational trees take advantage of nonlinear structure (trees) while maintaining the general idea of keeping attributes that contribute to primary key together (linked list).

Consider the following example

$$X \rightarrow Q, A$$

In previous method, a linked list of these three nodes is created. Nodes containing Q and A store address of X, but in relational trees, a linked list of primary key (i.e. X) is created. X has a pointer pointing to Q and a pointer pointing to A.

Following is the comparison between note structures of current (Table 3) and previous approach (Table 4).

7 Conclusion

It is concluded that we can represent a relational schema using relational tree. It is also expected that the results of the relational tree for representing relational schema will be faster than the previous method of representation which was representation

Table 4 Linked list node structure

attribute_name
attribute_type
determiner nodeid
Determinerofthisnode1
Determinerofthisnode2
Determinerofthisnode3
Determinerofthisnode4
keyattribute
ptrtonext

using a singly linked list [1]. Using relational tree to represent a relational with the functional dependencies overcomes the flaws of representation done using linked list. It gives the solution to children to determiner linking system in the previous approach by providing a new node structure for representation. Normalization algorithms can be applied, to normalize the given schema.

Thus, it is concluded with the expectation that relational tree performs better than tool proposed in [1] and *Micro* [4].

8   Future Scope

1. Applying efficient normalization algorithms on the relational tree, we can overcome the disadvantages of manual normalization.
2. Automation of normalization has the great industry application, and it reduces the cost of manual normalization.

References

1. Dhabe, P.S., Y.V. Dongare, S.V. Deshmukh. 2010. Representation of a database relation and it's functional dependencies using singly linked list data structure for automation of normalization. IJERIA 3 (No. II): 179–198.
2. Codd, E.F. 1970. A relational model of data for large shared data banks. *Communications of the ACM* 13 (6): 377–387.
3. Hetch, C. Stephen. 1998. US Patent 5778375—Database normalizing system.
4. Du, H., L. Wery. 1999. Micro: a normalization tool for relational database designers. Journal of Network and Computer Application 22: 215–232.
5. Yazici, A., Ziya, K. 2007. JMathNorm: a database normalization tool using mathematica. In Proceedings of International Conference on Computational Science, 186–193.
6. Bahmani, A., M. Naghibzadeh, B. Bahmani. 2008. Automatic database normalization and primary key generation. Niagara Falls Canada IEEE.
7. Maier, D. 1988. *The Theory of Relational Databases*. Rockville, MD: Computer Science Press.

8. Georgiev, Nikolay. 2008. A web-based environment for learning normalization of relational database schemata. Masters Thesis, Umea University, Sweden.
9. Kung, Hsiang-Jui, Hui-Lien, Tung. 2006. A web-based tool to enhance teaching/Learning database normalization. In Proceeding of International Conference of Southern Association for Information System.
10. Dhabe, P.S., Dongare, Y.V., Deshmukh, S.V. 2011. RDBNorma A semi-automated tool for relational database schema normalization up to third normal form. IJDMS 3 (No.1).

Real Drowsiness Detection Using Viola–Jones Algorithm in Tensorflow

Aseem Patil

Abstract Physical security and personal protection characteristics are developed to avoid lethargy. In the automotive industry, this is a significant technical challenge. Driving during grogginess is a key reason, particularly nowadays, behind road accidents. If drowsy, the risk of collapsing may be greater than in an alert state. There is therefore a important aid in preventing accidents using assistive technologies to monitor the driver's alertness level. In this publication, the driver uses visual characteristics and intoxication identification with an alcohol sensor for the identification of insobriety. Driver bleakness also depends on the driver's head, nose and nose classification in real-time driving either on a deserted road or a road full of traffic.

Keywords Drowsiness detection · Viola–Jones algorithm · Facial recognition · Eye detection · EAR · Fatigue · Image capturing · Landmark points

1 Introduction

Drowsiness causes the consciousness to decrease and is triggered by the absence of exercise or weariness. The person driving the vehicle crashes due to grogginess control of the car which could lead to serious injuries, which could divert him/her from the route. This is where Viola–Jones algorithm comes into the picture. The Viola–Jones algorithm is a commonly used object detection system. The primary feature of this algorithm is that it is slow to train but quick to detect. This algorithm is based on Haar-function filters, so multiplications will not be implemented. The integrated picture can be calculated with only four points for the Haar extractors. Detection takes place in the window of surveillance. The screen length is minimum and maximum, and a common stage length is selected for each volume. A range of cascade-connected classifiers is stored in each face recognition filter (from the collection of N filters). Each classifier examines a linear subset of the identification panel and determines whether the screen appears like a face. If so, the next evaluation

A. Patil (✉)
Department of Electronics Engineering, Vishwakarma Institute of Technology, Pune, India
e-mail: aseem.patil16@vit.edu

© Springer Nature Singapore Pte Ltd. 2020   317
D. Swain et al. (eds.), *Machine Learning and Information Processing*,
Advances in Intelligent Systems and Computing 1101,
https://doi.org/10.1007/978-981-15-1884-3_30

will be implemented. The screen provides a favorable response, and the picture is acknowledged if all classifiers offer a favorable response. Otherwise, the next filter ran from the set of N filters. The Haar function extractors (weak classifiers) are made in each of the classifier already present in the set. Each Haar characteristic is a weighted sum of 2-D integrals connected to each other in tiny linear zones. The weights can be ±1.

The sleepiness/drowsiness of the rider is the main factors that cause injuries, according to WHASTA operational stats. However, over the years, India faces serious repercussions for road safety, with the growing pace of every single individual trying to own their personal vehicle followed by an extension. In India, there has been a 5% increase in the complete percentage of route fatalities from 580,400 in 2013 to 611,414 in 2018.

2 Drowsiness Detection Techniques

If vehicle generation prevents or at the least warns of driving force fatigue, what are the signs and symptoms that may be detected by means of the motive force? There are numerous classes of technologies that can detect driver fatigue, according to investigate. The primary is the usage of cameras to display the behavior of a person. This process includes keeping track of their eye aspect ratio, yawning, the head position and a diffusion of various other things. The subsequent of those technologies is the popularity of voice. The voice of someone can often provide clues as to how worn-out they are. Specific rationalization of the underlying techniques is used for the detection of drowsiness are as follows:

- Electrocardiogram and electroencephalogram
- Local binary patterns
- Steering wheel motion
- Visual detection
- Eye blinking-based approach.

2.1 Electrocardiogram and Electroencephalogram

Many scientists have examined the preceding physiological measurement signals for lethargy and somnolence, and they are, namely electrocardiogram (ECG) and electroencephalogram (EEG). The heart rate (HR) also differs considerably between various phases of drowsiness, such as alertness and fatigue. Heart frequency can thus also be used to identify drowsiness effectively by using ECG measurements. For the HRV measurement of drowsiness, the short (LF) and large (HF) ranges between 0.06–0.19 and 0.18–0.8 were also used by other people. In order to measure drowsiness, electroencephalograms (EEG) are the most common physiological signals (Fig. 1).

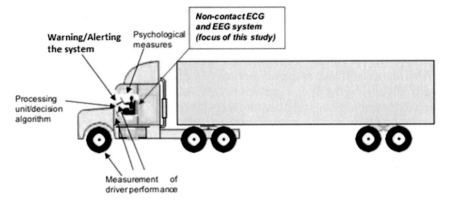

Fig. 1 Schematic of sensing system integration for driver drowsiness and alertness system using EEG

2.2 Local Binary Patterns

The concern in machine vision and image processing accelerated in local binary patterns. LBP efficiently sums up local pixel models as a non-parametric technique by linking each pixel to its neighboring pixel vertices. The most significant characteristics of LBP are their sensitivity and flexibility in computer to monotonic modifications in lighting. This method is used mostly to detect stimuli on the body such as joy, melancholy and thrill. In the drowsiness assessment process, LBP is used in the vehicle face detection, dividing the picture into four quadrants and separating the upper and lower parts.

2.3 Steering Wheel Motion

It is a large-use vehicle-based method for detecting driver drowsiness rate, measured by the steering wheel angle detector. A control sensor on the steering column is used to measure the driver's control posture [1]. Monitoring the motions of the steering wheels is a well-documented way to detect driver fatigue and drowsiness. Current steering wheel motions surveillance techniques are prohibitive for regular use due to elevated execution cost and the need for a complicated change for the re-configuration. These constraints limit the possibility to detect lethargy by means of steering wheels to experimental and simulator configurations.

When drowsy, the amount of nano adjustments on the steering wheel reduces in comparison with ordinary steering. The research teams presumed that only small steering wheel motions of 0.6°–5.1° were required to modify the horizontal stance within the lance to eliminate the effects of column modifications. By turning the steering wheel in small increments, drivers are constantly assessing the situation ahead and applying small, smooth, steering adjustments to correct for small road bumps and crosswords.

Therefore, it is possible to determine the driver's somnolence state based on small SWMs and thus provide an alert if necessary. In a simulated environment, light side winds pushing the car to the right side of the road were added along a curved road to create lateral position variations and force the drivers to make corrective SWMs.

2.4 Visual Detection

The most common feature of a vision detector model is electromagnetic or near-infrared LEDs to illuminate conductor pupils, which then are tracked through surveillance systems. Computer algorithms are used to determine drowsiness for the binding frequency and length. In the case of indications of drooling, for example, glare in eye and abrupt neck spikes, the camera mechanism can also monitor face and eye stance.

2.5 Eye Blinking-Based Approach

The blinking rate and duration of the eye closure are measured in this eye to detect the drowsiness of the driver. Because at that time, when the driver felt sleepy, his/her eye blinking and his/her eyelid gaze differed from normal situations so that drowsiness could be easily detected [2]. And a remotely positioned camera is used in this sort of system for collecting recordings and machine imaging methods to frames to evaluate the completion rate of the head, nose and ears [3]. Drowsiness of the rider can be identified with these deeper visual lenses and the blinding hoard. Any indications that sway the face, lower the ears or nodes continuously, can be monitored by the device installed in a special spot in the vehicle.

In many applications such as eye gaze tracking, iris detection, video conferencing, auto-stereoscopic displays, face detection and face recognition, eye detection is required. This paper uses color and morphological image processing to propose a novel technique for eye detection.

3 Implementation for the System

To understand the system more effectively, we shall have a look at the block diagram we have made for the design of the system (Fig. 2).

In the proposed system, we shall crop the image using an 8 MP USB camera that shall be mounted on a vehicles front view. We shall use facial detection to crop only

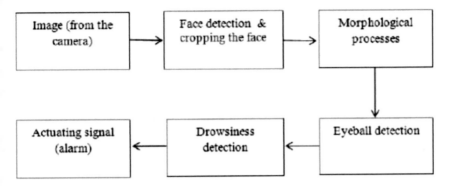

Fig. 2 Block diagram of the above mentioned system

the face and remove any other unnecessary details from the cropped photo. We shall use Viola–Jones algorithm in detecting the eyeball stats away from the entire facial image. Using the eye aspect ratio, the system will be able to determine the strong homogeneous coordinates and will store them in a basic excel file. From there, we shall implement the process and get the required result.

3.1 Pseudo Code

if ear < Threshold: # EAR Threshold COUNTER += 1
if ear < Threshold:

   DBAR+=10

if ear> Threshold:

   DBAR=0

if COUNTER > 2 : # Blink Detection if ear > Threshold:

   TOTAL +=1 COUNTER = 0

if DBAR > TDBAR:# Sleep Detection DEVENT+=1

If EAR drops below the set threshold in any case and remains for at least 1 s, it will be detected as blink, and COUNTER will store blink no value [2]. If further, EAR remains below the threshold for more than 3, it is considered to be sleep and will be displayed on somnolence scale, and no drowsiness events will be stored in the DEVENT variable. This is the logic we are going to use to implement the above-proposed system. This can only be done by Viola–Jones algorithm.

3.2 Image Capturing

This is the system's first and initial phase. For the present owner, the configuration is done and optimized. The driver's head position is the primary tracking phase. When the driver's face is effectively positioned, processing the specified image is straightforward, and the driver's actual mental status is also recognized [4]. The camera used is of an intensity of 8 MP. We can implement the face detection using the proposed method, which is Viola–Jones algorithm. The steps for doing so are as follows:

(1) To build a simple prescreening filter with template features for the first stage in the cascade.
(2) Encode the most basic differences between the template-based faces.
(3) Match the training set and the training/demonstration test set.
4) Draw the final decision on the positive training set in the first place and the negative training set in the second place.
(5) Read the image source and remove the ROI from the image face source. ROI is the image of the sub and must be smaller than the image of the target.
(6) Do standardized cross-correlation and find ROI and target image peak coordinates. Calculate and display the standardized cross-correlation as a surface plot. Here, Viola–Jones algorithm plays a vital role.
(7) Find the full inverted images. The complete distortion or transfer of pictures relies on how the top is located in the cross-correlation matrix and on whether the pictures are sized and positioned.
(8) Check that the face of the target image is extracted. Figure 3 where the face matches exactly within the target image.

The setup of the system includes two major processes: (i) extracting the driver's facial expressions and detecting their head movement and (ii) collecting different samples of eyes open or closed. This information gained is further processed to analyze the current state of the driver (Fig. 4).

From the 68 facial landmark points, we can assume the distances parallel to each other for each eyelid and the eye aspect ratio that holds the facial landmark points in visualization [5]. After applying Viola–Jones algorithm, we get the image to be transferred into a spider web-like structure that connects each point with each other (Fig. 6).

Imagine we have a tendency to have an interest into coaching a model that is able to localize solely the landmarks of the left and right eye [6]. To do this, we have to edit the image and vision computational annotations by choosing solely the relevant points. This can be done by calling the slice_xml () function that makes a replacement xml-file with solely the chosen landmark points of the respective person.

The final occurrence comes when the eye aspect ratio is measured for each eye that helps us to determine the accuracy and precision of the measurement taken when the person was drowsy. We shall scale a graph of the measures in the future before predicting the number of hits and the percentage of hits for each person used as a data point in the data set made.

Fig. 3 Facial detection and recognition

From the given figure (Fig. 5), the points that are shown red represent the 68 facial landmark points that are situated on the face of a human being. The facial landmark points are basically the strong points out of the points that are white in color. We extract the features based on the highest rate of the EAR. If the EAR for a certain point is greater than another facial landmark point, we shall neglect such points. In the given figure, we could identify a total of 68 facial landmark points, out of which 42 stood out. Using these 42 points, we found the average of the distances between the ear and the eye of the individual in his normal state and compared this average to the average of the same individuals' fatigue state. This helped us in getting progressive and real readings.

3.3 Alert Stage

If the driver is found in the abnormal state which could either be in a trance state or drunk, the system activates the alarm. The alarm that is rung can be in the form of music, vibrator placed under the seat or a normal alarm noise.

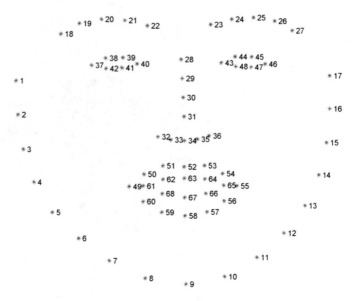

Fig. 4 The data set visualizes the following 68 facial landmark points of a higher resolution that is related to the HOG descriptor

This is measured using the steering angle sensor and is a widely used vehicle-based measurement for detecting the level of drowsiness in the driver. The driving activity of the driver is evaluated using an angle sensor attached to the steering wheel. If you are drowsy, your steering gear reduces the amount of nano adjustments relative to ordinary steering measurements [3]. We found that drivers who were sleeping poorly had smaller turns on the steering wheel than ordinary riders. We considered only small steering wheel movements (between 0.4° and 6°) necessary for adjusting the side stance inside the row to eliminate this impact of lanes. In particular, the directional conduct is then drive-like (e.g., driving practice), and drivers usually evaluate and use simple, smooth, motor modifications to fix little obstacles or crosswinds by rotating the steering wheel in slight increase based upon the challenge. In addition, the riding performance is usually determined by drivers.

3.4 Viola–Jones' Algorithm Significance

Paul Viola and Michael Jones presented a fast and robust face detection method that at the time of release is fifteen times faster than any technique with 95% accuracy at about 17 fps. The algorithm of detection Viola and Jones is used as the basis of our design. Since we know that all human faces have some similarities, we used this concept as a hair feature to detect face in image. Algorithm looks for a face's specific hair feature if the algorithm found passes the candidate to the next stage.

Fig. 5 Analysis of the detected face by using Viola–Jones algorithm

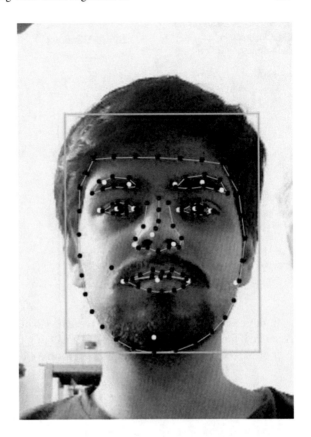

The candidate is not a complete image here but only a rectangular part of this image, known as a sub-window, is 24 * 24 pixels in size. Check the entire image with this window algorithm. In this method, we will be using the cascading method from Viola–Jones algorithm.

Using stage cascading, the false candidate can be eliminated quickly. If the first stage was not passed, the cascade eliminates the candidate. If it passed, it is more complicated than sending it to the next stage. If a candidate has passed the entire stage, a face will be detected. This makes Viola–Jones algorithm a discrete and different form of dividing the good from the wrong and helps in maintaining the accuracy percentage of the system (Fig. 6).

4 Calculating the EAR (Eye Aspect Ratio)

Each vision is represented by $6(x, y)$ points from the top angle of the mirror (as if you were looking at the individual), then in the chronological position operating around the mirror (Fig. 7).

Fig. 6 Using cascade as one
of the methods in
Viola–Jones algorithm

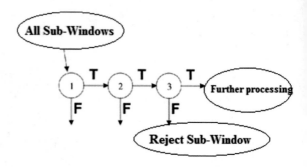

Fig. 7 The distances of the
homogenous coordinates that
are needed to calculate the
eye aspect ratio (EAR),
where p_1–p_6 are 2D facial
landmark locations

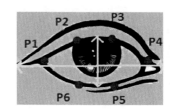

In our case of drowsiness detector, we will monitor the ratio of the attention to visualize if the value falls but does not rise again, implying that the person has closed their eyes. The subsequent formula is used for the calculation of EAR.

$$\text{EAR} = \frac{\|p_2 - p_6\| + \|p_3 - p_5\|}{2\|p_1 - p_4\|} \tag{1}$$

From the following Eq. (1), the numerator calculates the range between the vertical eye landmarks, while the denominator calculates the range between inverted eye landmarks, properly balancing the denominator because of only one set, but two sequences of vertical numbers exist. As the eye is accessible, the aspect ratio is almost continuous but falls to null rapidly when the eye is blinked as an involuntary action.

5 Results

Under normal and regular conditions, the mean of the percentage of hits gives us the accuracy of the system. It turns out to be 93.37%. By Table 1, we can conclude that the accuracy that we needed is more than expected.

Table 2 shows the result table of the different hits observed at circumstances in real life with a cabby and a person wearing spectacles. It is quite hard to understand whether the person is drowsy or not when he/she is wearing spectacles. That is why the accuracy of the driver with glasses is lesser than average.

Table 1 Under normal and regular conditions, the mean of the percentage of hits gives us the accuracy of the system. It turns out to be 93.37%. By the following table we can conclude that the accuracy that we needed is more than expected

Test	Number of observations	Number of hits	Percentage of hits
Yawn detection	170	143	84.11
Front nodding	200	184	92.0
Assent of the head to the right	200	190	95.0
Assent of the head to the left	200	191	95.5
Distraction to the right	200	184	92.0
Distraction to the left	200	193	96.5
Blink detection	200	197	98.5

Table 2 The following table shows the result table of the different hits observed at circumstances in real life with a cabby and a person wearing spectacles. It is quite hard to understand whether the person is drowsy or not when he/she is wearing spectacles. That is why the accuracy of the driver with glasses is lesser than average

Test	Number of observations	Number of hits	Percentage of hits
Driver with a cab	1400	1295	92.5
Driver with glasses	1400	1183	85.5

From the tables obtained, we have plotted four sample graphs randomly. We find that the accuracy is in its nominal state and that there may be variation in steering angle when a person is drowsy as every person has his/her own energy level till they get to their extreme level of drowsiness condition (Fig. 8).

6 Conclusions

The paper has suggested a scheme to help a driver by assisting his/her condition avoid significant collisions usually provoked by drowsiness. The driver status is determined using image processing algorithms. A buzzer and a horn warn the user when they are in the drowsy circumstances. The location of the driver's face is determined with the regard to the center of propulsion, and the present driver position is determined. High focal camera of excellent precision captures the motion of the body. The scheme provides an additional tracking function. If the rider yawns more often, then the alarm is still triggered. The detector detects whether or not the rider is drunk or drowsy. For correct precision, the sensor should be distanced from the rider. The warning, which can be in audio or vibration form, can encourage the rider to securely achieve the desired state of mind. The scheme can add to the detection of the actual driver's

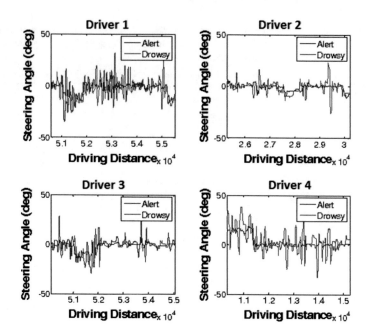

Fig. 8 Output results obtained by testing the code with four drivers. The result we go was 93.37% accurate, and this was only possible because we took real drowsy people as our test samples. The red line shows drowsiness condition, and the blue line shows alertness condition in that person

condition and the frequency of road accidents although further advanced research is still required.

The advanced driver help system that has been designed is used to observe the temporary state of driver. It may be used to check if the driver is drunk or not. This technique will be enforced in vehicles in real world to administer provision to take live video feed of driver when he/she is on the road either driving or stuck in traffic.

References

1. Saito, Yuichi, Makoto, Itoh, Toshiyuki, Inagaki. 2016 March 21. Driver assistance system with a dual control scheme: effectiveness of distinctive driver sleepiness and preventing lane departure accidents. *IEEE Transactions on Human-Machine Systems*, 4–8.
2. Sari, Nila Novita, Yo-Ping, Haung. 2016. A two-stage intelligent model to extract options from PPG for somnolence detection. In *2016 International Conference on System Science and Engineering (ICSSE) National Chi Nan University*, July 7–9, 2016.
3. Tran, D., E. Tadesse, W. Sheng, Y. Sun, M. Liu, S. Zhang. 2016. A driver help framework supported driver sleepiness detection. In *The sixth Annual IEEE International Conference on Cyber Technology in Automation, management and Intelligent Systems*, June 19-22, 2016, 7–9.
4. Aleksandar, Oge, Marques, Borko, Furht "Design and Implementation of a Driver somnolence Detection System A practical Approach.

5. Anjali, K.U., Athiramol K. Thampi, Athira, Vijayaraman, Franiya M. Francis, Jeffy N. James, Bindhu K. Rajan. 2002. Real-time nonintrusive observation and detection of eye blinking visible of accident interference thanks to drowsiness. In *2016 International Conference on Circuit, Power and Computing Technologies[ICCPCT]*.
6. Ahmed, J., Jain–Ping, Li, S. Ahmed Khan, R. Ahmed Shaikh. Eye Behavior Based Mostly Sleepiness Detection System, 6–8.

Smart Sound System Controller for Restricted Zones

Ranjana Jadhav, Rahul Dhotre, Hrishikesh Gaikwad, Vinit Bolla, Suresh Hamilpure and Aditya Unavane

Abstract Noise pollution is a major aspect and has become a serious issue. The scale or range of noise pollution is increasing rapidly day by day. According to the statistical data, the acceptable range of noise is 40–45 dB. But during festivals like Diwali and Ganesh festival, it is observed to be around 90–95 dB. This rate of growth of pollution is very high which leads to environment problem and disturbance to ecosystem. To bring it under control, it is necessary to take appropriate actions against the causes of noise pollution. In existing systems, basic monitoring of the noise is done which is not sufficient to control and reduce the problem of noise pollution. Hence, we are introducing a system which will overcome this issue and will not only help us to monitor the level of sound but also provide a way to control that sound-emitting device using IoT. This system provides a fast and automated way for taking action against major sound-polluting devices in specific areas. Thus, with the help of our system, the noise pollution can be detected as well as controlled to remain in specified limits.

Keywords Noise pollution · IoT · Sensors · RF transmitter–receiver · Monitoring and controlling system

R. Jadhav · R. Dhotre (✉) · H. Gaikwad · V. Bolla · S. Hamilpure · A. Unavane
Department of Information Technology, Vishwakarma Institute of Technology, Pune, India
e-mail: rahuldhotre777.77@gmail.com

R. Jadhav
e-mail: ranjanajadhav26@gmail.com

H. Gaikwad
e-mail: hrishikesh.r.gaikwad@gmail.com

V. Bolla
e-mail: vinitbolla10@gmail.com

S. Hamilpure
e-mail: sureshhamilpurefc@gmail.com

A. Unavane
e-mail: adityaunavane98@gmail.com

© Springer Nature Singapore Pte Ltd. 2020
D. Swain et al. (eds.), *Machine Learning and Information Processing*,
Advances in Intelligent Systems and Computing 1101,
https://doi.org/10.1007/978-981-15-1884-3_31

1 Introduction

Nowadays, the entire world is moving toward the era of technology and smart devices. These upcoming and rapidly developing technologies are making human life more and more comfortable [1]. All the systems around us are becoming smart and automated. Internet of Things is the main backbone of majority of the smart systems. The high amount of use of Internet and the human interaction with the machines have increased the requirements and importance of IoT on a large scale. IoT has contributed in many ways for the development of smart and automated devices and systems. The availability of low-cost sensors that can share information easily is the major part toward the success and popularity of IoT [2].

Along with the developments in the technologies and rise in population, environmental problems have also been increasing on a large scale. Pollution is one of the leading problems among all. Any type of pollution is not only harmful for the nature but also very much harmful for humans. Sound pollution is one such type of pollution [3]. There are various researches done on the problem of sound pollution detection, and various rules have been imposed for controlling sound pollution. But the existing model requires a large amount of manual work of collecting data from the pollution spot and then imposing rules for its control [4]. Our proposed model "Smart Sound Control System Controller" is a smart system which not only detects the sound pollution but also controls it in a complete automated manner. "Smart Sound Controller" makes use of sound sensors to detect the sound system that is causing sound pollution in the restricted zones and stops them at that particular instance. This paper gives the description of designing and working of our smart sound system controller using IoT [5].

2 Motivation

The main motivation for the system comes from festivals like Diwali, Shiv Jayanti, Ambedkar Jayanti, Moharam, Navratri, Ganesh Utsav, etc. [6]. The noise level increases up to 80–100 dB during evening hours against the normal level of 75 dB. Due to this unwanted sound, not only the small children and the older people get irritated but also the young people get rid of this thing which caused many problems [7]. This system will help to control that unwanted sound by using the RF transmitter and receiver and IoT concepts such as noise increases in the restricted zones the sound sensor will be able to sense it and send a signal to the controller board so that it will be able to keep track of that data accordingly and send the RF signals to the RF receiver [8]. All the sensor data will be uploaded to the web server for keeping track of that data.

3 Literature Survey

Smart sound control system is very beneficial for India by making use of this system in restricted areas such as hospitals, schools, and co-operate areas. Our system provides means to overcome the disadvantages of the existing system.

Sound pollution monitoring system includes device like Arduino UNO which is primarily based on ATmega328 microcontroller and helps to make analysis of the sound and the air pollution in particular region. If there is any increase in pollution, it notifies people through android application to take appropriate action [9].

Air and sound pollution monitoring system based on IoT consisting of air and sound level sensor is used to sense the live pollution of air and noise in that region and transmit that data through Raspberry Pi over an application. That allows the authorities to take required action against this noise and air pollution [10].

Air and sound pollution monitoring system using technology of IoT which includes some sensors, namely MQ7, MQ6, MQ135, and LM35, to measure the amount of carbon monoxide, cooking fumes, carbon dioxide, smoke, and temperature, respectively, in the environment. And it also uses sound level sensor just to detect the sound and air pollution. The output of the system obtained is in digital form so that the authorities can take appropriate action against it [11].

IoT-based air and sound pollution monitoring system uses the sound level sensor and gas sensor to detect the noise in restricted areas and measure the amount of toxic gases in the air to report this data to authorities to take appropriate action against the pollution [12]. The existing model is an embedded device which monitors the levels of noise pollution and carbon dioxide in the atmosphere and helps in making the environment intelligent as well as interactive with the objects using wireless communication. The existing system model (Fig. 1) is more adaptive and distributive in nature in monitoring the environmental parameters. The proposed architecture is a 4-tier model in which each module functions for noise and air pollution monitoring. The 4 tiers of proposed model consist of the environment in tier 1, sensor devices in tier 2, tier 3 for sensor data acquisition, and decision making and intelligent environment in tier 4 [13].

4 Proposed System

Most of the silent/restricted zones, while ensuring that the necessary facilities are available to its users, also seeks to provide an atmosphere conducive enough for research, study, and assimilation. However, noise is a major hindrance in achieving such conducive arrangements in our silent/restricted zone. Therefore, measures need to be taken to eliminate this problem. The proposed model device is divided into two parts: one for the restricted zone that is shown in Fig. 2 and second for the sound system that is connected between the sound system and amplifier which is shown in Fig. 3.

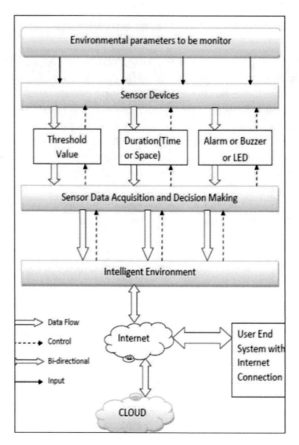

Fig. 1 Existing system model

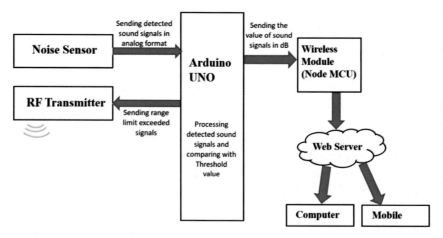

Fig. 2 Proposed model 1

Fig. 3 Proposed model 2

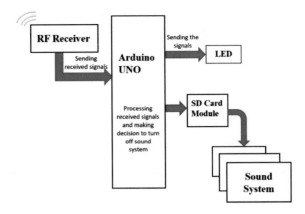

The proposed system monitors the sound and noise signals by using the noise sensor and sends the data to the arguing UNO board. After that data can be tracked by the Arduino. Until the sound is in range between 0 and 75 dB, no action will be taken. But when the sound range increases above 75 dB, the device sends the RF signals by using RF transmitter. Also, the data monitored by sound system is sent to the web server by using the Wi-Fi module.

4.1 Working of Proposed Model

The proposed system model is mainly used to control the noise pollution created by the sound/DJ system. To control the sound level in DJ sound systems, we have used two systems. One system will be on the hospital (or at any restricted zone), and the other system will be connected to the sound-emitting device like Dolby. In the first system which is there on the hospital, we have used NodeMCU, RF transmitter, and sound sensor (shown in Fig. 2). In the second system which will be connected to the sound-emitting device, we have mainly used Arduino UNO module and RF receiver (shown in Fig. 3). We have used SD card adapter as sound-emitting device.

When the sound-emitting device comes into the 1 km range of the hospital, sound sensor on the system on the hospital will start sensing the sound and will transfer those signals to the Arduino. The Arduino will then send the signals in decibels to the NodeMCU. At the same time, when the signals are received, the level of this sensed sound is checked by the NodeMCU. If the level of the sensed sound goes above 75 dB, then with the help of the RF transmitter, it will send the radio frequency signals to all the systems connected to all the sound-emitting devices in the 1 km range of the hospital so that we can reduce the sound level of all the sound-emitting devices by 50–80%. Now, suppose if you do not want to reduce the sound level of all the sound-emitting devices by 50% then you can also disable all the sound-emitting devices by sending stop signals which will switch off the power supply of sound in sound-emitting device. And now the sound-emitting device will remain disabled until the device goes out of the restricted zone [1].

4.2 Flow of the System

The most important thing before playing a DJ/sound system on roads or at any public places is to issue a No Objection Certificate (NOC) from the police inspector of that area. Without the NOC from police, it is illegal to play the DJ/sound system. Hence, when the sound system owner or the person who wishes to play the DJ comes to the police station, the police officer will complete his registration process by his approval and give him the NOC along with the controller device system shown in Fig. 3. Only if the controller device is connected to the DJ/sound system, the owner will be granted with NOC. If not, then he has no legal permission to play his sound system.

Once the sound system is played and the allotted time of permission for that registration expires, the sound system owner has to return the controller device back to the police station. To ensure the security of the controller device, police inspector can charge some deposit money from the sound system owner while issuing it to him. When he comes back to return it, after verifying its condition, the deposit money can be returned back as it is (Fig. 4).

4.3 Working of the Protocols

The communication between the transmitter and receiver is done using a simple transmission protocol. This protocol works for sending the messages in the form of packets from transmitter to desired receiver. This protocol mainly shows proper working when there are multiple receivers for one transmitter (Fig. 5).

Hence, the data packets sent by the transmitter have a fixed format which helps in identifying the desired transmitter to which the data has to be sent.

Each receiver has a specific ID by which it is recognized. When the message is sent by the transmitter, it has the receiver ID and then the corresponding message in the packet. Receiver ID is matched with all receivers in the range, and then, the one whose ID matches accepts the packet.

As shown in Fig. 6, the first 8 bits (1 byte) are reserved for the receiver ID and the next 8 bits (1 byte) are reserved for the message. So according to our working, if any system in the range is detected with sound signals greater than threshold value, the transmitter transmits the packet with the receiver ID of that specific receiver and the message to stop the sound system. On receiving the packet, corresponding action is taken by the receiver-side device.

Fig. 4 System workflow

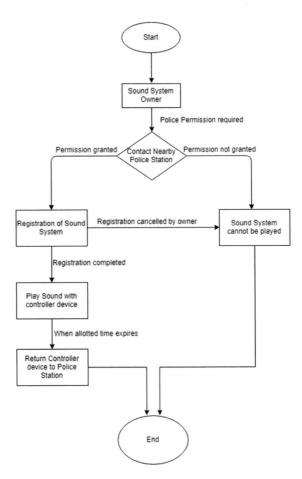

4.4 Devices Used

xcluma Sound Sensor Module Sound Detection Module: This is a sound sensor module sound detection module main chip, i.e., Lm393, electret microphone. It has 3 sets of servo horns and fittings. It can be used for sonic lamp with photosensitive sensors which acts as a sound and light alarm, and it can also be used in the occasion of voice control and sound detection circuit boards' output switch value.

RF Module Ask Wireless Transmitter Receiver Pair 433/434 MHz Compatible with Arduino, Raspberry Pi: It has an operating voltage of 3.6–5.5 V (5V recommended). It has receiver sensitivity as high as −110 dBm. It has high quality and compatible with HT12E-HT12D packed in anti-static packet to protect during handling and shipping. It has low power and good range (200–300 m).

ESP8266: The ESP8266 is a low-cost Wi-Fi microchip with full TCP/IP stack and microcontroller capability produced by Shanghai-based Chinese manufacturer

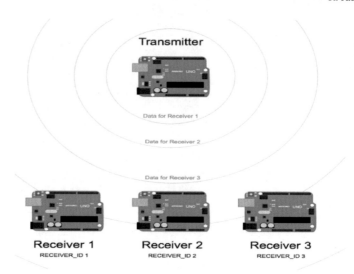

Fig. 5 System setup

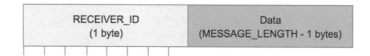

Fig. 6 Data packet format

Espressif Systems. Its memory size is 32 KiB instruction and 80 KiB user data. It has default CPU speed of 80 MHz or 160 MHz.

3–24 V Piezo Electronic Tone Buzzer Alarm 95DB Continuous Sound: Its alarm diameter is 22 mm/0.86″. Its alarm height is 10 mm/0.39″. It has two mounting holes of distance 30 mm/1.18″. It has two wires of length 90 mm/3.54″. Its buzzer type is piezoelectric. Sound pressure level is 95 dB. Rated voltage is 12 V DC. Its operating voltage is 3–24 V. Maximum current rating is 10 mA. Its frequency is 3900 ± 500 Hz.

Arduino UNO: The Arduino UNO is an open-source microcontroller board based on the microchip ATmega328P microcontroller and developed by Arduino.cc. The board is equipped with sets of digital and analog input/output pins that may be interfaced to various expansion boards and other circuits. Its CPU is microchip AVR (8-bit). It has SRAM memory. Its storage is flash EEPROM.

Micro TF Card Memory Shield Module SPI Micro Storage Card Adapter For Arduino: The module (micro-SD card adapter) is a micro-SD card reader module, and the SPI interface via the file system driver, microcontroller system to complete the micro-SD card read and write files. Arduino users can directly use the Arduino IDE which comes with an SD card to complete the library card initialization and read-write.

Techleads NodeMCU ESP8266 CP2102 NodeMCU Wi-Fi Serial Wireless Module: It is inbuilt with micro-USB, flash, and reset switches and is easy to program. It is Arduino compatible, and it works great with the latest Arduino IDE.

4.5 Experimentation

(a) This system is placed in restricted zone for detecting sound (Fig. 7).

The sound level sensor is used to detect the signals or loud noise and send the signal from transmitter to receiver at the sound system to control the sound level.

(b) This circuit is placed on SOUND SYSTEM for controlling the sound after monitoring the loud noise (Fig. 8).

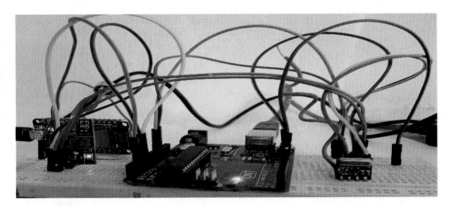

Fig. 7 Model 1 circuit

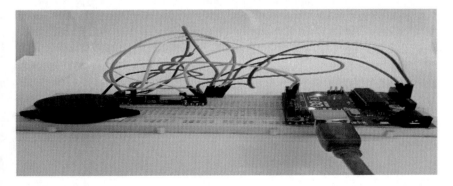

Fig. 8 Model 2 circuit

Table 1 Observed sound signals

Time (in s)	Sound signals (in dB)
1	7
2	11
3	22
4	48
5	69
6	65
7	52
8	70
9	95
10	7
15	45
16	60
17	83
18	7
23	15
24	37
25	48

Receiver at the sound system gets signal from transmitter end and controls the sound if it crosses the limit.

(c) The data is monitored, and data is sent to server.

The sound signals detected by the sound sensor and that particular instance of time are shown in Table 1. The level of the sound coming from the sound system that is detected by the sound sensor in analog format is converted into digital format, i.e., in decibels (dB), and displayed at the wireless module through the NodeMCU.

Graph 9 gives a clear graphical representation of exactly how the system works. If the detected sound range is less than 75 dB, there is no action taken by the system, but once the sound signal is detected above the threshold value, i.e., 75 dB, the system is stopped by sometime. In the above graph, we can see, at the time instance of 9 s, the sound range went to 95 dB. Here, the entire working of the system takes place and the sound system is stopped. As the system is stopped, the sound sensor can sense only the default values until the sound system is in stopped state. When the system starts again, the same procedure is repeated. The next limit is exceeded at time instance of 17 s with 83 dB of sound level. Hence, again the sound system is stopped.

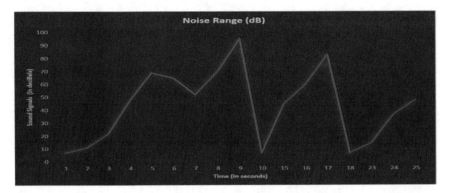

Fig. 9 Graphical representation of Table 1

5 Result

The proposed system is designed to overcome problem of noise in restricted area. Our system helps us to control the noisy environment, unlike monitoring which is in existing system. Instead of notifying authorities to take action, our system does it automatically if it detects the noise.

6 Conclusion

With the help of this smart sound control system, we can automatically control level of sound when a high-noise-making system like loudspeaker enters in restricted zones like hospitals, schools, etc. After installing this system, it can continuously track and control the sound level within given range around restricted zones. Additionally, this system has advantages like easy to use as well as less expensive in comparison with the other systems developed till date for controlling sound pollution in restricted zones. This helps our system to contribute toward digital India, Skill India.

References

1. Chaitanya, Kulkarni, Kulkarni, Shruti, Bhopale, Siddhi, M.M. Raste. 2017 February. Sound and air pollution monitoring system. *International Journal of Scientific & Engineering Research* 8 (Issue 2).
2. Joshi, Lalit Mohan. 2017 November. Research paper on IOT based air and sound pollution monitoring system. *International Journal of Computer Applications* 178 (No. 7), (0975 – 8887).
3. Sai, Palaghat Yaswanth. 2017 March. An IoT based automated noise and air pollution monitoring system. *International Journal of Advanced Research in Computer and Communication Engineering (IJARCCE)* 6 (Issue 3).

4. Sumithra, A., J. Jane Ida, K. Karthika, S. Gavaskar. 2016 March. A smart environmental monitoring system using internet of things. *International Journal of Scientific Engineering and Applied Science (IJSEAS)* 2 (Issue-3).
5. Guthi, Anjaiah. 2016 July. Implementation of an efficient noise and air pollution monitoring system using internet of things. *International Journal of Advanced Research in Computer and Communication Engineering* 5 (Issue 7).
6. Kaur, Navreetinder, Rita, Mahajan, Deepak, Bagai. 2016 June. Air quality monitoring system based on Arduino microcontroller. *International Journal Innovative Research in Science, Engineering and Technology (IJIRSET)* 5 (Issue 6).
7. Sai Chandana, P., K. Sreelekha, A. Muni Likith Reddy, M. Anil Kumar Reddy, R. Senthamilselvan. 2017 March. IOT air and sound pollution monitoring system. *International Journal on Applications in Engineering and Technology* 3 (Issue 1).
8. Al-Ali, A.R., Imran, Zualkernan, Fadi, Aloul. 2010 October. A mobile GPRS-sensors array for air pollution monitoring. *IEEE Sensors Journal* 10 (10).
9. Sharma, Anushka, Vaishnavi, Varshney, Roopank, Maheshwari, Upasana, Pandey. 2018 March. IOT based air and sound pollution monitoring system. *International Research Journal of Engineering and Technology* 5 (Issue 3).
10. Pan, Meng-Shiuan, Yu-Chee, Tseng. 2013. ZigBee Wireless Sensor Networks and Their Applications, vol. 12, issue 3.. Department of Computer Science, National Chiao Tung University Hsin-Chu, 30010, Taiwan
11. Singh, Arushi, Divya, Pathak, Prachi, Pandit, Shruti, Patil, Priti. C. Golar. 2018. IOT based air and sound pollution monitoring system. *International Journal of Advanced Research in Electrical, Electronics and Instrumentation Engineering* 4 (Issue-3).
12. Deshmukh, Sarika, Saurabh, Surendran, M.P. Sardey. 2019 April. Air and sound pollution monitoring system using IoT. *International Journal on Recent and Innovation Trends in Computing and Communication* 7 (Issue IV).
13. Akinkaude, Shadrach Tunde, Kowawole, Peter Fasae. 2018 March. A Survey of Noise Pollution in Ado-Ekiti Metropolis Using Mobile Phone, vol. 4, Issue 11. Science Technology Department, Science Research Publishing.

Predictive Analysis of Co-seismic Rock Fall Hazard in Hualien County Taiwan

Aadityan Sridharan and Sundararaman Gopalan

Abstract Rock fall hazards pose a significant danger to human lives. Being the most abundant among the slope failures in an earthquake event, rock falls are one of the most destructive co-seismic events. A recent earthquake in Taiwan (M_w 6.1) on April 18, 2019, has been analyzed, and artificial accelerograms were generated using SeismoArtif software. In preserving the site properties, the Chi-Chi earthquake accelerogram was used to match the spectral envelope. Data of rock fall during earthquake in the Zhongbu cross island highway in the Hualien County was collected from a dash-cam recording of the event. This rock fall was modeled in 2-D using the 'Rockfall' software by Rocscience, and the number of rocks with respect to the rotational energy of the modeled rock was studied. The artificial accelerogram was used as an input to the predictive model, and the results predicted the Newmark's displacements. It was found that the predicted displacement values were significant enough to trigger the rock fall but the topography as observed by simulation has aided in the propagation of the rock fall.

Keywords Co-seismic rock fall · Newmark's method · SeismoArtif · Rockfall 2019

1 Introduction

Co-seismic rock falls are the most common type of landslides that affect large communities in mountainous regions in the event of an earthquake. Apart from posing a threat to human lives, they are capable of instant property damage. There have been incidents of houses being crushed by dislodged boulders during earthquake events

A. Sridharan (✉)
Department of Physics, Amrita Vishwa Vidyapeetham, Amritapuri, India
e-mail: aadityans@am.amrita.edu

S. Gopalan
Department of Electronics and Communication Engineering, Amrita Vishwa Vidyapeetham, Amritapuri, India
e-mail: sundar@am.amrita.edu

© Springer Nature Singapore Pte Ltd. 2020   343
D. Swain et al. (eds.), *Machine Learning and Information Processing*,
Advances in Intelligent Systems and Computing 1101,
https://doi.org/10.1007/978-981-15-1884-3_32

[1]. Predictive models that can identify regions prone to rock fall could reduce the societal risk for communities in mountainous regions that are tectonically active [2–4]. With the advancement in computational techniques, predictive analyses have become more reliable for hazard prediction and modeling [5].

Many rock fall hazard modeling techniques have been proposed in the literature [6, 7]. Generally, initial field conditions that cause rock fall such as the exact topography are difficult to be assessed. Therefore, reliable predictive models have to employ high resolution digital elevation model (DEM) and probabilistic methods in their analysis [8]. Some of the recent models employ the values of acceleration obtained from accelerogram to determine and predict earthquake-induced hazard [9]. The article by Jibson in 2007 discusses models that can be used to assess seismic landslide hazard [10]. The models were based on 2270 individual acceleration records from 30 worldwide earthquakes. While the proposed models give a global perspective to assess earthquake-induced slope failures, region-specific prediction requires careful selection of accelerogram records. The nature of seismic coefficient required for prediction varies among the models, and some usual terms are peak ground acceleration (PGA), static horizontal coefficient, peak ground velocity (PGV) and others [11, 12]. Introducing terms that are closely correlated with the dependent terms in the model has been carried out very carefully by various models [13].

In this study, we present a two-phase analysis of a very recent rock fall event in Taiwan. The event was reported on April 18, 2019, simultaneously along an earthquake of magnitude (M_W 6.1). The field observations were recorded in a dash-cam video of a car that nearly missed a massive rock fall. Analysis presented here explains the possible reason for such a rock fall event triggered by the earthquake event.

1.1 Study Area and Seismicity

Taiwan is situated in the Pacific Ring of Fire on convergent boundary of the 'Eurasian plate' and the 'Philippines sea plate.' Devastating earthquakes such as the 1999 Chi-Chi earthquake of moment magnitude M_W 7.6 have occurred in this country [14, 15]. Recently, on April 18, 2019, there was a strong tremor due to an earthquake of moment magnitude M_W 6.1 with epicenter at 23° 59′ 20.38″ N and 121° 41′ 34.81″ E off the eastern coast of Taiwan. An intensity map of the earthquake generated by United States Geological Survey (USGS) is shown in Fig. 1. There were many rock falls reported all over the Hualien County on the eastern coast of Taiwan. The event considered for the analysis here was along the Zhongbu cross island highway situated at 24° 12′ 12.77″ N and 121° 26′ 57.19″ E. The rock fall has been simulated, and the results were correlated with Newmark's displacement model.

Topography of the Hualien County of Taiwan comprises steep slopes generally varying from 45° to 85° [17]. The highways in this region are covered by tunnels at various stretches. Due to the steep topography and slope cuts along the highway, this region is among the most vulnerable areas prone to rock fall in the country of Taiwan.

Fig. 1 Intensity map
generated by USGS post the
earthquake (*Photo Courtesy*
USGS [16])

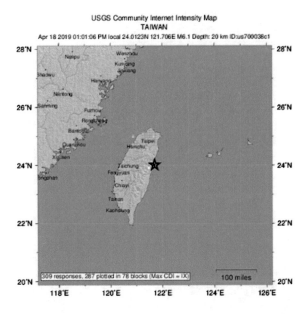

2 Methodology

There are two main stages to the analysis presented here; initially the rock fall location
information was collected from the dash-cam video posted after the event shown in
Fig. 2. As the quality of dash-cam video was not very good, the time frame was not
easy to extract. But the video does show that within 3 s of earthquake the rocks reach
the highway. The first stage of the analysis involves simulation of rock fall based on
slope design and other parameters using the 'Rockfall' software [18]. The second
stage involves the predictive model that is based on strong motion records generated
for this particular earthquake using SeismoArtif software [19]. SeismoArtif generates
artificial accelerograms based on magnitude, location and hypocentral distance of
an earthquake. Artificial accelerograms were processed to calculate the probability
of failure for the rock slope using predictive models. Finally, the simulation results
were validated by the predicted values.

2.1 Rockfall Software-Based Analysis

The topography of the slope was imported to the software from Google Earth as
a polyline that represents the slope. To make the simulation more realistic, a factor
called coefficient of restitution is input into the software, and coefficient of restitution
is the measure that can predict the impact of a rock fall [20]. Definition of coefficient
of restitution (R_c) can be given by

Fig. 2 Pre and event images of the Hualien County rock fall, Taiwan. Figure on the top (pre) was adapted from Google street view, in the bottom (event) the dash-cam video still that was recorded during the event

$$R_c = \text{Rebound velocity/Approach velocity} \qquad (1)$$

From the literature, it was found that a certain range of restitution values have been used in rock fall analysis for various cases in Taiwan [8, 21]. Coefficient of restitution has two components: normal (R_n) and tangential (R_t). Equation (1) will accordingly change for the two components, the R_n will have normal components of the velocities, and R_t will have tangential components, respectively. Typical values of the two components were carefully studied, and values that closely represent the characteristics of study area were selected for analysis ($R_n = 0.4$–0.7, $R_t = 0.5$–0.9) [21].

Around 53 rock fall paths were simulated by the software and mapped on a 2-D slope model. Along with the coefficient of restitution, evolution of kinetic energy of the rock mass is another factor that affects the impact. The variation of the rotational energy of the rocks in the simulation was calculated, and the values of number of rocks that correspond to the values of measured angular velocity were plotted.

2.2 Generating the Artificial Accelerogram Using SeismoArtif Platform

The distance of rock fall event from epicenter was calculated from Google Earth by drawing a polyline, and the depth of the earthquake was reported to be at 20 km by USGS [16]. Hence, the hypocentral distance was estimated to be 37.73 km. To generate the synthetic accelerogram in the SeismoArtif software, the site classification information is important, and the classification is based on shear wave velocity (V_S) at 30 m depth in the site. Based on V_S 30 values reported in Taiwan, the value in the region of rock fall is more than 760 ms^{-1} [22].

Generating the synthetic accelerogram requires preserving site properties, and variations in spectral accelerations are different for different sites. The natural period of the accelerogram changes with different site classes. SeismoArtif considers the site classification to produce the synthetic accelerogram; in our case, the site class was B, and corresponding characteristics of study area were used as inputs. Other inputs include the moment magnitude and the location of site with respect to the plate boundaries.

The tectonic location of the study area is modeled in the software as various options as either 'Inter-plate regimes,' 'Intra-plate regimes' and 'Regimes of active tectonic extension.' As mentioned earlier, Taiwan is an active tectonic region and is near the converging plate boundary, 'Regimes of active tectonic extension' option was used to generate the accelerogram in the software. 'Inter-plate regime' region was another option that represents Taiwan. Chi-Chi earthquake data was used for matching the spectral envelope. Figure 3 shows one of the generated accelerogram plots.

2.3 Predictive Model Based on Newmark's Methodology

The Newmark's model has been an intermediate model that can analyze earthquake-induced slope displacements. There are conventional pseudo static models and the

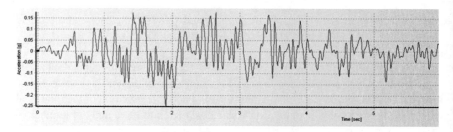

Fig. 3 Simulated synthetic accelerogram generated by SeismoArtif software [19]. Due to space constraints, only part of the accelerogram has been displayed

finite element-based detailed models that are less and more sophisticated, respectively. Also termed as rigid block analysis based on the permanent displacement of landslide block, it has been one of the most successful models to estimate co-seismic landslides [13]. Albeit being simplistic and with many approximations, this model has been used in seismic landslide hazard analysis on a region level [13]. Various statistical models have employed this analysis for predicting seismic slope failures [2, 10, 23]. Hseih and Lee in 2011 modified the earlier models for Taiwan based on Chi-Chi earthquake data. The equations are given by [24]

$$\text{Log } D_n = 1.782 \, \text{Log} I_a - 12.104 a_c + 1.764 + / - 0.671 \tag{2}$$

$$\text{Log} D_n = 1.756 \, \text{Log} I_a - 2.78 \, \text{Log} a_c - 2.728 + / - 0.658 \tag{3}$$

D_n is the Newmark's displacement in cm, I_a is the Arias intensity in m/s, and a_c is the critical acceleration in m/s^2.

The computed artificial accelerograms were used to compute all the input terms except a_c values mentioned above. A range of a_c values from 0.02 to 0.2 has been used for the analysis [24]. I_a values were averaged out of three artificial records (two horizontal and one vertical components of ground motion) generated from SeismoArtif and were found to be 0.218 m/s. From Eqs. (2) and (3), predicted displacement values were obtained based on input parameters as plotted in Fig. 4a, b.

3 Results and Discussion

It was found that average displacement values were 3.96 cm and 7.64 cm in case of model 1 and 2, respectively. Usually in field, predicted Newmark's displacement of few tens of centimeters corresponds to major failure of slope. The average values obtained here are significant enough to trigger the rock fall. The complete failure of the boulder however must have been due to the steep topography. Maximum slope in the areas of rock fall was found to be 60.8° calculated from the elevation profile. Variation of the slope was found to be in the range 41°–60.8°; this shows that the

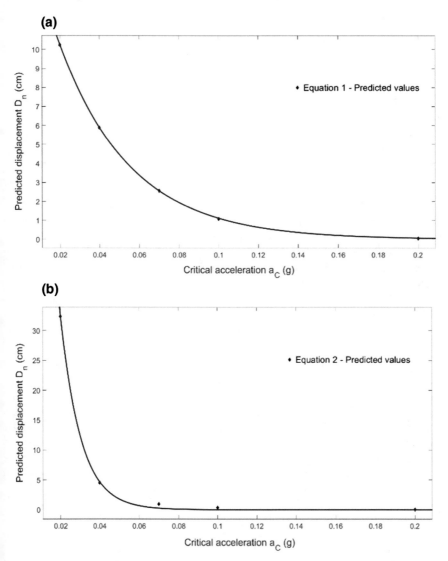

Fig. 4 Predicted displacement values in cm have been plotted against critical acceleration values. **a** The predicted values for Eq. (1) and **b** The predicted values for Eq. (2)

trigger caused by the earthquake and the topography aided the boulder to continue the roll. The rotational velocity distribution is shown in Fig. 5, and increased values are observed upon impact near the highway. From the plot out of the 53 simulated rock paths on an average, every five rock fall paths have an angular velocity of more than 10 rad/s. Given the approximate size of the boulder from the video still, it is safe to assume that this value could cause serious damage on impact.

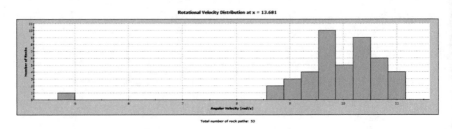

Fig. 5 Rotational/angular velocity along the simulated rock paths for 53 trials. Y-axis shows the number of rock paths that have corresponding angular velocity (rad/s)

Figure 6 shows the bounce height on slope for 95th percentile of rock paths (The simulations that are close to field observations). It can be seen here that bounce height has suddenly peaked on reaching 10 m distance from the fall. From the video still in Fig. 2, the boulder shown is approximately 3–4 m above the ground which is consistent with the bounce height plot which shows values around 8 m from the simulation.

From the simulated accelerogram, the I_a values have been found to be large values which suggest a moderate to high ground shaking scenario [13]. Typical values of I_a are consistent with the literature [4, 25] and suggests that these values can be more reliable in assessing the rock fall than the displacements alone. While the magnitude of the earthquake is more than above 6, there can be attenuation of kinetic energy of the ground motion at a distance of 37.73 km, which is the distance of the rock fall from the source. But higher I_a values on the other hand might signify increased ground shaking. For boulders that are already ready to roll, this increased shaking could have provided the initial trigger. The 95th percentile of simulations show close resemblance to field observations. These trials were used to assess the bounce height as well as angular velocity.

It has to be kept in mind that the simulated rock fall paths need not fully converge to the exact path that might be observed in the field. While most of the rock fall paths are showing similar characteristics as the available field observations, it is

Fig. 6 Simulated bounce height of the rock mass versus location of fall from initiation. The 95th percentile of graphs were used for plot

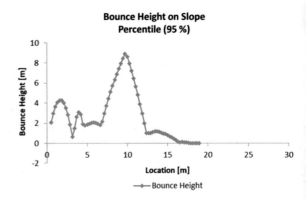

possible that the bounce height as mentioned earlier could be an over estimation. Albeit the simulated values have predicted close resemblance to the field observation, there might always be slight aberrations to the calculated values to the actual field conditions. Synthetic accelerograms generated have also been indicative of higher values of acceleration as expected in a near field condition (<100 km from epicenter). Though Chi-Chi earthquake accelerograms were used to mainly site amplification properties, there might be certain local site effects during the tremor that might not have been accounted for.

Recent techniques involve combining the Newmark's algorithm with rock fall simulations using Geographical Information System (GIS)-based analysis that have been proved reliable for rock fall analysis [26]. Future work will involve using the regional predictive capability of the Newmark's rigid block model to enhance the prediction to warn communities vulnerable to such rock fall hazards. Utilizing newer physical models such as Newmark's multi block analysis might aid in precision of the predicted displacement [27]. Early warning systems based on such new models combined with seismic hazard analysis [28, 29] and other parameters can improve the safety of travelers along these highways [30].

4 Conclusion

A combination of predictive analysis and simulation for rock fall hazard has been presented here. Moderate agreement between the two models shows the capability of extending such work for wider range of analysis. The resultant analysis has successfully correlated with the predictive algorithm to generate significant results. Field observations based on the video stills have provided crucial information that has aided in fruitful analysis of the rock fall. Highways along the Hualien County are prone to such rock fall risks and signboards have been placed in various places along the highway that warn travelers of imminent danger. Models such as the one presented above can be deployed with a combination of earthquake warning systems to early warn travelers in real-time scenarios.

References

1. Jacklitch, C.J. 2016. A geotechnical investigation of the 2013 fatal rockfall in Rockville, Utah. https://etd.ohiolink.edu/!etd.send_file?accession=kent1464978379&disposition=inline.
2. Ma, S., C. Xu. 2018. Assessment of co-seismic landslide hazard using the Newmark model and statistical analyses: a case study of the 2013 Lushan, China, Mw6.6 earthquake. *Natural Hazards*. https://doi.org/10.1007/s11069-018-3548-9.
3. Saygili, G., and E.M. Rathje. 2008. Empirical predictive models for earthquake-induced sliding displacements of slopes. *Journal of Geotechnical and Geoenvironmental Engineering* 134: 790–803. https://doi.org/10.1061/(ASCE)1090-0241(2008)134:6(790).

4. Saygili, G., and E.M. Rathje. 2009. Probabilistically based seismic landslide hazard maps: an application in Southern California. *Engineering Geology* 109: 183–194. https://doi.org/10.1016/j.enggeo.2009.08.004.
5. Fan, X., G. Scaringi, Q. Xu, W. Zhan, L. Dai, Y. Li, X. Pei, Q. Yang, R. Huang. 2018. Coseismic landslides triggered by the 8th August 2017 Ms 7.0 Jiuzhaigou earthquake (Sichuan, China): factors controlling their spatial distribution and implications for the seismogenic blind fault identification. *Landslides* 15, 967–983. https://doi.org/10.1007/s10346-018-0960-x.
6. Ii, F., P.D. Calcaterra, G. Pappalardo, P. Sebastiano, P. Zampelli, P.M. Fedi, S. Mineo. 2017. Analysis of rock masses belonging to the Apennine-Maghrebide Orogen by means of in situ and remote methodologies applied to rockfall risk assessment. www.fedoa.unina.it/11456/1/Mineo_Simone_XXIX.pdf.
7. Harp, E.L. 2002. Anomalous concentrations of seismically triggered rock falls in Pacoima Canyon: are they caused by highly susceptible slopes or local amplification of seismic shaking? *Bulletin of the Seismological Society of America* 92: 3180–3189. https://doi.org/10.1785/0120010171.
8. Ku, C.Y. 2014. A 3-D numerical model for assessing rockfall hazard. *Disaster Advances* 7: 73–77.
9. Du, W., and G. Wang. 2016. A one-step Newmark displacement model for probabilistic seismic slope displacement hazard analysis. *Engineering Geology* 205: 12–23. https://doi.org/10.1016/j.enggeo.2016.02.011.
10. Jibson, R.W. 2007. Regression models for estimating coseismic landslide displacement. *Engineering Geology* 91: 209–218. https://doi.org/10.1016/j.enggeo.2007.01.013.
11. Sepúlveda, S.A., W. Murphy, R.W. Jibson, and D.N. Petley. 2005. Seismically induced rock slope failures resulting from topographic amplification of strong ground motions: the case of Pacoima Canyon, California. *Engineering Geology* 80: 336–348. https://doi.org/10.1016/j.enggeo.2005.07.004.
12. Jibson, R.W. 2011. Methods for assessing the stability of slopes during earthquakes—a retrospective. *Engineering Geology* 122, 43–50.
13. Jibson, R.W., E.L. Harp, and J.A. Michael. 2000. A method for producing digital probabilistic seismic landslide hazard maps: an example from the Los Angeles, California, USA. *Engineering Geology* 58: 271–289.
14. Wang, G., A. Suemine, F. Zhang, Y. Hata, H. Fukuoka, T. Kamai. 2014. Some fluidized landslides triggered by the 2011 Tohoku Earthquake (Mw 9.0), Japan. *Geomorphology* 208, 11–21.
15. Wang, K.-L., and M.-L. Lin. 2010. Development of shallow seismic landslide potential map based on Newmark's displacement: the case study of Chi-Chi earthquake, Taiwan. *Environmental Earth Sciences* 60: 775–785. https://doi.org/10.1007/s12665-009-0215-.
16. USGS. 2019. Earthquake overview—United States Geological Survey. https://earthquake.usgs.gov/earthquakes/eventpage/us700038c1/executive.
17. Khazai, B., and N. Sitar. 2004. Evaluation of factors controlling earthquake-induced landslides caused by Chi-Chi earthquake and comparison with the Northridge and Loma Prieta events. *Engineering Geology* 71: 79–95.
18. https://www.rocscience.com/: Rockfall 2019—Rocscience. https://www.rocscience.com/help/rocfall/#t=rocfall%2FGetting_Started.htm.
19. Seismosoft. 2018. SeismoArtif, https://www.seismosoft.com/seismoartif.
20. Li, L.Ping., S. qu. Sun, S. cai. Li, Q. qing. Zhang, C. Hu, S. shuai. Shi. 2016. Coefficient of restitution and kinetic energy loss of rockfall impacts. *KSCE Journal of Civil Engineering* 20, 2297–2307. https://doi.org/10.1007/s12205-015-0221-7.
21. Wei, L.W., H. Chen, C.F. Lee, W.K. Huang, M.L. Lin, C.C. Chi, and H.H. Lin. 2014. The mechanism of rockfall disaster: a case study from Badouzih, Keelung, in northern Taiwan. *Engineering Geology* 183: 116–126. https://doi.org/10.1016/j.enggeo.2014.10.008.
22. Lee, C.T., and B.R. Tsai. 2008. Mapping Vs30 in Taiwan. *Terrestrial, Atmospheric and Oceanic Sciences* 19: 671–682. https://doi.org/10.3319/TAO.2008.19.6.671(PT).

23. Saade, A., G. Abou-Jaoude, and J. Wartman. 2016. Regional-scale co-seismic landslide assessment using limit equilibrium analysis. *Engineering Geology* 204: 53–64. https://doi.org/10.1016/j.enggeo.2016.02.004.
24. Hsieh, S.-Y., and C.-T. Lee. 2011. Empirical estimation of the Newmark displacement from the Arias intensity and critical acceleration. *Engineering Geology* 122: 34–42. https://doi.org/10.1016/j.enggeo.2010.12.006.
25. Chousianitis, K., V. Del Gaudio, N. Sabatakakis, K. Kavoura, G. Drakatos, G.D. Bathrellos, and H.D. Skilodimou. 2016. Assessment of earthquake-induced landslide hazard in Greece: from arias intensity to spatial distribution of slope resistance demand. *Bulletin of the Seismological Society of America* 106: 174–188. https://doi.org/10.1785/0120150172.
26. Yue, X., S. Wu, Y. Yin, J. Gao, and J. Zheng. 2018. Risk identification of seismic landslides by joint Newmark and RockFall analyst models: a case study of roads affected by the Jiuzhaigou earthquake. *International Journal of Disaster Risk Science* 9: 392–406. https://doi.org/10.1007/s13753-018-0182-9.
27. Song, J., Q. Fan, T. Feng, Z. Chen, J. Chen, and Y. Gao. 2019. A multi-block sliding approach to calculate the permanent seismic displacement of slopes. *Engineering Geology* 255: 48–58. https://doi.org/10.1016/j.enggeo.2019.04.012.
28. Sitaram, T.G., N. James, and S. Kolathayar. 2018. Comprehensive seismic zonation schemes for regions at different scales. *Springer*. https://doi.org/10.1007/978-3-319-89659-5.
29. Nirmala, V., K. Sreevalsa, S. Aadityan, R. Kaushik. 2016. An investigative study of seismic landslide hazards. https://doi.org/10.2991/rare-16.2016.60.
30. Ramesh, M.V., Vasudevan, N. 2012. The deployment of deep-earth sensor probes for landslide detection. *Landslides* 9, 457–474.

Abstractive Text Summarization and Unsupervised Text Classifier

Aditya, Akanksha Shrivastava and Saurabh Bilgaiyan

Abstract In this day and age, as the Internet gets increasingly cluttered for content, the comprehension of generated huge texts is growingly becoming a source of major inconvenience for a normal viewer, and classifying it is another gruesome task. Text summarization and classification prove to be a boon as it abridge and categorize the massive text into sizeable length without removing the key information within it and assign it to its genre. In this paper, the authors have tried to implement the abstractive type of summarization using sequence-to-sequence RNN. Using this model, the output perceived is a summary which is short, lossy, and whose length does not necessarily depend on source text length, and then the authors have used bidirectional LSTM to find the best one to assign a particular genre.

Keywords Text summarization · RNN · Sequence-to-sequence models · Lossy text · Text classification · Bidirectional LSTM · Machine learning

1 Introduction

Text summarization is the process of abridging very large texts into a concise and fluent summary while keeping the key information and overall meaning intact [1]. The area of implementation of text summarization is becoming increasingly popular because be it anything, a small institution, a hefty Web site, a business firm, all require shortening of the vast texts into readable summaries [2, 3]. It could be anywhere from machine translation to speech recognition, from image captioning to language identification, etc. [4].

Aditya · A. Shrivastava · S. Bilgaiyan (✉)
School of Computer Engineering, KIIT Deemed to be University, Bhubaneswar, Odisha, India
e-mail: saurabhbilgaiyan01@gmail.com

Aditya
e-mail: adikid1996@gmail.com

A. Shrivastava
e-mail: shivi.s98@gmail.com

© Springer Nature Singapore Pte Ltd. 2020
D. Swain et al. (eds.), *Machine Learning and Information Processing*,
Advances in Intelligent Systems and Computing 1101,
https://doi.org/10.1007/978-981-15-1884-3_33

Text summarization is of two types: extractive summarization and abstractive summarization [3]. In extractive summarization, important sections of the text are identified, and then a summary is generated which is a subset of the sentences from the original text [3]. It has been a widely researched subject, and people have extensive work in this area and have almost reached its maturity stage [1]. The world is now turning toward abstractive text summarization [1]. In contrast to extractive summarization, the abstractive approach involves understanding the intent and writes the summary in one's words, meaning rephrasing of words is used.

Abstractive summarization is further classified into structure-based, semantic-based, deep-learning with neural network-based and discourse and rhetoric-based structure-based summarization [1, 3].

In this paper, we have attempted Abstractive text summarization with attention mechanism on sequence-to-sequence model with LSTM. A question could be why not just use machine translation sequence-to-sequence model [5]. The problem with machine translation sequence-to-sequence model is that the target summary does not necessarily depend on source length [1, 5]. It varies according to the length of the source text. Also, the summary the authors got in sequence-to-sequence RNN model is lossy or so as to say the source text content is partially discarded and inexactly approximated to encode data [5].

Next, text classification is the process of assigning the text into one or better-defined categories. It is one of the most widely and extensively used natural language processing applications [6]. Some of the examples of text classification are understanding audience sentiment from social media, detection of spam and non-spam emails, auto tagging of customer queries, categorization of news articles into defined topics, health-related classification using social media in public health surveillance, language identification, etc. [6, 7]. There are several types of text classification methods which include (1) support vector machines, (2) Naive Bayes, (3) rough set-based classifier, (4) instantaneously trained neural networks, etc. [6, 7].

In this paper, the authors have implemented bidirectional LSTM to classify processed summary and assign its genre or category [6]. The authors have divided the content into the following parts: Sect. 2 contains the explanation modus operandi by the authors and the abstract mentioning of all previous work done in this particular topic. Section 3 contains an explanation of the working of attention mechanism in sequence-to-sequence RNN and the algorithm used to implement the model. Section 4 contains a brief description of the bidirectional RNN model implemented. Section 5 is used in this section to describe the code for Rogue's summary and showcase their model accuracy. Section 6 is describing how the program functions on the whole. Section 7 describes the conclusion and future improvements that could be implemented further to improve the current model used by the authors. References contains the mentioning of all the papers which were referred by the authors.

2 Related Works

The authors are proposing a model for text classification and summarization of huge text using sequence-to-sequence RNN and bidirectional LSTM. There are many researches already done in this field but the task was to merge both the broad scenarios and provide results at one singular place.

The paper also elucidates several explanations about text summarization. The work by Gupta, et al. is an overview in which they have clarified the facts about abstractive text summarization [1]. They have thoroughly demonstrated the broad categories of techniques of text summarization and have explained them in depth [2, 3]. The paper also infers similar work done by Islam et al. in which they used sequence-to-sequence RNN to translate Bangla text [4]. They have furthered their study LSTM and have worked to implement this technique in the Python framework [4]. The work in this paper is a largely inspired work from the paper of Nallapati et al. where they proposed the system to attention encoder–decoder RNN and achieved state-of-the-art performance and were able to address many critical problems in summarization [5]. Azmi et al. have implemented this to convert Arabic text using user granularity [8]. The text is segmented text topic wise and applied to an enhanced extractive summarizer inferred to as rules-based sentence reduction technique or RST-based extractive summarizer, and the length of the extracted summary was modified to control the size of predicted summary. The results were better in comparison with an abstractive model. In the work of Bhargava et al., the authors have implemented the sentiment infusion to implement abstractive text summarization [9]. The work done by Sahoo et al. implements sentence clustering using Markov clustering principle followed by sentence ranking [10]. The then top-ranked sentences from the clusters are worked through linguistic rules to form the summary. Text summarization was also implemented in the work of Negi et al. in which the authors have used NLP and ML to implement a mix of syntactic featuring to comprehend the writing style of incoming reports for supporting timely report monitoring and perceive authoritative statements [11].

On the other hand, work by Mirończuk et al. demonstrates the type of text classification techniques available and how can they be used [6]. Liu et al. have worked to implement bidirectional LSTM with attention mechanism and convolutional layer to classify text [7]. This is another approach for classifying text which provided better results than normal supervised techniques. The semantic text classification done by Altinel et al. demonstrates the advantages of semantic text classification over text classification which included extraction and using in-words latent relationship and textual semantic understanding, etc. [12]. The work also compares the performance comparison of knowledge-based approaches [13].

3   Summarization Algorithm

A.   Model for the Attention Mechanism in Sequence-to-sequence RNN

The sequence-to-sequence model consists of an encoder, a decoder, and an attention mechanism.

Encoder: It is a bidirectional long short-term memory (LSTM) layer that extracts information from the original text. It reads one word at a time and as it is an LSTM, it updates hidden state based on the current word and the words it has read before [1]. In short, all encoder does is it takes the input data and passes the last state of its recurrent layer as an initial state to the decoder's first layer (recurrent).

Decoder: It is unidirectional long short-term memory (LSTM) that summaries one word at a time. The decoder LSTM starts working once it gets signaled that the entire source text has been read. It creates a probability distribution over the next word using information from the encoder and previously written text [1]. In short, the decoder takes the last state of the encoder's last recurrent layer and uses it as its first recurrent layer's initial state.

Attention Mechanism: Through this mechanism, the decoder can access the intermediate hidden states in the encoder and use all that information to decide which word is next [3] (Fig. 1).

The algorithm is implemented in the following steps:

1. The input text is encoded, and the decoder is initialized with internal states of the encoder. The <start> token is passed as an input to the decoder.
2. The decoder is executed for one-time step with the internal states.
3. The output will be the probability for the next word. The word with the maximum probability will be selected.
4. The sampled word is passed as an input to the decoder in the next time step, and the internal states are updated with the current time step.

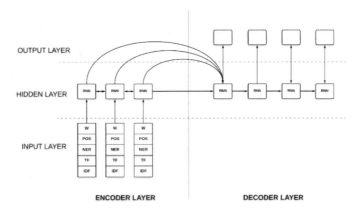

Fig. 1   Attention mechanism for sequence-to-sequence RNN

5. The steps 3–5 are iterated until <end> token is generated to the maximum length of the target sequence is hit.

4   Classifier Algorithm

The algorithm is implemented in the following steps:

1. First, the input layer is provided.
2. After adding the first embedding layer, a bidirectional LSTM layer is added.
3. The output layers are then added, i.e., Dense (activation = "ReLU"), Dropout, Dense (activation = "sigmoid") in consecutive order.
4. The optimizer used to train the model is Adam's optimizer and loss parameter is 'binary_crossentropy' (Fig. 2).

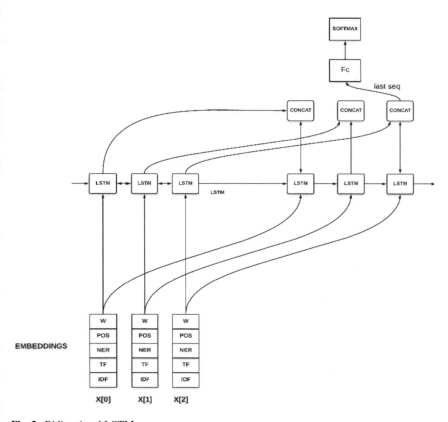

Fig. 2 Bidirectional LSTM

5   Experimental Studies

Performance Benchmarks The authors are using ROUGE summary to compare the efficiency between their model and presently extractive summarization model.

ROUGE is an acronym that stands for Recall-Oriented Understudy for Gisting Evaluation. It is actually a system of measurement for appraising the automatic summarization of texts. There are two score sets in ROUGE.

$$\text{Precision} = \frac{\text{number\_of\_overlapping\_words}}{\text{total\_words\_in\_reference\_summary}} \qquad (1)$$

$$\text{Recall} = \frac{\text{number\_of\_overlapping\_words}}{\text{total\_words\_in\_system\_summary}} \qquad (2)$$

```
Precision is: 0.38095238095238093
Recall is: 0.25
F Score is: 0.30188774460662915
Sum of ROUGE Score: 14.068020212266791
Average ROUGE Score = 0.3271632607503905
Count: 43
```

6   Results and Discussion

The authors in this manuscript have used amazon-fine-food-reviews dataset to train the sequence-to-sequence model of summarizer; then the outcome of the results of the model is used on classifier to give the genre as well.

The program works in the following manner:

For text summarizer:

```
Using TensorFlow backend.
```

The length of the reviews and the summary for an idea about the distribution of length of the text (Fig. 3).

```
Generating the proportion of the length of the
summaries:
0.9424907471335922
```

This means 94% of the summaries have length below 8.

So, we can fix the maximum length of the summary.

Preparing the text tokenizer:

```
% of rare words in vocabulary: 66.12339930151339
Total Coverage of rare words: 2.953684513790566
```

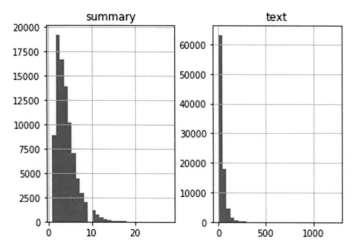

Fig. 3 Diagnostic plot to observe the distribution of length over the text

Preparing the summary tokenizer:

```
% of rare words in vocabulary: 78.12740675541863
Total Coverage of rare words: 5.3921899389571895
```

Model building:

```
Layer (type)                   Output Shape            Param #
Connected to
================================================================
=================================
input_1 (InputLayer)           (None, 30)              0


embedding (Embedding)          (None, 30, 100)         844000
input_1[0][0]


lstm (LSTM)                    [(None, 30, 300), (N    481200
embedding[0][0]
```

```
input_2 (InputLayer)              (None, None)            0


lstm_1 (LSTM)                     [(None, 30, 300), (N    721200
lstm[0][0]


embedding_1 (Embedding)           (None, None, 100)       198900
input_2[0][0]


lstm_2 (LSTM)                     [(None, 30, 300), (N    721200
lstm_1[0][0]


lstm_3 (LSTM)                     [(None, None, 300),     481200
embedding_1[0][0]
lstm_2[0][1]
lstm_2[0][2]


attention_layer (AttentionLayer   [(None, None, 300),     180300
lstm_2[0][0]
lstm_3[0][0]


concat_layer (Concatenate)        (None, None, 600)       0
lstm_3[0][0]
attention_layer[0][0]


time_distributed (TimeDistribut   (None, None, 1989)      1195389
concat_layer[0][0]
================================================================
=================================
Total params: 4,823,389
Trainable params: 4,823,389
Non-trainable params: 0


Train on 41346 samples, validate on 4588 samples
Epoch 1/50
41346/41346 [==============================] - 85s 2ms/sample -
loss: 2.8152 - val_loss: 2.5780
Epoch 2/50
41346/41346 [==============================] - 79s 2ms/sample -
loss: 2.4859 - val_loss: 2.4072
================================================================
…Epoch 19/50
41346/41346 [==============================] - 80s 2ms/sample -
loss: 1.7070 - val_loss: 2.0398
Epoch 00019: early stopping
```

In the following plotted, we can see the validation loss has incremented after 17th epoch and kept increasing for two epochs. So, we used early stopping at 19th epoch (Fig. 4).

Here, a few summaries generated by the model:

```
Review: gave caffeine shakes heart anxiety attack plus
tastes unbelievably bad stick coffee tea soda thanks
Original summary: hour
Predicted summary: not worth the money

Review: got great course good belgian chocolates better
Original summary: would like to give it stars but
Predicted summary: good

Review: one best flavored coffees tried usually like
flavored coffees one great serve company love
Original summary: delicious
Predicted summary: great coffee

Review: salt separate area pain makes hard regulate
salt putting like salt go ahead get product
Original summary: tastes ok packaging
Predicted summary: saltReview: really like product
super easy order online delivered much cheaper buying
gas station stocking good long drives
Original summary: turkey jerky is great
Predicted summary: great product

Review: best salad dressing delivered promptly
quantities last vidalia onion dressing compares made
oak hill farms sometimes find costco order front door
want even orders cut shipping costs
Original summary: my favorite salad dressing
```

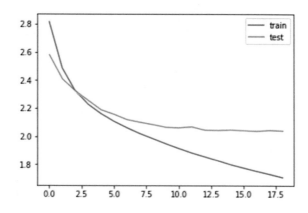

Fig. 4 Diagnostic plot to observe the behavior of the model

```
Predicted summary: great product
```

For Text Classifier:

```
Epoch 1/1
10500/10500 [==============================] - 48    s 4
ms/step - loss: 0.6389
Bidirectional-RNN, Word Embeddings 0.4318
```

7 Conclusion and Future Work

In this work, the authors applied the attention mechanism on RNN to find the best possible summary. The model's output was then transferred to the bidirectional LSTM classifier model to assign a category. The former model could be further improved by using beam strategy for decoding the test sequence. The accuracy could also be measured using BLEU score and pointer-generator networks and coverage mechanisms could be used to handle unique words. Then, the classifier model can also be improved by using hierarchical attention networks or using CNNs and RNNs with a greater number of layers.

References

1. Gupta, S., and S.K. Gupta. 2019. Abstractive summarization—an overview of the state of the art. *Expert Systems with Applications, Elsevier* 121 (1): 49–65.
2. Mahajani, A., V. Pandya, I. Maria, D. Sharma. 2019. A comprehensive survey on extractive and abstractive techniques for text summarization. In *Ambient Communications and Computer Systems (Part of the Advances in Intelligent Systems and Computing)*, vol. 904(1), 339–351. Berlin: Springer.
3. Gambhir, M., and V. Gupta. 2017. Recent automatic text summarization techniques: a survey. *Artificial Intelligence Review* 47 (1): 1–66.
4. Islam, S., S.S.S. Mousumi, S. Abujar, S.A. Hossain. 2019. Sequence-to-sequence Bangla sentence generation with LSTM recurrent neural networks. In: *International Conference on Pervasive Computing Advances and Applications—PerCAA 2019, Procedia Computer Science*, 152(1), 51–58. Amsterdam: Elsevier.
5. Nallapati, R., B. Zhou, C. Santos, Ç. Gulçehre, B. Xiang. 2011. Abstractive text summarization using sequence-to-sequence RNNs and beyond. In *The SIGNLL Conference on Computational Natural Language Learning (CoNLL)*, 1–12.
6. Mirończuk, M.M., and J. Protasiewicz. 2018. A recent overview of the state-of-the-art elements of text classification. *Expert Systems with Applications* 106 (1): 36–54.
7. Liu, G., and J. Guo. 2019. Bidirectional LSTM with attention mechanism and convolutional layer for text classification. *Neurocomputing* 337 (1): 325–338.
8. Azmi, A.M., and N.L. Altmani. 2018. An abstractive Arabic text summarizer with user controlled granularity. *Information Processing & Management* 54 (6): 903–921.
9. Bhargava, R., Y. Sharma, G. Sharma. 2016. ATSSI: abstractive text summarization using sentiment infusion. In *Twelfth International Multi-Conference on Information Processing*, 1–8.

10. Sahoo, D., A. Bhoi, R.C. Balabantaray. 2018. Hybrid approach to abstractive summarization. In *International Conference on Computational Intelligence and Data Science (ICCIDS 2018)*, 1–10.
11. Negi, K., A. Pavuri, L. Patel, C. Jain. 2019. A novel method for drug-adverse event extraction using machine learning. In *Informatics in Medicine Unlocked*, 1–6. Elsevier, In Press, Corrected Proof.
12. Altinel, B., and M.C. Ganiz. 2018. Semantic text classification—a survey of past and recent advances. *Information Processing and Management* 54 (6): 1129–1153.
13. Harish, B.S., B. Udayasri. 2014. Document classification: an approach using feature clustering. In *Recent Advances in Intelligent Informatics (Part of the Advances in Intelligent Systems and Computing book series (AISC))*, 235 (1), 163–173.

Image Classification Using Convolutional Neural Network (CNN) and Recurrent Neural Network (RNN): A Review

Patel Dhruv and Subham Naskar

Abstract With the advent of technologies, real-time data is essentially required for future development. Everyday, a huge amount of visual data is being collected, but to use it efficiently, we need to recognize, understand and arrange the visual data for a perfect approach. So, the neural network was introduced to find out patterns from images, a form of visual data as the neuron functionality in a human brain. It is biologically inspired programming approach to allow the machine to learn from observational data. Neural networks have provided solutions to several problems of image recognition, and it is actively utilized in the medical field due to its efficiency. This paper concentrates upon the use of RNN and CNN in the feature extraction of images and the challenges. The paper also presents a brief literature review of the neural networks like CNN and RNN.

Keywords Image classification · CNN · RNN · Feature extraction

1 Introduction

With more and more visual data accumulation, every day, the use and application of automated image classification are required. Automated image classification can make the image search procedures more efficient, and the whole process will be beneficiary for the visual understanding of medical security and military applications [1]. Here is an example concluding the above context. The image can be deferred into different tags like 'temple,' 'river,' 'boat' and 'person.' Similarly, these logical reasoning results may vary to different people since we all have different opinions. As shown in Fig. 1, there are several different aspects using which the images are classified as the concept of the image.

P. Dhruv · S. Naskar (✉)
School of Computer Engineering, KIIT, Bhubaneswar, India
e-mail: subhamnaskar671@gmail.com

P. Dhruv
e-mail: dhruvpatidar2912@gmail.com; 1605377@kiit.ac.in

© Springer Nature Singapore Pte Ltd. 2020
D. Swain et al. (eds.), *Machine Learning and Information Processing*,
Advances in Intelligent Systems and Computing 1101,
https://doi.org/10.1007/978-981-15-1884-3_34

Fig. 1 Image shows the sideview of Varanasi [1]. *Source* cleartrip

Supervised convolutional models have completely changed the way computer vision and machine learning was before. Nowadays, the available large supervised datasets [2] and the training models use hardware acceleration and graphics processing units (GPUs) [3]. CNN succeeded in overcoming the traditional low-level vision features with complement classifiers [4]. CNN has shown capabilities which are equal to the performance of the neurons present in the cortex of the human brain [5].

2 Image Analysis and Classification

The fundamental errands of visual data analysis, image classification came to existence only to cease the gap between the computer and human vision. It differentiates the image into a category based on the content and type of perception. Image classification uses pixel-wise segmentation which identifies parts of the image and tries to find out where they belong, and it relies on object detection and classification. Image classification plays an important role in medical operations, therapy estimate, surgical emulation, anatomical structure review and the treatment strategy. Image classification could be done manually by manually pointing out the boundaries with the related regions of interest (ROI) which requires the intervention of experts. Classification can be done in two ways.

1. Semantic classification—It studies the image and classifies the pixels of the image to meaningful objects which correspond to real-world categories. The pixels belonging to different objects need to be treated and grouped separately, and hence, it performs pixel-wise classification.
2. Instance classification—It redetects the same instance of the object. In this process, the system focuses specifically on the similar look-alike of the object, and if the objects belonging to the same class looks different, then those objects will be ignored since the process is based on instances.

Effects of clusters on image segmentation: Expert intervention is required to determine the optimal image cluster number in each image since image segmentation utilizing clustering requires prior knowledge to determine the location centroid. Utilization of metaheuristic optimization search algorithm can, however, be a probable solution for 'fuzzy clustering.' Fuzzy clustering is a type of clustering in which the datapoints can be a part of more than one cluster, and only for its robust ambiguity features, it can retain more information than other segmentation methods. Fuzzy c-means clustering (FCM) is one of the most famous fuzzy clusterings. FCM-based artificial vee colony [6], FCM-based differential equation [7], FCM-based ant colony optimization [8], FCM-based genetic algorithm, FCM-based harmony search algorithm, FCM-based firefly algorithm and FCM-based on image segmentation suffer from noise. The challenges related to the FCM are as follows: FCM is sensitive to the process of initiating center cluster and outlier noise. It helps in predicting the number of clusters. Hence, noise-sensitive images, the sensitivity of initializing cluster centers and the unpredictable number of clusters present in the dataset have been observed as the main challenges of fuzzy clustering.

3   Overview of CNN

In deep learning, convolution neural network (CNN) comes out as one of the significant topics which help us in the recognition of faces, detection of an object and classification of images [9]. CNN consists of many layers. These layers are fully connected layers like neurons in a biological brain. Neurons help to carry the 'message' from one cell to another cell. One neuron is surrounded by many other neurons.

CNN also works on this approach. One layer is connected with many other layers, so the output of one layer becomes an input for many other layers those are directly connected to that particular layer [10]. CNN takes input in the form of audio signals, images or any structured samples [11].

Working of CNN: As in Fig. 2, CNNs consist of three layers, the input layer, the hidden layer (fully connected layers) and the output layer. Each neuron of a given layer is connected to every single neuron of the subsequent layer [12]. Input layer takes 2D images in the form of array of pixels, and to do this, the input images are represented in the form of pixel values. Main goal is to extract the features from

Fig. 2 Image showing different layers of CNN

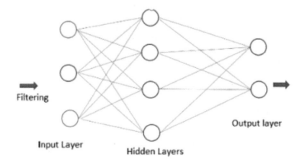

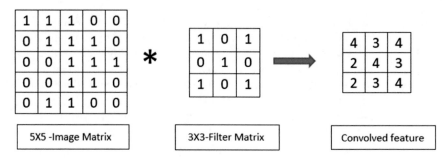

Fig. 3 Calculate 'feature matrix' using image matrix and filter matrix

image, and it can be done by analyzing the weights of filters and helps to find out the connection between the path and neurons [12].

As in Fig. 3, CNNs extract features in three processes:

(1) For getting local features, apply set of weights.
(2) Apply multiple features to get some more characteristics.
(3) Spatially share parameters to each and every filter.

These processes help us to extract any features from the image, e.g., the image below shows that by applying, 3×3 filter matrix on 5×5 image matrix with stride value 1 gives us convolved feature. It is known as 'convolution.' Convolution layer uses different features and performs many operations on image such operations used for detecting edges, sharpening images and blurring it [9].

Anyone can control the behavior of convolution by changing the number of filters and the size of filters. But there are many other parameters as in Fig. 4:

(1) *Stride*: Number of filter matrix shifts over input matrix is called 'stride.' If stride is two, then filter matrix shifts [9] two pixels over the input matrix, and here, we get smaller convolution matrix than we got before. Whenever applying stride, its value should be low because low stride value gives us better resolution per each filtered image due to high overlapping of images happens in receptive field.

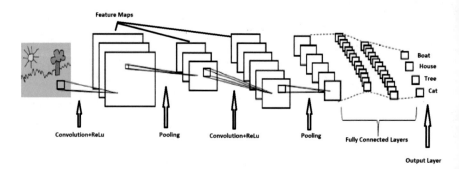

Fig. 4 Functionality of CNN [13]. *Source* https://medium.com

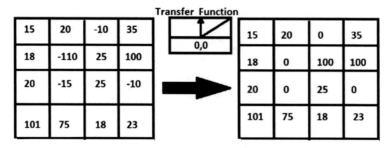

Fig. 5 Working of ReLU function [13] *Source* https://medium.com

On contrary, higher stride value gives us low resolution per each filtered image and also low overlapping in receptive field (Fig. 5).

(2) *Padding*: Padding is useful when filter does not perfectly fit image. There are two options:

Zero padding: It means to pad picture with zeros and Valid padding: It means drop that part of the image where filter does not fit.

(3) Receptive field: A neuron in a downstream layer that is connected to its respective input layers is termed as receptive field [12].

(4) NonLinearity (ReLU): ReLU stands for rectified linear unit, and it is used for nonlinear function which produces the output $f(y) = \max(0, y)$. Nonlinearity is introduced in our convolution network by ReLU.

(5) Pooling: Pooling is used to reduce the dimensionality and preserve the spatial invariants when image is too big. It reduces the computation complexity. There are three different types of pooling:

1. Max Pooling: It takes maximum value from patch or feature map.
2. Average Pooling: It takes an average of all values from feature map.
3. Sum Pooling: It takes sum of all values from feature map.

4 Overview of RNN

We, human beings have the ability to think and deduce conclusion on any topic. However, for inferring anything, we should know the context beforehand. If the context is known, we do not need to start everything from scratch every time we try to make a conclusion. This cannot be done by traditional neural networks, and it is the main problem of the traditional neural network. Traditional neural networks cannot remember past things, so they cannot be able to do things like text prediction, language translation which are heavily dependent on the previous context. Few major problems often occur while implementing traditional neural network:

It does not work for long-term dependencies. Count does not preserve order, and no parameter sharing occurs.

372
P. Dhruv and S. Naskar

RNN persists information being a network consisting of loops. The chain-like structure of the RNN can state that it is related to lists and sequences [14].

Working of RNN: RNN takes vector x_t as input and produces vector h_t as output as in Fig. 6. However, this vector produced as output is not only influenced by the input we give in real time but also on the input history that is fed before. Every time, when the step is called the RNN, class updates internal state. The single hidden vector h being present can be concluded as the simplest case [15].

This RNN's parameters are the three matrices:

W_{hh} Matrix based on the 'previous hidden state'
W_{xh} Matrix based on the 'current input'
W_{hy} Matrix based between 'hidden state and output' (Fig. 7).

The 'hidden state update' notation is

$$h_t = \tanh(w_{hh}h_{t-1} + W_{xh}x_t) \qquad (1)$$

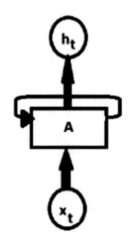

Fig. 6 RNN LOOP *Source* https://medium.com

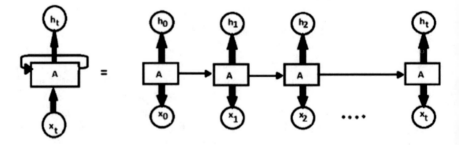

Fig. 7 An unrolled RNN *Source* https://medium.com

where tan*h* is applied with the elements consecutively.
Calculation for the 'present state' is as follows:

$$h_t = f(h_{t-1}, x_t) \tag{2}$$

h_t 'the new state'
h_{t-1} 'the previous state'
x_t 'the current input.'

Now, we have the previous input state apart from current input state because the transformations are applied on the previous input by the input neuron. Each and every one of these inputs is termed as time stamp. Taking tanh as the activation function, W_{hh} as the weight of recurrent neuron and W_{xh} as the weight of input neuron, the equation for state time can be written as

$$h_t = \tanh(w_{hh}h_{t-1} + W_{xh}x_t) \tag{3}$$

After getting current state, output state is calculated as

$$y_t = w_{hy}h_t \tag{4}$$

Back-propagation in a Recurrent Neural Network: Recurrent neural network uses back-propagation to update the weights of the neuron to learn better. During forward propagation, the inputs move forward and pass through each layer to calculate output state. In case of back-propagation, we go back and change the weights of the neuron to make it learn in a better way.

Problems faced by RNN algorithm: *Vanishing gradient problem*: As we back propagate to lower layers, gradient often gets smaller. Eventually, it causes weights to never change at lower layers. Pictorially, vanishing gradient problem can be shown as in Fig. 8:

Consider the output $y^{<3>}$, it is mainly affected by the input sequences near $y^{<3>}$ like $x^{<1>}, x^{<2>}$, etc., but, the output $y^{<T_y>}$ is not affected by input sequences like $x^{<1>}$,

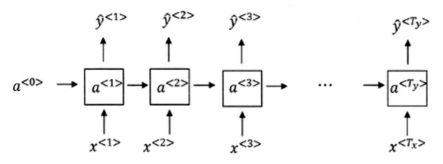

Fig. 8 Vanishing gradient

$x^{<2>}$, $x^{<3>}$, etc., which is not desired and gives wrong output [16]. LSTM is used to solve this issue of long-term dependency.

Working principle and basic calculations of a neural network: The output of the hidden neurons is a part of the next neural network. The next neural network receives the output of the previous hidden neuron as an input. The neural network passes the hidden state info back to itself, and then, a new time-step is received. 'Activation function' is added to introduce nonlinearity. Sigmoid function transforms linear output to a value between 0 and 1. With four layers of 0.2 value, the output will be 0.2^4, and then, most of the layers will perish with values nearing to 0. This would result in 'gradient vanishing.' The network with the value 0 will not be able to learn and be functional anymore. As a result of this problem, the low-level layers will vanish. But the low-level features are fundamentals of high-level features, so the vanished gradients will not only affect the output, but it is also a serious issue. The lower layers are important because they learn the simple patterns used for building up the network. A ReLU activation function has been introduced to avoid the vanishing gradient problem. In ReLU, only negative inputs end up being 0, so the backpropagated error can be ruled out whenever there is a negative input. In leaky ReLU, if the input is smaller than 0, the output will be between the input and 0.01. Here, even if the activation function is negative, the inputs will not be 0. Hence, the neural network can be reactivated and reused, thereby improving the overall performance (Fig. 9).

x_t = input, h_{t-1} = previous hidden state, c_{t-1} = cell state, c_t = output of memory pipeline, h_t = hidden state output.

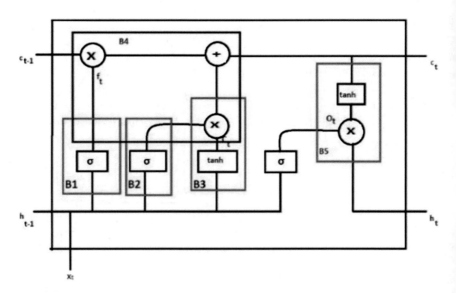

Fig. 9 LSTM memory block (one neuron) *Source* https://colah/github.io

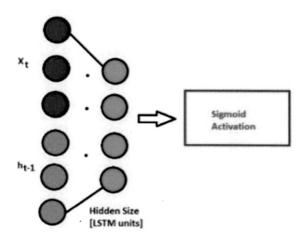

Fig. 10 Concatenated layers of neuron
Source https://colah/ github.io

The layers are concatenated and then passed through a linear layer. The output constitutes the hidden size, number of LSTM units present. In Fig. 10, the output enters the sigmoid function to translate the output between 0 and 1.
f_t = forget gate.
Box 1 is responsible on how much to keep from the old memory

$$f_t = \sigma\left(W_f \cdot [h_{t-1}, x_t] + b_f\right) \tag{5}$$

Value 0 means = forget all the memory
Value 1 means = keep all the memory
Value 0.5 means = keep a partial part of the memory.

Box 2 is responsible on how much to add to new memory

$$I_t = (\sigma\left(W_i \cdot [h_{t-1}, x_t] + b_i\right) \tag{6}$$

Value 0 means = not to add anything from the returned values
Value 1 means = allow all the values that have been returned
Value 0.5 means = partially add the returned values.

'Generating new memory'
Box 3: Here, new memory is received, then, it is passed through the tan h activation function, and the new memory is now controlled by the previous gate i_t.
Box 2: Determines how much to take from the new memory generated by Box 3.

$$\vec{c}_t = \tanh(W_c \cdot (h_{t-1}, x_t) + b_c) \tag{7}$$

Box 4: Determines what is required and what needs to be deleted from the old memory.

$$c_t = f_t * c_{t-1} + i * c_t. \tag{8}$$

Now, the new memory is constructed. Box 5 is responsible for the purpose.

$$o_t = \sigma\left(W_o\left[h_{t-1}, x_t\right] + b_o\right) \tag{9}$$

Value of 0 implies = not to output anything from the new memory
Value of 1 implies = to output everything from the new memory
Value 0.5 implies = to output some of the new memory.

The output is calculated basing on the memory, and the gate calculated previously controls it.

$$h_t = o_t * \tanh(C_t) \tag{10}$$

The memory state is passed through a tanh activation function to generate a new hidden state in Box 5.

The sigmoid function located previously to Box 5 controls how much to output [17].

5 Literature Survey

This is a literature survey of RNN and CNN. A detailed review of the architectures of the different neural networks and their function is given here.

5.1 Literature Survey on CNN

1. *LeNet*

This was recognized as the first successful work in application of convolution neural network. In 1990, the work was done by LeCun et al. [18] and later improvised in 1998 [2]. They worked on 'handwritten digit recognition' by using 'ConvNets.' It is applied in reading zip codes, house numbers, digits, etc. Due to the unavailability of machines with high computational power during the period, it took too much time to implement CNN.

2. *AlexNet*

Krizhevsky et al. [19] introduced this feature. They are famous for making convolution network more popular in the field of computer vision. AlexNet is quite similar to LeNet though, the major difference between AlexNet and LeNet is that, in LeNet, there is alternation between Conv and pooling layer, while in AlexNet, all Conv layers put together. AlexNet is much bigger and deeper than LeNet. It won the ILSVRC (2012) competitions [20].

3. *GoogleNet*

Szegedy et al. developed the architecture [21]. It earned reputation at ILSVRC competition (2014). A newly developed architecture called Inception (v1). By using this architecture, they can improve the utilization of the computing resources in the network. GoogLeNet comprises 22 layers of Inception modules but with lesser number of parameters as compared to AlexNet. Many improvements had been done on Inception-v1 lately. The most significant addition was the introduction of batch normalization which formed Inception-v2 by Ioffe et al. [22]. With some more modern additions, the architecture is called as Inception-v3 [23].

4. *VGGNet*

Andrew Zisserman and Karen Simonyan introduced VGGNet by analyzing the depth factor in a ConvNet, where the remaining parameters were held the same. By doing this, they were successful in extracting a large number of parameters in the network. The layers still consisted of '3 × 3' convolution filters. The innovation gave community to the new accurate version of architecture that is better than ConvNet.

5. *Inception V4*

Szegedy et al. introduced an architecture [24], in 2015. It was an extended version of the GoogleNet. When the Inception modules (by Szegedy et al. [21]) were trained with residual connections (introduced by He et al. [25]), the training process was accelerated. The network produced a state-of-the-art performance in the ILSVRC challenge and won the contest in 2015.

6. *ResNet*

'Residual learning framework' was introduced by Kaiming et al. The layers learnt residual functions from the inputs received and avoided unreferenced functions. Optimization of residual networks made the usage of ResNet more superior adding more accuracy to it. This network is expensive to evaluate because a large number of parameters are present. The removal of the first fully connected layers, however, did not affect the performance [25].

Difference between architectures of different neural networks:

AlexNet: The first famous convolutional neural network (CNN) was AlexNet. In AlexNet, convolutional layers are put together, performing max pooling at each layer and after that dense layers at the back. It is deeper than LeNet.
GoogleNet: GoogleNet comprised of Inception modules which gives more efficient result when used with convolutional neural network. It is very helpful to reduce cost for computation. GoogleNet comprises 22 layers of Inception modules.

VGGNet: VGGNet introduced by analyzing deep factor in ConvNet. Each filter has size '3 × 3.' With the help of this network, one can extract very large parameter of network. It is deeper than AlexNet. It consists of 16 convolutional layers.

Inception V4: This architecture is extended version of GoogleNet. In which, Inception modules are trained with residual connections to increase training speed.

ResNet: This architecture is similar to Inception V4. Layers learnt residual functions from the inputs and avoided unreferenced functions. This network is expensive to evaluate because a large number of parameters are present.

5.2   Literature Survey on RNN

In modern days, RNN is used in many different applications. Using natural language processing with RNN is recently used in many applications.

1. *RNN For Voice Activity Detection*:

In 2013, IEEE International Conference on Acoustic Speech and Signal processing (ICASS), Thad Hughes and Keir Mierle presented RNN for VAD. They have used Gaussian mixture models (GMM) and hand-tuned state machines (SM) for temporal smoothing purpose [26].

2. *Multi-lingual acoustic models using distributed deep neural networks*:

In 2013, IEEE International Conference on Acoustic Speech and Signal processing (ICASS), multi-lingual mode was proposed by Heighold and Vanhoucke using 'deep learning.' It was found to be useful in solving difficulties arising in data and decreasing the performance gap of resource rich and resource scarce [27].

3. *Multi-frame deep neural networks for acoustic modeling*:

Multi-frame deep neural network for acoustic modeling was introduced by Vincent Vanhoucke, Matthieu Devin, Georg Heighold in 2013, IEEE International Conference on Acoustic, Speech and Signal processing (ICASS). Compared with Gaussian mixture (GM), this model was costlier for real-time applications. In performance, it is same as the frame synchronous model. This model reduced computation cost of neural network-related issues up to 4× times [28].

4. *ASR error detection using recurrent neural network language model and complementary ASR*:

In 2014, IEEE International Conference of Acoustic, Speech and Signal processing (ICASS), automatic speech recognition (ASR) is proposed by Yun Lei, Wen Wang, Jing Zheng and YikCheung. This model is very helpful for the computer to understand human spoken language. This model locates problem occurring in utterance and then manages it [29] (Tables 1 and 2).

Table 1 Literature survey of CNN

Year	Author	Contribution
1990	LeCun et al.	Handwritten digit recognition with a back-propagation network [18]
2012	Krizhevsky et al.	Imagenet classification with deep convolutional neural networks [19]
2014	Szegedy et al.	Going deeper with convolutions [21]
2014	Simonyan and Zisserman	Analyze depth factor in ConvNet
2015	He et al.	Deep residual learning for image recognition [25]
2015	Szegedy et al.	Accelerate training of inception modules [24]

Table 2 Literature survey of RNN

Year	Author	Contribution
2013	Hughes and Mierle	Use of Gaussian mixture models (GMMs) and hand-tuned state machines (SM) was done for temporal smoothing purpose [26]
2013	Heigold and Vanhoucke	It was found to be useful in solving difficulties arising in data and decreasing the performance gap of resource rich and resource scarce [27]
2013	Vanhoucke et al.	This model was costlier than Gaussian mixture (GM), for real-time applications. In performance, it is same as the frame synchronous model, succesfully reduces the computational costs of these neural networks up to 4X times [28]
2014	Tam et al.	This model is very helpful for the computer to understand human spoken language. This model locates problem occurring in utterance and then manages it [29]

6　Conclusion

We discussed the process of image classification using CNN and RNN, their basic working principles and terminologies in this paper. A brief review on the architecture of some primitive models of the neural networks and their corresponding differences has also been presented in the literature survey. Convolutional neural networks (CNNs) and recurrent neural networks (RNNs) are gradually occupying an important position in the field of image classification, text prediction and are extensively used by data scientists for their easiness in usability and satisfactory minimal contingent outcomes. The neural networks can be used to provide better architecture for efficient memory management, reduce gradient exploding problems.

References

1. Image shows the sideview of Varanasi, and is downloaded from https://www.cleartrip.com/activities/Varanasi/ganga-aarti-tour-4-hours. on August 16, 2019.
2. LeCun, Y., L. Bottou, Y. Bengio, and P. Haffner. 1998. Gradient-based learning applied to document recognition. *Proceedings of the IEEE* 86: 2278–2324.
3. Yamins, D, H. Hong, C. Cadieu, and J.J. Dicarlo. 2013. Hierarchical modular optimization of convolutional networks achieves representations similar to Macaque IT and human ventral stream. In *27th Annual Conference on Neural Information Processing Systems, NIPS 2013,* December 5–10, 2013, Lake Tahoe, NV, United States, 2013. Air Force Office of Scientific Research (AFOSR), Amazon.com, Facebook, Google, Microsoft Research.
4. Hochreiter, S., and J. Schmidhuber. Long short-term memory. *Neural computation*.
5. Understanding a 3D CNN and its Uses. https://missinglink.ai/guides/neural-network-concepts. on 16 August, 2019.
6. Ouadfel, S., and S. Meshoul. 2012. Handling fuzzy image clustering with a modified ABC algorithm. *International Journal of Intelligent Systems and Applications* 4: 65.
7. Das, S., and A. Konar. 2009. Automatic image pixel clustering with an improved differential evolution. *Applied Soft Computing* 9: 226–236.
8. Yu, Z., W. Yu, R. Zou, and S. Yu. 2009. On ACO-based fuzzy clustering for image segmentation. In *Advances in Neural Networks–ISNN 2009*, 717–726. Berlin, Heidelberg: Springer.
9. Prabhu. 2018. *Understanding of Convolutional Neural Network (CNN)—Deep Learning*. https://medium.com/@RaghavPrabhu/understanding-of-convolutional-neural-network-cnn-deep-learning-99760835f148. March 4, 2018.
10. Wikipedia. 2019. *Convolution Neural Network*. https://en.wikipedia.org/wiki/Convolutional_neural_network. July 9, 2019.
11. Oruganti, Ram Manohar. 2016. Image description using deep neural networks. Thesis, Rochester Institute of Technology.
12. Ava Soleimany. 2019. MIT 6.S191: Convolutional Neural Networks. https://www.youtube.com/watch?v=HHVZJ7kGI0&list=PLtBwnjQRUrwp5__7C0oIVt26ZgjG9NI&index=3. on February 11, 2019.
13. Functionality of CNN. https://medium.com/@RaghavPrabhu/understanding-of-convolutional-neural-networkcnn-deep-learning-99760835f148. on August 16, 2019.
14. Banerjee, Survo. 2018. An Introduction to Recurrent Neural Networks. on May 23, 2018.
15. Gupta, Dishashree. 2017. Fundamentals of Deep Learning—Introduction to Recurrent Neural Networks.
16. Thapliyal, Manish. 2018. Vanishing Gradients in RNN.
17. Sammani, Fawaz. 2019. Applied Deep Learning with Pytorch published.
18. LeCun, B.B., J.S. Denker, D. Henderson, R.E. Howard, W. Hubbard, and L.D. Jackel. 1990. Handwritten digit recognition with a back-propagation network. In *NIPS*.
19. Krizhevsky, Alex I. Sutskever, and G.E. Hinton. 2012. Imagenet classification with deep convolutional neural networks. In *NIPS*.
20. Zeiler, M., Taylor, G., and Fergus, R. 2011. Adaptive deconvolutional networks or mid and high-level feature learning. In *ICCV.*
21. Szegedy, C., W. Liu, Y. Jia, P. Sermanet, S. Reed. 2014. Going deeper with convolutions. In *CVPR.*
22. Ioffe, S., and C. Szegedy. 2015. Batch normalization: accelerating deep network training by reducing internal covariate shift. In *Proceedings of the 32nd ICML*.
23. Szegedy, C., V. Vanhoucke, S. Ioffe, J. Shlens, and Z. Wojna. 2015. Rethinking the inception architecture for computer vision. arXiv:1512.
24. Christian Szegedy, Sergey Ioffe, Vincent Vanhoucke. 2016. Inception-v4, Inception-ResNet and the impact of residual connections on learning. arXiv:1602.07261.
25. Hinton, G.E., N. Srivastava, A. Krizhevsky, I. Sutskever, and R.R. Salakhutdinov. 2012. Improving neural networks by preventing co-adaptation of feature detectors. Preprint at arXiv:1207.0580.

26. Thad Hughes, and Keir Mierle. 2013. RNN for voice activity detection published. In *IEEE International Conference on Acoustics, Speech and Signal Processing*.
27. Heigold, G.V., A. Vanhoucke, P. Senior, M. Nguyen, M. Ranzato, and Devin J. Dean. 2013. Multilingual acoustic models using distributed deep neural networks. In *IEEE International Conference on Acoustics, Speech and Signal Processing*.
28. Vanhoucke, Vincent, Matthieu Devi, and Georg Heigold. 2013. Multiframe deep neural networks for acoustic modeling. In *IEEE International Conference on Acoustics, Speech and Signal Processing*.
29. Yik-Cheung Tam, Yun Lei, Jing Zheng, and Wen Wang. 2014. ASR error detection using recurrent neural network language model and complementary ASR. In *IEEE (ICASSP)*.

Analysis of Steganography Using Patch-Based Texture Synthesis

Kalyani O. Akhade, Subhash V. Pingle and Ashok D. Vidhate

Abstract Steganography is one of the techniques to hide secret information. Texture synthesis is a process to get large image from small piece of image without disturbing image texture. This paper focuses on patch-based texture synthesis. Patch-based texture synthesis process is a way to achieve steganography. This is performed by embedding message into image meanwhile not disturbing image properties and appearance. Message hiding is done by spreading secret information throughout the image. The distortion present in image is very fine sharp and person trying to interfere this cannot able to obtain required information. A secret message is stored in the form of patches which uses texture synthesis. At the receiver end, opposite action carried out to obtain original secret message and texture image. Using patch-based texture synthesis process, embedding capacity increased as well as image quality.

Keywords Steganography · Data embedding · Image processing · Texture synthesis

1 Introduction

A technique which is used for hiding communication is nothing but a steganography; this process inserts hidden matter in non-recognizable cover medium so as not in view of a meddler's doubt. Steganographic applications are used in secret communication between two users. Generally, cover medium used for steganography is any digital media. However, this paper emphasize on image as cover medium and text message embedding. This method does not use fixed-size cover image because

K. O. Akhade (✉) · S. V. Pingle
CSE Department, SKNSCOE, Solapur University, Solapur, India
e-mail: akhade.kalyani@gmail.com

S. V. Pingle
e-mail: subhash.pingle@sknscoe.ac.in

A. D. Vidhate
E&TC Department, SMSMPITR, Solapur University, Solapur, India
e-mail: vidhate.a.d@gmail.com

© Springer Nature Singapore Pte Ltd. 2020
D. Swain et al. (eds.), *Machine Learning and Information Processing*,
Advances in Intelligent Systems and Computing 1101,
https://doi.org/10.1007/978-981-15-1884-3_35

insertion capacity relates to size of texture image. For fixed-size image, more insertion of data gives less quality image appearance and that can be easily identified by attacker. So, in order to avoid such problem, patch-based algorithm firstly resizes the existing image as per requirements, and then, message is inserted to image. Insertion of message is achieved by texture synthesis process. Texture synthesis can be done by two approaches either pixel-based texture synthesis or patch-based texture synthesis. We make use of patch-based texture synthesis. Patch is a small unit of image. Image considers for texture synthesis is any digital image taken by camera or pictorially drawn [1–3].

In specific, this algorithm conceals the source texture picture as opposed to using a current cover picture to conceal messages and inserts hidden messages by patch-based texture synthesis process. Patch-based texture synthesis process allows to extract hidden messages and source texture from synthetic texture. In the literature of texture synthesis, steganography using patch-based algorithm has always been submitted to the best of our understanding. Test comes out has confirmed that this method gives different numbers of implanting capacities; it gives embedded image which is not distorted. Ideally, it is seen that there is very less chances to break this steganographic approach. This method can stand up to an RS steganalysis attack.

2 Related Work

H. Otori and S. Kariyama have given one approach for data embedded texture synthesis. This technique has major drawback such that it gives some sort of error during recovery of the message. The method gives an idea of joining pixel-based texture synthesis with data coding. The embedding of data takes place in the form of dot pattern, and the capacity of embedding depends on the number of data patterns to be used [4, 5]. K. Xu has given one approach as feature-aligned shape texturing. The approach pastes the source texture to form the whole image. The image formed is of uniform source texture so that the quality of the image is enhanced. One of the drawbacks such that during paste of texture, the neighbouring pixel required to check from time to time [6].

Efros and Freeman have given image quilting methodology. In image quilting, patch stitching performed by overlapping the neighbouring patch, and due to that, at neighbouring side, similarity between adjacent gets increased. So, the minimum error obtained through the overlapped region. Optimum boundary utilization is obtained by image quilting [7]. Z. Ni et al. given algorithm for recovery of the secret message hidden on to image texture. Two types of algorithms are present, the first is forward data hiding, and the second is reversible data hiding algorithm. Generally, reversible data hiding algorithm is used in which during recovery distortion is less [8].

3 Methodology

The proposed methodology consists of top–down approach. Figure 1 shows break-down structure of complete methodology. We make use of some terms such as patch. Patch is small unit of resized image where patch size is provided by user. Patch structure is shown in Fig. 2a. Kernel region is central part of patch, and remaining area with depth Pd is boundary region. Figure 2b shows structure of source texture which divided into number of kernel blocks [9]. Expanding boundaries of each kernel block source patch is created. Number of source patches can be calculated with Eq. (1). Once source patches are ready, index table is generated which stores location of source patches, and also, same table helps in retrieving source texture. Hidden message is inserted to embeddable patches. All these patches are stitched each other by index table. Index table gives the location of the patches where to stick them. When patches are stitched as per index table, the obtained image is composite image. The composite image with secret message and source texture is called stage synthetic texture as shown in Fig. 3. The capacity of data embedding is product of bit per patch (BPP) with number of embeddable patches EP_n.

$$\text{Total source patches} = [S_h/K_h] * [S_w/K_w]. \qquad (1)$$

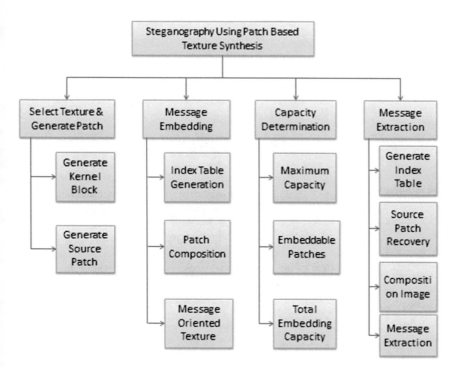

Fig. 1 Steganography process

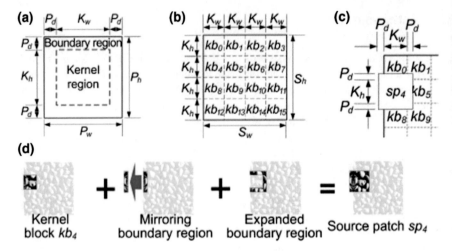

Fig. 2 Elements of steganography process. **a** patch structure; **b** source texture kernel structure; **c** overlapping process; **d** expansion by mirroring source patch

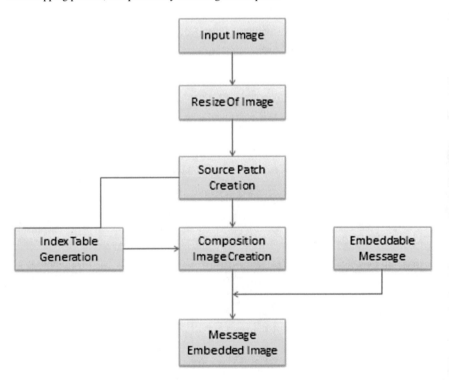

Fig. 3 Message embedding procedure

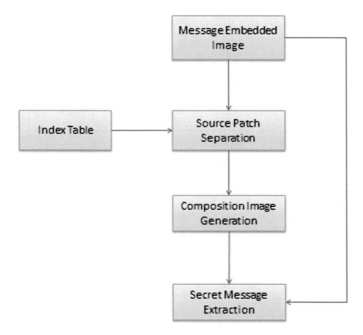

Fig. 4 Message extracting procedure

$$TC = BPP \times EP_n = BPP \times (TP_n - SP_n). \qquad (2)$$

At receiver side, index table is generated which is exactly same as that of generated at transmitter side. Each kernel is retrieved by referring to index table and ready to organize source texture as per order, hence recovering the recaptured input texture which is able be precisely copy of source texture. Within the third process, as per the index table, composite image is formed. Finally, message extraction and authentication are done (Fig. 4).

4 Result and Discussion

Net beans IDE 8.0.2 and JDK 1.8 were used to implement the scheme. The effect of variations of the steganography finds out by using portable desktop computer system. The system outcome shows improvement in the embedding capacity as BPP obtains its optimum value. Table 4.1 shows results on different embedding capacities for different size source texture for $P_h * P_w = 58 * 58$, $P_d = 8$, $T_h * T_w = 500 * 500$. The size of source texture is proportional to the embedding capacity. Large size source texture has less embedding capacity. As source texture size is large, source patches are more which ultimately reduce embeddable patches, so total embedding

Table 1 Increasing embedding capacity with respect to BPP

Source width × source height	Source Patch n	Embeddable patch n	TC (5BPP)	TC (10BPP)	TC (BPPmax)
128 × 128	9	87	435	870	1229
250 × 250	35	61	305	610	1518
300 × 300	51	45	225	450	1584

capacity decreases. To calculate embedding capacity given by this algorithm, we make use of following equations.

$$\text{Total Embedding Capacity} = \text{Bits per patch} * \text{Total embeddable patches.} \tag{3}$$

$$\text{Total Embeddable Patches} = \text{Total Patches} - \text{source Patches.} \tag{4}$$

$$\text{Total Patches} = [((T_h - P_h)/(P_h - P_d)) + 1] * [((T_w - P_w)/(P_w - P_d)) + 1]. \tag{5}$$

$$\text{BPP}_{\max} = \left[\log_2(S_w - P_w + 1) * (S_h - P_h + 1)\right]. \tag{6}$$

where

$\text{BPP}_{\max}$	Bits per patch maximum embedding capacity
T_h	Height of synthetic texture
T_w	Width of synthetic texture
P_h	Height of patch
P_w	Width of patch
S_h	Height of source texture
S_w	Width of source texture
P_d	Depth of patch (Table 1)

Variation and robustness of the steganographic outcome with optimum values are as per Figs. 5 and 6; Table 2.

The accuracy of image is defined by using mean squared error of the overlapped area (MSEO). It is the analysis in which the overlapping in between source patches has been checked. The MSEO is always positive quantity. The MSEO value is inversely proportional to image quality. The value of MSEO resembles good quality of image. Texture synthesis procedure gives the MSEO. Table specifies the MSEO with the embedding capacity for the source textures. This has been calculated for patch size having equal height with width of 58 and patch depth of 8. Image quality is compared using mean squared error of overlapped area. For peanut, source texture in case of 5BPP MSEO is 2837, and for 10BPP, MSEO is 2913 which shows slightly higher image quality (Table 3).

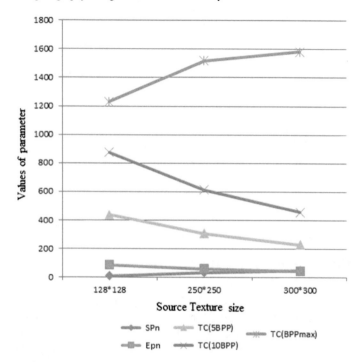

Fig. 5 Output of embedding capacity

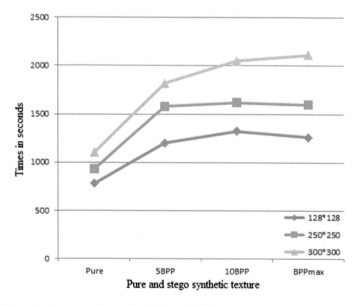

Fig. 6 Schematic for speed of execution in interval of time

Table 2 Execution speed variation with respect to BPP (second)

Patch width × patch height	Original	Five BPP	Ten BPP	BPP_{max}
128 ×128	780	1200	1325	1262
250 ×250	931	1580	1622	1600
300 ×300	1100	1820	2050	2112

Table 3 Performance of MSEO

	Original	5BPP	10BPP
Ganache	1116	1125	1204
Peanuts	2832	2837	2913
Metal	8763	8768	8919
Rope net	6842	6846	6790

5   Conclusion and Future Work

Steganography using patch-based texture synthesis produces a large stego synthetic texture having secret messages. This method embeds secret message and also retrieves embedded message, making possible a second round of texture synthesis if needed. This algorithm is secure and stands against an RS steganalysis attack.

One possible future study is to expand this method to support other kinds of texture synthesis approaches to improve the image quality of the synthetic textures. Another possible study would be to combine other steganography approaches to increase the embedding capacities, and also, this study can be extended to other digital media. Hence, for better security by enhanced coding technique, steganography is useful to avoid stag attack.

References

1. Johnson, N.F., and S. Jajodia. 1998. Exploring steganography: seeing the unseen. *Computer* 31: 26–34.
2. Provos, N., and P. Honeyman. 2003. Hide and seek: an introduction to steganography. *IEEE Security and Privacy* 1: 32–44.
3. Akhade, K.O., S.V. Pingle, and A.D. Vidhate. 2019. Steganography using texture synthesis—a review. *IJIRCCE* 7: 1798–1800.
4. Otori, H., and S. Kuriyama. 2007. Data-embeddable texture synthesis. In *Proceedings of the 8th International Symposium on Smart Graphics,* Kyoto, Japan.
5. Otori, H., and S. Kuriyama. 2009. Texture synthesis for mobile data communications. *IEEE Computer Graphics and Applications* 29.
6. Xu, K., D. Cohen-Or, L. Liu, H. Zhang, S. Zhou, and Xiong, Y. 2009. Feature-aligned shape texturing. *ACM Transactions on Graphics* 28.
7. Freeman, W.T., and A.A. Efros. 2001. Image quilting for texture synthesis and transfer. In *Proceedings of the 28th Annual Conference on Computer Graphics and Interactive Techniques.*

8. Ni, Z., Y.-Q. Shi, N. Ansari, and W. Su. 2001. Reversible data hiding. *IEEE Transactions on Circuits and Systems* 16.
9. Wu, Kuo-Chen., and Chung-Ming. Wang. 2015. Steganography using reversible texture synthesis. *IEEE Transactions on Image Processing* 24.

Non-invasive Approach for Disease Diagnosis

Anita Gade and V. Vijayabaskar

Abstract The human respiration involves inhalation and exhalation of atmospheric air. The exhaled breath contains around 3000 organic compounds (VOCs) which are linked to human metabolic and biochemical processes occurring inside the body. The electronic nose plays a vital role in identification of various diseases by breath sample signature analysis. Electronic nose is a non-invasive technology for disease monitoring with systematic integration of sensor array and artificial intelligence. The blood glucose monitoring methods currently used in medical field are invasive and uncomfortable to patients due to needle pricking for blood sample collection. Due to which the need for non-invasive and alternative screening/diagnostic technologies is anticipated. The lung cancer is globally one of the cancer types with high fatal ratio. The diagnostic techniques like sputum cytology, radiology and tomography are not accessible to worldwide population for screening. The non-invasive techniques like mass spectrometry and gas chromatography require skilled technicians and are not portable and economical for wide range of population. The exhaled human breath contains biomarkers for several diseases so non-invasive methods will definitely be advantageous over the existing invasive and expensive methods. The review briefs the modern collaborative machine-learning approach with non-invasive, economical and easy to use framework for disease detection and diagnosis.

Keywords Diabetes · E-nose · Exhaled breath · Gas chromatography · Lung cancer · Non-invasive

A. Gade (✉) · V. Vijayabaskar
Sathyabama Institute of Science and Technology, Chennai, India
e-mail: anita.gade@yahoo.co.in

V. Vijayabaskar
e-mail: v_vijaybaskar@yahoo.co.in

© Springer Nature Singapore Pte Ltd. 2020
D. Swain et al. (eds.), *Machine Learning and Information Processing*,
Advances in Intelligent Systems and Computing 1101,
https://doi.org/10.1007/978-981-15-1884-3_36

1 Introduction

1.1 E-nose—Sensor Technologies

In case of illness conditions, the metabolic biochemical organic processes inside the body are changed, resulting into the formation of new/altered chemicals. These altered composition of body fluids leads to formation of volatile organic compounds in human exhaled breath pattern [1]. These organic breath patterns are the biomarkers linked to specific diseases and play a vital role in advance detection and screening of illnesses. According to WHO published report, the non-contingent diseases (NCD) indicate annual worldwide death ratio of around 70%. Diabetes, cardio-vascular, respiratory illnesses and cancer cover most of the NCD fatalities. The global list of major illnesses causing deaths indicates lung cancer and diabetes mellitus on sixth and seventh ranks. These two major illnesses are reviewed, studied, analysed and characterized in this review paper using electronic nose techniques. Globally, 430 million populations are diagnosed with diabetes with 1.6 million death ratios and the percentage is increasing in upcoming years. The diabetes-related ailments like renal failure, visual impairment and obesity are also rising in upcoming years. The diabetes individuals are more prone to diseases in comparison with normally fit persons. Modern advanced blood glucose level detection methods like fasting plasma glucose test, oral glucose tolerance test are simple and economical with very few drops of blood sample requirement for diabetes diagnosis [2]. But these methods involve of needle pricking for blood sample collection so it is painful and invasive method especially in case of patients which requires continuous monitoring. Diabetes is a silent killer and cancer is a direct killer. Amongst the various types of cancers, the lung cancer is having maximum number of fatalities around the globe. Usually, the cancer is diagnosed and detected with clinical analysis of blood and urine sample tests. The diagnostic techniques like X-rays, computed tomography (CT), fibre-optic endoscopy (FOE), ultrasonography (USG) and positron emission tomography (PET) scan are invasive, expensive and more time slot consumption with requirement of specialized trained expertise. Due to these factors, the large-scale population coverage and feasibility are affected which leads to demand and need of economical, non-invasive and user-friendly diagnostic technologies—nose combines the VOC sensors with artificial intelligence and neural networks ensuring non-invasive technology for detection of targeted VOC's from exhaled human breath for early detection and diagnostics of various diseases [3]. This review paper focusses on study and analysis of existing E-nose techniques and its applications in health industry for disease diagnostics and detection.

E-nose tool consists of gas sensors as basic building blocks. E-nose sensors should have characteristics like high accuracy, high sensitivity, quick response time, stability, reproducibility, less sensitivity for humidity, low operating temperature, ease for data processing, easy calibration, smaller in size and economical. Since last five decades, various types of sensors like field, optical gas sensor, polymer, metal oxide semiconductor (MOS), effect transistor (FET) and electrochemical catalytic

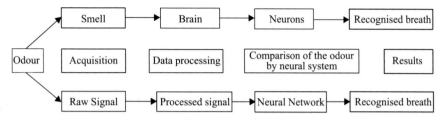

Fig. 1 E-nose principle

and piezoelectric sensors had been explored. The main focus of researchers is sensing logic and fabrication methodologies to detect variety of bandwidths of concentrations of gases for achieving user-friendly sensors with high sensitivity and low cost. In optical transduction gas sensor technique, the detection of gases is based on change in absorption properties, refractive index, luminescence, scattering, reflectivity and optical path length. The optical absorption and luminescence techniques are based on changes in light intensity and wavelength once it is passed through gases. The surface plasmon resonance (SPR) is the most effective technique in gas sensing. Even if the optical gas sensors are excellent in sensitivity and cost effectiveness but due to its size and signal conditioning, it is rated less based on fabrication feature on E-nose. The piezoelectric gas sensors used in some of the E-nose systems are surface acoustic wave (SAW) device and the quartz crystal microbalance (QCM). In quartz crystal microbalance (QCM), a piezoelectric disk is coated with semiconductor metal oxide or gas sensitive polymer and travelling wave resonant frequency is captured. In line with this, surface acoustic wave (SAW) gas sensor captures the propagation frequency changes via piezoelectric substrate-coated sensing material due to gas. The salient features of piezoelectric gas sensors include compactness, less power consumption, robust, quick response time and higher sensitivity. Due to the operational capability at higher frequencies and complexity in circuits, the signal-to-noise ratio for piezoelectric sensors is lower leading to less effective element for electronic nose system. The commercial electronic nose systems consists of polymer composites with number of reversible physical and chemical properties and these composites can operate at room temperature with higher sensitivity, accuracy and selectivity (Fig. 1).

2 Breath Analysis

The VOCs in individual's breath will fluctuate; still the basic centre of breath which is available in each individual remains same. The basic breath analysing techniques for bifurcation of inhaled or exhaled breath biomarkers are:

- Gas chromatography-based techniques, combined with a type of mass spectroscopy.

- Electronic investigation using sensors as electronic noses [4].

The E-nose shows demonstrated functional capability of diagnosis in diagnosis of diabetes, renal diseases and lung malignancy. By using real-time and canister analysis, the study of the presence of 108 organic compounds in human breath is detected.

2.1 Breath Background Correction Importance

It is important to note that exhaled breath is exposed to atmosphere containing exogenous contaminants. The standardization in collection of breath sample is important and it should be captured for a fixed duration of time in controlled environment to avoid the contamination in sample [5]. The fruitful approach would be subtraction of exogenous concentrations from the exhaled air from which alveolar gradient can be calculated. The VOC origin can be correctly captured with above method whether it is metabolic or external. So background correction of it is utmost importance for creation of normal concentration ranges of human VOS's with reference to physiological and sociological variables like age, ethnicity and smoking status [6].

3 State of the Art

3.1 Diabetes

There are four fundamental demonstrative techniques for monitoring and diagnosing diabetes mellitus:

- Glycosylated haemoglobin test,
- Oral glucose tolerance test,
- VOC comprehensive test,
- Gene diagnosis.

The blood glucose detection techniques which are available in market are dependent on laboratories and hand held glucometer devices for sugar-level analysis [7]. The testing methods involve needle pricking for blood sample complex operation along with involvement of high cost. The exhaled breath contains involvement of different natural mix gases in VOC biomarkers. But $(CH_3)_2CO$ fixation is viewed as a dependable pointer of diabetes since it is discharged from skin and its fixation is with h-hydroxy butyric corrosive [8]. Analysis of $(CH_3)_2CO$ in individual's skin foreseen as a screening test for ketoacidosis and also can be utilized as diabetes control.

3.2   Lung Cancer

The exhaled breath biomarkers diagnosed with lung disease and analysis from various research groups is still not in a decisive and conclusive diagnostic stage due to unstable natural gases mixture in the captured samples [9].

With the period of customized prescriptions requesting quick be spoken analysis and treatment, this region of atomic diagnostics is starting to see an upsurge in biotechnological headway. Here, we talk about late enhancements and headings in the improvement of breath VOC examination and finding stages that offer the potential for illness biomarker revelation and ailment guess. Actually, so as to analyse the lung malignancy, low portion chest computed tomography check is effective method.

3.3   Childhood Asthma

There are presently around 7 million youngsters around the world suffering from asthma, the WHO evaluates that asthma is the most widely recognized non transmittable infection in kids. This fast ascent has been seen in numerous different populaces; however, asthma commonness is not falling all over the place [10]. While over conclusion of youth asthma unquestionably added to the "asthma plague", almost certainly, there was an authentic ascent in asthma predominance related with a move towards "westernized living". The important factors for which they are mostly proned to be change in microbial exposures, air quality and action levels.

4   Literature Survey

The human nose and brain are systematically interconnected for detection and distinguishing number of chemicals leading to recognition of food freshness, perfumes, strong odours and leakage of gases [11]. The complex volatile organic compounds with the help of neuron receptors send the signals for pre-processing to brain through nerve impulse. The hypothalamus and olfactory cortex store the signals and analyse it in the brain in conclusive steps. On similar lines, electronic nose technique contains sensor array, data acquisition system, signal processing unit, data storage unit and AI neural network for detection and screening of chemical vapours. Until 1980, the variety of chemical sensors using electrical, thermal, mass and optical transducer principles was available due to the scarcity of advanced electronics components and advanced computational tools. The world's first E-nose was developed by Persaud and Dodd, using a chemical sensor array and pattern recognition system for gas discrimination. Further similar research is continued at Hitachi Research Laboratory, Japan by Ikegami et al. at Warwick University, United Kingdom in 1982. E-nose as "an instrument which comprises an array of electronic chemical sensors with partial sensitivity and an appropriate pattern recognition system capable of recognizing simple or complex odours" is defined by Gardner and Bartlett in 1994. Since the last decade, researchers working on E-nose have been focusing on developing highly sensitive and selective sensors, a compact electronic module (electronic circuits for amplifying signals, signal conditioning, and analog-to-digital (A/D) converters) for better feature extraction with reduced redundancy as well as classifier (primarily neural network) for learning and validation process as pattern recognition algorithm to discriminate the chemical of interest. Several E-noses have been developed by combining nonspecific sensor array and machine learning, for different applications such as biomedical, healthcare, safety, food industry, chemical industry and pharmaceutical [12].

Table 1 shows previous work on medical E-nose with different techniques.

5 Future Directions in E-Nose Technology Research

The major challenges faced in Electronic nose technology during disease diagnostics are briefed below.

5.1 Respiratory Disease

It is reasonable to check whether the old style graphical yield of the E-nose fits a change in a graphical capacity that would streamline its substance and make it effectively comprehensible [13]. Such rearrangements would make the scientists progressively sure about the down to earth utilization of the method and afterward increasingly inspired to investigate its possibilities.

As of now E-nose has been projected as an indicative system for authentication and diagnostic strategies based on individual symptoms.

The progressing evaluation of BP of individual sicknesses, just somewhat, might be interpreted in clinical practice because of three significant confinements of chemometrics:

- Age-related ordinary benchmarks are not accessible, in spite of the fact that age-related contrasts are in all respects likely. This restrains the translation and routine utilization of the BP as an indicative device.
- It is obscure to what degree the deliberate BP changes with momentary introduction to contamination. In this way, a similar subject may have diverse BP relying upon whether the expirate is gathered following seven days end at home, or a work day.
- Both intense and ceaseless maladies are never disengaged in grown-ups and, significantly more, in the old [14].

Table 1 Literature survey

S. No.	Journal and year	Study	Technique	Design based on technology study	Main finding
1	Iain et al. Elsevier 2019	Offline breath analysis	Capture and storage of breath	Canisters and tedlar bags breath collection apparatus	Tedlar bags ideal for breath sampling
2	Judit et al. Elsevier 2019	Chronic disease—asthma, COPD, obstructive sleep apnea, interstitial lung diseases	Breath analysis	Case study	Environmental and metabolic factors shape the breath print
3	Paul Brinkman Elsevier 2019	Prediction of respiratory diseases	VOC	Case study	It's not a one-size-fits-all method regarding measurements, analysis and reporting which hamper affirmation of Electronic nose findings
4	Fan et al. Elsevier 2019	Breath analysis	VOC	Case study	Breath printing based diagnosis
5	Crucitti et al. Elsevier 2019	Lung cancer	E-nose	Case study	Limited size of populations analysed
6	Arnaldo et al. Elsevier 2019	Lung cancer diagnosis, diabetes	GC-MS	MOS detection, SAW detection and diagnosis units	Very expensive and not portable
7	Stefania et al. Elsevier 2019	Childhood asthma	Breath analysis	GC-MS for identification of biomarkers relevant to a particular illness	(1) No standard test for asthma (2) inflammatory components of asthma
8	Corrado et al. Elsevier 2019	Data analysis		Feature extraction and classification algorithm	Limited size of populations analysed

(continued)

Table 1 (continued)

S. No.	Journal and year	Study	Technique	Design based on technology study	Main finding
9	Giorgio et al. Elsevier 2019	Experts opinions	Breath printing	Approach	Exhaled breath generates a huge, significant library of breath prints
10	Raffaele et al. Elsevier 2019	Non-neoplastic chronic	Breath printing	Case study	Poorly suitable for rapid and large-scale application
11	Xidong et al. Journal 2019	Acetone sniffer (diabetes detection)	Acetone detection	YSZ-based gas sensor with mixed potential with Sm2-xSrxNiO$_4$ (x = 0.4, 0.6 and 0.8) -SEs synthesis by sol–gel method	Low detection limit of 300 ppb to acetone (-1.8 mV)
12	Ayten et al. IEEE conf. 2018	Lung cancer	Breath	8 Metal oxide semiconductor gas sensors and 14 quartz crystal microbalance sensors	
13	Dongmin et al. IEEE Trans. 2018	Odour analysis	Odour	Tedlar gas sampling bag and an airtight box	Airway inflammation sensitivity 73.79%
14	Alphus Sensors 2018	Gastrointestinal diseases	VOC	WOLF E-nose: eight electrochemical sensors, two non-dispersive infrared optical devices and one photoionization detector	Unique VOC profile signatures are indication of complex mixtures of specific biomarker metabolites with unique properties
15	Jiang et al. Journal 2018	Diabetes (acetone recognition)	Acetone	Gas chromatography (Usage of aluminium foil bag to collect the breath acetone samples of 25 number of Type 2 diabetes patients and 44 number of healthy people)	Low feasible

5.2 Heart Failure

In spite of the fact that VOC examination is still in the fundamental stages, it is without a doubt a standout amongst the most creative and intriguing methodologies in heart failure investigation. For ensuring and validating the results, accessibility of cutting edge innovations like genome sequencing, medication is switching to a platform with precise and customized treatment with reference to individual's health risk profile analysis, the treatment choices are tailored to a characteristics of patient [15]. Later on, with the advancement of the diagnostics equipment's performing VOC investigation, we would most likely inhale into our gadget serenely at home, and the results of the examination would be conveyed to the medicine specialists.

5.3 Chronic Liver Diseases

Albeit extremely encouraging, VOC examination in liver illnesses is influenced by the nonattendance of adequate perceptions permitting an authoritative assessment of its clinical criticalness. Undoubtedly, bigger longitudinal examinations with outer approval ought to be given so as to ensure the sufficient quality of proof and bolster the estimation of VOC investigation in clinical practice. Endeavours ought to be additionally made to institutionalize examination methods: most by far of studies depended on costly and tedious investigations yet this clearly makes related outcomes inadequately appropriate for fast and huge scale pertinence [16]. To this point, more noteworthy and well-organized research endeavours ought to be coordinated towards the enlistment of more extensive partners of patients in well-organized and result explicit investigations, including legitimate follow-up periods. Regardless of breathomics speaking to an engaging and promising methodology in finding new bits of knowledge into asthma pathobiology, a few confinements must be considered. In case of paediatric patients, it is a challenge to collect exhaled breath since there are various variable factors like rate of breathing, work out, natural temperature and moistness. Information handling and investigation are of additionally testing. Distinctive factual calculations (administered or unsupervised) can be utilized, contingent upon the examination point.

6 Conclusions

The investigation of breath is non-invasive and test sample collection can be simple and without any harm to individual without production of any noise. As of now, breathed out breath tests can be estimated predominantly by gas chromatography and an electronic nose. Gas chromatography and electronic nose are the best practises in this field with optimal third-generation capability and flexibility. Gas chromatography results can be with high level of systemized reciprocal capability and accuracy but with more space and cost with verification and validation of results by experts.

References

1. Guo, Dongmin, David Zhang, Naimin Li, Lei Zhang, and Jianhua Yang. 2018. A novel breath analysis system based on electronic olfaction. *IEEE Transactions on Biomedical Engineering*.
2. White, Iain R., and Stephen J. Fowler. 2019. Capturing and storing exhaled breath for offline analysis. In *Breath analysis*. Amsterdam: Elsevier.
3. Pennazza, Giorgio and Marco Santonico. 2019. Breathprinting roadmap based on experts' opinions. In *Breath analysis*. Amsterdam: Elsevier.
4. Gao, Fan, Xusheng Zhang, Min Wang, and Ping Wang. 2019. Breathprinting based diagnosis, selected case study: GCMS and E-nose collaborative approach. In *Breath analysis*. Amsterdam: Elsevier.
5. D'Amico, Arnaldo, Giuseppe Ferri, and Alessandro Zompanti. 2019. Sensor systems for breathprinting: a review of the current technologies for exhaled breath analysis based on a sensor array with the aim of integrating them in a standard and shared procedure. In *Breath analysis*. Amsterdam: Elsevier.
6. Brinkman, Paul. 2019. Breathprinting based diagnosis, selected case study: U-BIOPRED project. In *Breath analysis*. Amsterdam: Elsevier.
7. Jiang, Y.C., M.J. Sun, and R.J. Xiong. 2018. Design of a noninvasive diabetes detector based on acetone recognition. In *Journal of Physics*.
8. Zeng, Xianglong, Haiquan Chen, Yuan Luo, and Wenbin Ye. 2016. Automated diabetic retinopathy detection based on binocular siamese-like convolutional neural network. *IEEE Access* 4.
9. Tirzite, Madara, Māris Bukovskis, Gunta Strazda, Normunds Jurka, and Immanuels Taivans. 2019. Detection of lung cancer with electronic nose and logistic regression analysis. *Journal of Breath Research*.
10. La Grutta, Stefania, Giuliana Ferrante, and Steve Turner. 2019. Breathprinting in childhood asthma. In *Breath analysis*. Amsterdam: Elsevier.
11. Aguilar, Venustiano Soancatl, Octavio Martinez Manzanera, Deborah A. Sival, Natasha M. Maurits, and Jos B.T.M. Roerdink. 2018 Distinguishing patients with a coordination disorder from healthy controls using local features of movement trajectories during the finger-to-nose test. *IEEE Transactions on Biomedical Engineering*.
12. Subramaniam, N. Siva, C.S. Bawden, H. Waldvogel, R.M.L. Faull, G.S. Howarth, and R.G. Snell. 2018. Emergence of breath testing as a new non-invasive diagnostic modality for neurodegenerative diseases. *Brain Research*. Elsevier.
13. Pako, Judit, Helga Kiss, and Ildiko Horvath. 2019. Breathprinting-based diagnosis: case study: respiratory diseases. In *Breath analysis*. Amsterdam: Elsevier.
14. Le Maout, P., J-L. Wojkiewiczy, N. Redony, C. Lahuec, F. Seguin, L. Dupont, A. Pudz, and S. Mikhaylov. 2018. From drifting polyaniline sensor to accurate sensor array for breath analysis.In *IEEE*.
15. Zhang, Wentian, Taoping Liu, Miao Zhang, Yi Zhang, Huiqi Li, Maiken Ueland, Shari L. Forbes, X. Rosalind Wang, and Steven W. Su. 2018. NOSE: a new fast response electronic nose health monitoring system. In *IEEE*.
16. Incalzi, Raffaele Antonelli, Antonio De Vincentis, and Claudio Pedone. 2019. Breathprinting-based diagnosis, selected case study: nonneoplastic chronic diseases. In *Breath analysis*. Amsterdam: Elsevier.

Alleviation of Safety Measures in Underground Coal Mines Using Wireless Sensor Network: A Survey

Raghunath Rout, Jitendra Pramanik, Sameer Kumar Das
and Abhaya Kumar Samal

Abstract Mining is one of the oldest endeavors regarded of mankind along with agriculture. To ensure personal safety of miners and maximizing the mining process improvements in safety measures is a must. Also wireless sensor networks play a vital role for monitoring the mining environment. This paper describes a survey of the improvements in safety measures in underground coal mines using the wireless sensor network finding the unexplored areas which needs improvement. To ensure quick rescue of miner's, ground station plays a vital role. In view of all the safety measures taken together with the development of technology it is feasible to apply some of the technologies with modifications. This paper surveys the need for extended research in wireless sensor network and applications to monitor the UG mine environment effectively.

Keywords Underground mining · Sensor system · Sensor node

1 Introduction

Mining is the process of obtaining coal or other minerals from a mine. Several activities are carried out within an underground mine. The workers who are indulged in mining are exposed to several callous situations such as roof fall, mine fires, release of toxic gases, etc. They face several unsuitable conditions for communication and monitoring of systems. Safe mining depends on several environmental factors and

R. Rout (✉)
Department of CSE, DRIEMS Autonomous Engineering College, Cuttack, India
e-mail: raghunathrout_78@yahoo.co.in

J. Pramanik
Centurion University of Technology and Management, Bhubaneswar, Odisha, India

S. K. Das
Department of CSE, Arihant Junior College, Ganjam, India

A. K. Samal
Department of CSE, Trident Academy of Technology, Bhubaneswar, India

© Springer Nature Singapore Pte Ltd. 2020
D. Swain et al. (eds.), *Machine Learning and Information Processing*,
Advances in Intelligent Systems and Computing 1101,
https://doi.org/10.1007/978-981-15-1884-3_37

conditions of mines, like airflow, surrounding temperature, humidity, dust and gases present. To undergo safe coal mining, the basic requirement is to regularly monitor the level of gases like methane, oxygen, carbon dioxide, carbon monoxide, etc., monitoring air pressure and detecting mine fires. To combat the problem of toxic gases and overcome the traditional mining hazards, WSNs are widely deployed in multidisciplinary areas where human intervention is not possible.

2   UG Mines: An Insight

Earth reserves abundant amount of coal and mineral deposit underneath. Most of the coal companies try to extract profound deposit of coal safely by different methods. Columns of coal are left to support the ground in room and pillar mines. Then they are often taken out and the mine is left to disintegrate that is called subsidence. In long-wall mining, mechanical shearers strip the coal from the mines. Support structures that allow the shearers' get admission to the mine are eventually removed and the mine collapses.

2.1   Potential Impact of Underground Mining

It reasons subsidence as the land above it starts sinking and mines crumple which results in causing severe damage to buildings.

It lowers the water table altering flow of the groundwater and streams. Several gallons of water are pumped out each year. Only a small portion of it is used by the corporation or by the nearby villages; the rest of it is wasted.

Coal mining produces additionally greenhouse fuel emissions.

3   Sensor Nodes and Types

There are several key elements that build a WSN node shown in Fig. 1.

3.1   Low Power Embedded Processor

The job of a WSN node consists of processing of both locally sensed data and also information shared by other sensor nodes. Embedded processors face several challenges regards to computational energy (e.g. several of the units used presently in doing research and development of an eight-bit 16-MHz processor) Due to the

Fig. 1 Basic building block
of a sensor node

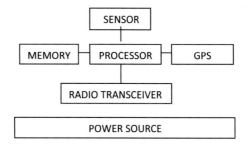

limitations of such processors, gadgets typically run specialized module based on embedded working structures, which include Tiny OS.

3.2 Memory

Two kinds of memory are most widely used in WSN nodes; they are on-chip memory of microcontroller and FLASH memory. Generally, off-chip RAM is rarely used. The portions of memory and storage on an embedded WSN device are frequently controlled generally with the aid of financial issues, and also are probably to improve over time.

3.3 Radio Transceiver

The primary task of a radio transceiver is to transmit and receive data through sensor nodes. It operates in different modes like the on/off mode. Wireless sensor nodes consist of a low-charge, concise range wi-fi radio (<100 m). Due to their limited capability sensor nodes undergo gradual improvement in upgrades in cost, spectral efficiency, tunability, and immunity to noise, fading, and interference, etc. Radio communication takes the maximum energy intensive operation in a WSN device, and hence the radio must use different energy proficient ways like sleep and wake-up modes.

3.4 Sensors

Due to limitations of bandwidth and energy constraint, wireless sensor nodes assist least data rate sensing. For this reason, there is a requirement of several sensor nodes on board [1]. These sensors are application dependent, it may include sensors like pressure sensor, humidity sensor, light sensor, temperature sensors, accelerometers, chemical sensors, etc.

3.5 Geo-positioning System

For getting an optimized output of sensor nodes, sensor locations must be predefined. This is possible by preconfiguring the sensor places for deployment; however, this may only be feasible in constraint deployments. Sensor nodes are deployed in an ad hoc manner for outside operations. The data of these nodes are effortlessly obtained via satellite-based GPS. In most of the application, a small portion of the nodes is geared up with GPS functionality because of environmental and monetary constraints. Here, other nodes obtain their location by using network localization algorithms.

4 Systematic Review Process

Numerical reviews relating to the safety in UG mines have been carried out (Andrey somonov et al.). To our understanding, no such systematic literature review (SLR) has been conducted. The papers [2] studied the sensor data by measuring the analog voltage values by studying the close characteristics of gases. This paper focussed more on the technological progresses by monitoring the environment but this paper did not make a SLR and studied inadequately in an unplanned manner. In this paper, we present a SLR on the wide application areas of WSN in UG mines.

4.1 Review Result

The SLR was performed using four methods of identifying, selecting and verifying by making crucial studies of the abstract, the full text and keywords and finding their relevance to SLR (Table 1).

5 Research Challenges Faced in WSN

5.1 Energy

Energy performance of nodes is essential for sustainability of the network in UG mines. One of the important factors for WSN design is power. We incorporate energy competent algorithms to increase the longevity of network and outline the hardware that uses energy intelligently. The duty cycle is one of the important concepts for energy competence [3] when the nodes in a network are continuously in the function of receiving or sending information, they use minimum energy resource. For this

Table 1 Crucial study of systematic review of processes

Factor	Writer	Paper id	Problem addressed	Year	Approach	Restrictions	Verification
Mine atmosphere	Meng lei et al.	1	Environmental effects of coal mining	2009	Cumulative effects of underground coal mining	Data and assessment method	Yes
	Siva prasad dontala et al.	2	Environmental Mitigational measures of corporate coal mining	2015	Studying Mitigational measures	Lack of validation and performance analysis of proposed system	No
	Robert J. Timko et al.	3	Studying the atmospheric samples in UG mines	2006	Gas Sampling methods	Lack of implementation	Yes
Gas sensing	Andrey somonov et al.	4	Gas sensing using wireless sensor network	2012	Differential gas measurement approach	Accuracy of gas detected is not considered	No
	Qin Xianli et al.	5	Monitoring the coal mine gases	2011	Zig Bee module	Paper needs to focus more on sensitivity and accuracy of detection systems	Yes
	Ajay k singh and jaywardhan kumar	6	Utilization of methane emissions in coal mines for generating clean energy	2015	Using the extraction technologies of methane like CMM, VAM and CMM	Policy initiative needed to exploitation of gas	No
	Andrey somonov et al.	7	Methane leak detection using WSN	2012	Used WSN nodes (with embedded Catalytic sensors)	WSNs cannot be used for long duration due to high power consumption	No

(continued)

Table 1 (continued)

Factor	Writer	Paper id	Problem addressed	Year	Approach	Restrictions	Verification
Coal mine safety	MO LI Yunhao Liu	8	UG coal mine monitoring	2009	Structure aware self-adaptive sensor system	Lack of mechanisms to detect holes	Yes
	Sinan UGUZ	9	Monitoring of structural parameters using WSN	2016	Antenna design for horizontal and vertical communication using Zigbee approach	A lot to be specified of antenna design	no
	Valdo Heniques et al.	10	Ambient characteristics of mining environment	2016	Zigbee wireless module used for data communication	Xbee module for smart mesh network to be addressed	Yes

purpose, they stay passive within the duration when they do now not exchange data, namely they bounce to the sleep mode.

5.2 Communication and Routing

Communicating using WSN in UG mines finds research challenge; a more prominent challenge is designing a low price petite sensor node. Recent sensors that are developed are examples that are suitable to fulfill the objective. A more detailed study can envision into the development of low price sensors by peeping into the recent developments using MEMS.

5.3 Security

Among several research issues, security is a vital concern in UG wireless sensors. As in WSN, data travel wirelessly through the air and wireless signals are open to everyone; thus, anyone can monitor and participate as well, in communication. Mostly, nodes in WSN operate in ISB license-free band. Security becomes very important to prevent malicious attacks, unauthorized access and denial of service DoS attacks. More about security requirements is found in [4–6].

5.4 Efficient OS

Sensor nodes in coal mining contain very limited amount of memory and processing power, so an efficient and small size OS need to solve this issue. The operating system of sensor node must be capable of providing basic resource management, memory management and should be less complex as compared to traditional OS. TinyOS, Nano-Q plus and mantis operating system are specially designed for WSN but still need for improvement.

5.5 Hardware and Software Issues

For the reducing size of sensor nodes, limited amount of resource like memory, processing speed and energy is also an issue for WSN. Normally, sensor node has microcontroller, sensor, transceiver and power backup. Sensor collects or sense data and sends to microcontroller for processing. Microcontroller performs necessary operation and transmits via transceiver to sink node. Microcontroller is responsible for basic WSN functions and communication protocols. For storage data, flash memory is suggested to use in nodes because flash memory is cheap and fast. In order to save power microcontroller should work in three state sleeps, active and idle.

6 Scope of WSN in UG Mining

6.1 Cognitive Sensing

For automatic intelligent sensing of the environment, cognitive sensor networks are used in localized procedure by deploying a huge number of sensor nodes in a particular location. Managing these huge number of sensor nodes is a vital task and is a challenge for researchers for which researchers are moving for bio-inspired sensing and networking [7].

Two accepted an example that best describes cognitive sensing is quorum sensing and swarm intelligence: Swarm intelligence is urbanized using AI for studying the cooperative behavior of decentralized in self-organized systems. Example of quorum sensing is bio-inspired sensing and networking. Quorum sensing is the capability of coordination bacteria to communicate and behavior via signaling molecules.

6.2 Spectrum Management

As we realize that most applications require sensor nodes that utilize minimum power and its requirement is increasing day by day, we are able to envision a future wherein wi-fi devices, which includes wireless keyboards, smartphone headsets, condition monitoring sensors, etc. But the recognition of these common tools guides to extended interference and blockage within as well as among networks. Out of many approaches introspective radios and MACs that have manifold frequency looms to make use of numerous frequencies for parallel communication. A standard solution for WSNs is a self-adaptive spectrum management [SAS] is supplied by [7] which can be effortlessly incorporated together with the prevailing only frequency.

6.3 Underwater Acoustic Sensor System

Underwater acoustic sensors collect the hydrological information like temperature, pressure, PH of water, etc., and find wide applications supervising environment pollution and disaster intuition. A complete survey in underwater sensor networks is presented in [8]. Sensors deployed in UG mines are prone to water sprouts. So a need arises to communicate among sensor nodes and coordinate information reducing the delay and transmitting information to the base station. Underwater sensor networks face several challenges compared to terrestrial sensors. Distributed underwater clustering scheme (DUCS) is routing protocol that shrivels the hands-on routing message exchange and does no longer use flooding approach. DUCS additionally makes use of data aggregation to eliminate redundant facts.

6.4 Coordination in Heterogeneous Network

The principal obstacle of most of the sensor nodes is the power constraint in coordination with other networks. Sensor networks are very useful in the domain of health tracking, natural world habitat tracking, monitoring the forest fire, building controls, etc. Sensor nodes are deployed in UG mines to supervise the information of other sensor nodes, and must be communicable. This can be achieved by linking wireless sensor nodes with the prevailing infrastructure such as WWW, LAN or any privatized network [4].

6.5 Hole Problem

Sensor nodes deployed in UG mines may find research challenges related to holes, i.e. sometimes, it may happen that the nodes that need to transmit data because of node failures cannot transmit or it may not sense the correct data. Holes normally breach communication within the sensor nodes.

7 Conclusion

The intrinsic behavior of WSN makes them deployable in a multitude of conditions. Sensor nodes have the capacity to be ubiquitous, at homes and workplaces, forests, battleground, catastrophic regions, or even for underwater explorations. This research survey the application areas wherein WSNs have been deployed inclusive of site visitor's surveillance, military sensing, target monitoring, environment tracking, and healthcare tracking and underground coal mining. It additionally surveys that WSN to be broadly deployed in UG coal mines for secure and safe mining. WSNs may be deployed for several applications related to various fields such as undersea exploration using acoustic sensor devices, sensing based on cyber physical systems, real-time programs, introspective sensing, safety seclusion control and spectrum control, etc. These application domains are being researched substantially by numerous researchers throughout the enterprise and intellectual areas.

References

1. Tiwari, A., P. Ballal, and F.L. Lewis. 2007. Energy efficient wireless sensor network design and implementation for condition based maintenance. *ACM Transactions on Sensor Networks* 3 (1): 17.
2. Henriques, V., and R.Malekian. 2016.Mine safety system using wireless sensor network. *IEEE Access* 4:3511–3521.
3. Anastasi, G., M. Conti, M. Di Francesco, and A. Apassarella. 2009. Energy conservation in wireless sensor network. A survey. *Ad Hoc Networks* 7 (3): 537–568.
4. Arms, S.W., C.P. Townsend, and M.J. Hamel. 2001. Validation of remotely powered and interrogated sensing networks for composite cure monitoring. In *International Conference on Composite Engineering* 8(8):7–11.
5. Fernandes, LL. 2007. Introduction to Wireless Sensor Networks Report, University of Trento.
6. Zia, T. 2008. A security framework for wireless sensor networks. In *International Conference on Intelligent Sensors*.
7. Yang, Guang Zhong. 2008. Cognitive sensing for body sensor networks. sensors. In *IEEE* 26–29.
8. Akyildiz, I.F., D. Pompili, and T. Melodia. 2005. Underwater acoustic sensor networks research challenges. *Ad Hoc Networks* 3:257–279.

Automation of Reliability Warranty Report Using SAS Software for Data Analysis

Priya Urkude and Sangita Lade

Abstract For successive growth of modern organization, it is necessary to implement technique which can understand the complex analytical problem and segregate it for collaborative outcome. For this approach, process automation is the key for development. As such modern programming techniques help to develop new software to implement automation of business processes. Data analysis proves to be important mean for overall development of the business process by achieving streamline operations. To overcome the existing capability and improve the efficiency for the repetitive task, automations play a huge role. Hence, the outcome of this publication is to widen the area of automation with JMP tool focusing on statistical data analysis. Moving ahead with this analyzed data helps data analyst and reliability team to interpret data and translate it for firm decision making. Therefore, this publication discusses the application of JMP tools and their customization using scripting language and graph builder for effective communication. JMP Add-Ins provides complete workflow for building dashboards, data visualization, and statistical analysis.

Keywords Reliability analysis · JMP tool · Statistical analysis · Warranty report · Visualization

1 Introduction

JMP has a functionality for data preparation using techniques like scripting and recoding of data. Saving an analysis of task so that it can be used on the same or different data tables is an important step in creating an analysis workflow that is efficient and repeatable. The JMP scripting language (JSL) can be used along with the drag and drop tools available in JMP. Dashboard and Application Builder are to quickly build and deploy dashboard and application. Designing report with data filters, parameterized application can be created using JMP to analyze results, including interactive HTML. Multiple dashboards can be created by using JMP which

P. Urkude · S. Lade (✉)
Computer Science Engineering, Vishwakarma Institute of Technology, Pune, India
e-mail: sangita.lade@vit.edu

© Springer Nature Singapore Pte Ltd. 2020   413
D. Swain et al. (eds.), *Machine Learning and Information Processing*,
Advances in Intelligent Systems and Computing 1101,
https://doi.org/10.1007/978-981-15-1884-3_38

makes the analysis easier and faster. In manufacturing industries, product reliability strongly influences business process. Reliability tools in JMP help you to prevent the failures and improve warranty performance. By using reliability tools in JMP, outliers can easily have identified and actual model can be predicted. JMP helps to find important design faults, pinpoint defects in product design and processes. Reliability prediction is used to identify the system requirement perform analysis using appropriate statistical distribution.

Reliability testing is often used to make decisions in the form of warranty returns. Warranty database is created for financial reporting purposes, but more it provides reliable information about product design. When the product is released to market, reliability engineers provide the warrant cost of the product after few months. There is difference between prediction and product design models. These differences provide failures modes. Using JMP for early detection of reliability, issues are implemented in software and help the company financially. The reliability process, recent addition to SAS software, provides reliability tools and data analysis as well as recurrence data analysis. The reliability process provides all the failure data for uncensored data.

The purpose of the JMP application and Application Builder is to provide the following capabilities to do reliability analysis:

1. JMP is a lightweight process to organize objects into building blocks for application.
2. JMP helps to integrate with JSL debugger.
3. Reliability analysis is useful in prediction of the future, so that improvement action can be taken in advance to support end customers.
4. Identify and correct the causes of failures that do occur despite the efforts to prevent them.
5. Determine ways of coping with failures that do occur, if their causes have not been corrected.
6. Apply methods for estimating the likely reliability of new designs, and for analyzing reliability data.
7. Reliability analysis provides insights about performance of product across different geographic locations.

1.1   Related Work

For proposed work, we consider two methods to process work:

1. The usage of JMP tool to perform reliability data analysis
2. Statistical driver analysis on reliability data
3. How reliability forecasting helps to predict the future failures.

1. JMP tool to perform reliability data analysis

JMP is a scripting language. Scripting language is designed to automate tasks that usually involve interacting with an external program like SQL query, R or python program. Complex application program provides some sort of program that lets user to automate the repetitive tasks. Function-based language which is based on function filled with arguments to accomplish tasks.

Visual data analysis can be performed by using JMP which gives the standard visualization. Instead of using spreadsheet which is not enables to access information but the JMP makes the data table to access and link result statistics with graphical review. JMP allows to transfer the excel sheet in data shaping, preprocessing, and importing. In JMP, we can perform data mining, data modeling, and visualization functionality. By using JMP, creation of analytical applications can be performed which can be shared with user for interaction. Using JMP you can easily make the business decision through visualization. JMP functionality allows to import the packages of R, SAS, and excel. To uncover the hidden data and encapsulates the insight business solution, JMP provides wide functionality. JMP is open source which can be available for everyone who wants to perform analysis on dataset by using graphical representation and statistical result.

Functionality of JMP allows to visualize and performs statistical analysis of the data using different mathematical distribution. Data tables, bar charts, statistical report, and visualization help JMP user to execute analysis. Reliability engineers and data analysts are using JMP to uncover patterns and trend. JMP Functionality:

- Interactive visualization helps to explore datasets and dependence between data
- Find matching data patterns of different variations
- Encapsulate and enquire large amounts of data
- Build automated statistical view to forecast result
- Perform custom application build and report.

The JMP scripting language (JSL) helps to create the automatic process for business process. The large collection of statistical functions used for statistical analysis and can be augmented in SAS and R. Following application is provided by JMP application and Application Builder:

1. A lightweight structure to standardize objects into building blocks for application.
2. Summarize to remove the side effects.
3. Links to categorization mechanism.
4. Integration of.jsl script.

2. Statistical driver analysis on reliability data

Statistical is best practice to deal with real data. The purpose of this paper is to provide how to use JMP statistical software to automate the manual process and create a dashboard. Master file creation as input file for statistical driver analysis. Calculate estimated in-service date by using build month and fail month. Working with data tables is to customize the graphs and calculations at one click. Statistical results not provide analysis of variance (ANOVA), but on predicted failure rate. Consider whole factors for prediction. Understand the confounding factors not an assignable cause analysis tool but will steer the direction of the investigation, understanding the characteristics that are significantly influencing predicted failure rate.

Statistical models help to encapsulate trends and patterns, which helps to take the best decision for organization development. Developing models in JMP include a statistical platform to build data-driven decision. JMP provides a different way to fit the linear and non-linear models. JMP model provides platform to construct model terms. JMP includes advanced statistical techniques: Principal components, clustering, predictive modeling, screening, and multivariate methods. Exploring meaningful results from data so that missing values or outliers can be easily removed. JMP provides many multivariate methodologies with automated functionality.

- The statistical tool is used to dig into an identified issue, using fields that describe the engine to determine what characteristics are significantly influencing the failure rate.
- The tool is highly dependent on what data is available. Users should be aware that the most influential drivers may not be in their dataset.
- Highly dependent on what data is available.
- Utilizing SAS JMP and scripting.

3. Reliability forecasting helps to predict the future analysis

Reliability analysis identifies issues quicker, process will need to be streamlining to include forward-looking process. Reliability analysis used in prediction of the future. So that improvement action can be taken in advance to support end customers [2]. Process provides meaningful information about the product reliability before launching into market. It provides insights about performance of product across different location. Reliability analysis can be done in following way: Deep-dive analysis, monitoring, and finding out the emerging issues in reliability. While performing data analysis, the reliability engineers attempt to predict analysis of the product. To perform analysis in statistical tool:

1. Gather product data
2. Select a distribution that will fit for the product of the life
3. Evaluates the parameters that will fit the distribution of data
4. Implement analysis and generate plot
5. Display result that estimates the product life, such as failure and probability.

Warranty performance and failure prevention are two most essential reasons to implement advanced automated methods to predict the performance of new product. JMP helps to find out the defects in processes, it also helps to identify outliers. JMP distribution functionality is used to make reliability lifetime prediction of product or components used in engine manufacturing companies. Using life distribution analysis, we can be able to plot the different parametric distribution.

1.2   Contributions of This Work

In this paper, we demonstrate JMP scripting language—An easy way to regenerate reports in JMP to capture and reuse scripts that are automatically generated. Often the next step in automating a workflow is to run these scripts on specific datasets. The JMP scripting language (JSL) provides a set of functions which create interactive user interface. JMP enables user by providing different analytical functionality and linking other software packages, which allow user to work with data efficiently. JMP is having great functionality for interaction and visualization. Dependency and relationship can be easily handled in JMP. Quality and improvement of product can be driven through JMP statistical report. JMP focuses on how to make things better to understand which help to take business decisions and complete the work. Warranty report is used to reduce cost of product and meet the customer expectations. Forecasting result of reliability can be helpful to analyze the warranty cost to predict the future.

Quality of report can be improved through predicting build volume and forecasting time period of product. Explore the undefined patterns, which help to discover result from any angle of data and perform visualization from one set to other set of data. Graph is used to refine interactively to build a new graph. By using graph, we can emerge as you click and drag variables. This paper covers basic JSL syntax and structure and creating dialogs to capture file names and inputs using column selection boxes, and other elements such as check boxes, sliders, and text boxes.

2   Proposed Work

Proposed work is based on the life data analysis using some statistical driver's. By using some of the statistical tool, we can describe the failure mode in components. Reliability analysis used to develop product decisions through warranty report.

Warranty report is used to make financial decisions but helps to execute the reliability testing. Method to calculate the early faults in the product and component help to detect reliability problems and help to take business decisions to minimize the cost of production. As soon as reliability analysis is performed on product, statistical report helps to forecast the warranty cost of product. To identify the reliability data, most of the organizations are deploying modern statistical tool. Sensor and smart chips in a product can be easily installed to execute rate data product. By using reliability testing, we can get the patterns which help to figure out the problems. Statistical methods using JMP software play an important role in reliability analysis. To deploy new statistical methods is the need of engineers to meet customer satisfaction. Today, we are having the best and needed statistical tool for implementation (Fig. 1).

The proposed process incorporates information from distinct database. The process of mapping data is from different database into a consolidated one by using JMP functionality.

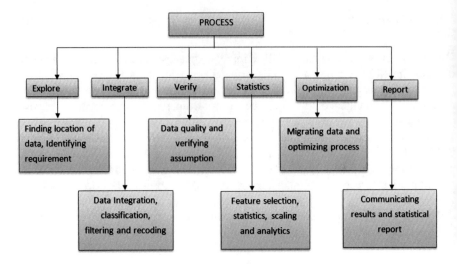

Fig. 1 Data flow diagram for data recoding and analysis

4 ways to automate the warranty workflow:

1. Download data
2. Merge and recode data
3. Analyze data and create graphs
4. Export graphs and generate summary reports.

1. **Download Data**

Finding relevant information to perform analysis, need to download data from multiple databases. Fetching data from different data sources is time consuming. Industries have lack of documentation and insufficient amount of data to process and difficult to handle large amount of data in tool like Excel. So, the JMP add-in created to download data from different data sources, as it takes less time to download data. JMP add-in helps to remove missing values from the data table and able to fetch the standardized dataset in JMP. Query builder provides interactive user interface to pull data from ODBC databases. Database allows organization to provide massive amount of data but databases are organized for efficient storage and transactions. Data analysis can be performed on different datasets. To make a significant work process, JMP has functionality to join multiple data tables. We can use SQL and other tools to join them (Fig. 2).

Steps for connection to database to fetch data from JMP Add-in:

Step 1: Open administrative tool
Step 2: Click on the ODBC data sources
Step 3: Create a DSN name for database connection

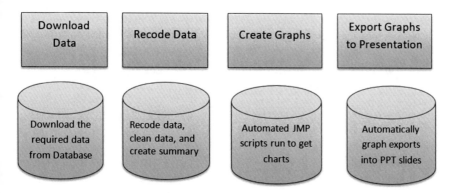

Fig. 2 Proposed system to perform the reliability data analysis on the dataset

Step 4: Click on finish and provide login information and server address
Step 5: Provide the same ODBC connection in JMP Add-In to fetch data directly from database without manual intervention.

2. Merge and Recode Data

Once a dataset is ready, an analyst can able to perform statistical analysis. Recoding and cleaning are performed on dataset. To perform this process on Excel is tedious work but JMP is having a functionality to perform the recoding of data in one click. Once the dataset is ready to use, analyst can be abled to perform downstream analysis. Most of the data are in unstructured format but the visualization will not be possible on these types of data. Therfore, data transformation is the foremost thing to do visualization. In unstructured dataset most of the claims for suppliers are missing, JMP helps to link the claims with suppliers. Another problem of data cleaning is the data which we get from database is heterogeneous. This occurs due to the human manual intervention in the log file while updating the data in the database. Time required to prepare data for data analysis is more, but by using JMP, we can easily prepare data, faster and reliable one. No matter how you process the data cleaning, JMP automates the process. Customer we required a visualization, where the user interface can be easily be created using JMP. Process to perform Recode: **Select column → COL → Recode**.

3. Analyze and Create Graph

JMP helps to explore the visualization of graphs interactively so that reliability engineers can easily perform analysis which will helps to understand the problems and take appropriate decision for improvement in product. Graphs provide better understanding of our datasets. Effective graph transformation is available in JMP tool. The purpose of analysis is not only to create interactive graphs and statistics, but to provide better understanding of dataset to take business decisions. Graphs display the dependancies between the variables used in datasets. It shows the distribution to predict, or graphically display significant effect. JMP help to communicate with analysis results through dashboards. Patterns and trends in datasets are not easily available in spreadsheet. Visualization tools help to discover statistical functionality effective which make discoveries. JMP graphs can be enhanced by using graph builder properties.

4. Export Graph and Generate Summary Report

To save the report as text, image, HTML, or PowerPoint. To create an automation system for work which is used to automatically update the graphs in JMP. When.jsl script runs, the latest graph will update automatically.

Communicate results: After completion of analysis, visualization is best way to communicate the results. Functionality of JMP gives you the tools to share the graphs. Following information is displayed by the graphs (Fig. 3):

- Product failure date.
- How the number of claims increased with prediction RPH increase as well during this time.

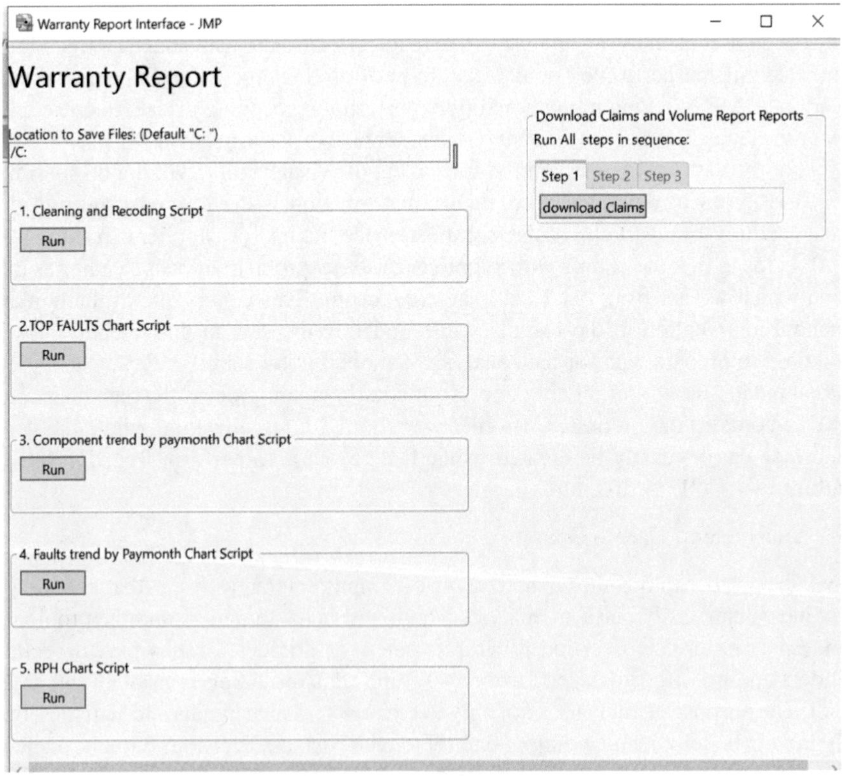

Fig. 3 Data dashboard in JMP to perform reliability analysis

Fig. 4 Final output of the dashboards to analyze top faults in the product

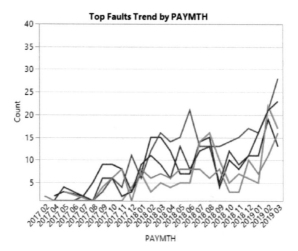

Interactive JMP Dashboard for real-time analysis:
Sometimes, we may need to compare real time with historical trends. In order to perform this, we have developed customized dashboard which allows user to run stored process.

JMP provides statistics, predictive modeling, and data mining. Using JMP, not only modeling is done quickly, JMP easily compares and contrasts models built using different techniques, and generates code in C, Python, JavaScript or SAS that can be deployed to new cases. JMP helps to perform regression on the data in better way than other techniques and easily compares with competing models. JMP reduces dimensionality and processes unstructured data which generate reports and can be easily consumed by other predictive modeling. If user wants to describe, predict or explain the problems, JMP provides intrinsic modeling between visualization and modeling. JMP provides a different functionality tool that makes easier for a reliability engineers from obtaining and manipulating data to visualizing clearly (Fig. 4).

3 Conclusion

JMP is expeditiously developing statistical and visualization tool for the reliability analyst. By using JMP scripting language, JMP offers a validated statistical analysis and integrated tool to extend the demand of the report. This paper has demonstrated how to monitor organizational business process, as business process is on-going and recurring work. JMP scripting language is used to automate the business process, which help to reduce analysis time. Through simple points and clicks, data analysts and reliability engineers can visualize the products relationships. JMP automation work will be used for statistical modeling, report building, and text processing.

Analysis through JMP provides interactive and non-static. JMP promote continuous improvement and best practices through custom interface. JMP deliver standardized warranty agreements for suppliers. Visualization of JMP report serves as a basis for competent discussion and business decision. JSL speeds up the work to get results and possible to correlate the data using script. JMP configures the work and provides consistent graph output. JMP helps to stabilize and optimize process without loss of time. Warranty report and reliability analysis reports are the two most important reasons to understand the new product and current product. By using JSL profiler, we can tune a scripts performance and iteratively optimize code for peak efficiency and speed.

Automatic exporting of graphs provides:

- Easy to organize the report
- Retains data interactivity
- Improve cycle times
- Standardized and transparent reporting
- Save countless hours of manual processing
- Improvements in efficiency, scalability, and best practices
- Statistical and graphical elements presented using JMP report
  Save scripts to regenerate analysis reports without writing any code.

References

1. Wu, S., and M. Xie. 2008. Warranty cost analysis for non-repairable services products. *International Journal of System.*
2. Thomas, MU. 2006. *Reliability and warranties: Method for product development and quality improvement.* Taylor and Francis Group.
3. Amar, R., J. Eagen, and J. Stasko. 2005. Low-level components of analytics activity in information visualization. *IEEE Information Visualization (InfoVis).*
4. Blischke, W.R., D.N.P. Murh. 1991. Product warranty management—I. A taxonomy for warranty policies. *European Journal of Operational Research.*
5. Hill, Eric. 2011. JMP 9 add-ins. Taking visualization SAS data to new heights. In *SAS Global Forum 2011 Conference.* Cary, NC: SAS Institute Inc.
6. Murthy, D.N.P. 2006. *Product warranty and reliability, annals of operation research,* 133–146.
7. Pecht, M.G. 2006. Establishing a relationship between warranty and reliability. *IEEE Publication.*
8. Ascher, H., and H. Feingold. 1984. *Repairable systems reliability.* New York: Marcel Dekker Inc.
9. Deshmukh, Amol, Jeff McDonald. Real-time market monitoring using SAS BI tools (1835–2014).
10. Isenberg, P., D. Fisher, M. Morris, K. Inkpen, and M. Czerwinski. 2010. An exploratory study of co-located collaborative visual analytics around a tabletop display. In Proceedings of the IEEE Visual Analytics Science and Technology (VAST), 179–186.

11. Nelson, W. 1988. Graphical analysis of system repair data. *Journal of Quality Technology* 20 (1): 24–35.
12. Cao, Jin Hua, Yan Hong Wu. 1988. Reliability analysis of a multistate repairable system with a replaceable repair facility. *Acta Mathematical Applications Sinica* (English Ser.) 4: 113–121.

A Critical Analysis of Brokering Services (Scheduling as Service)

Abdulrahman Mohammed Hussein Obaid, Santosh Kumar Pani and Prasant Kumar Pattnaik

Abstract Cloud of things integrates cloud computing and Internet of things. The cloud broker acts as an interface between the cloud service providers and service consumer. The objective of this paper is to provide an in-depth knowledge on various aspects of a cloud broker. It aims to focus on scheduling as a service provided as a brokering service. This paper will provide initial background knowledge to researchers on cloud brokering services and platform to explore the open research issues related to scheduling as a service in cloud of things.

Keywords Brokering · Cloud of things · Min-Min algorithm · Scheduling as service · Brokering services

1 Introduction

Nowadays, with rapid growth of connected devices across networks leads to the necessity of analyzing and storing huge volume of data. Cloud of things integrates both of the cloud computing and Internet of things [19]. In recent past, cloud of things is gaining popularity due to its ability to handle massive data that came from things (smart watches, smart phones, smart vehicles, laptops, home appliances, etc.) in an effective manner in the cloud. The cloud brokerage service is an intermediary between cloud service providers and service consumer to conduct the negotiation process, ensuring the good performance of both [21]. Cloud broker creates an interface to facilitate communication between the service consumer and service providers. A simple example of cloud brokerage (as an intermediation service) is price negotiation, which helps the cloud consumer to select the desired cloud service [19].

The paper is organized as following. The basic concepts and background of brokering services are presented in section two. Task scheduling as a brokering service

A. M. H. Obaid (✉) · S. K. Pani · P. K. Pattnaik
School of Computer Engineering, Kalinga Institute of Industrial Technology (KIIT-DU), Bhubaneswar, Odisha 751024, India
e-mail: Obaid.eng@gmail.com

© Springer Nature Singapore Pte Ltd. 2020
D. Swain et al. (eds.), *Machine Learning and Information Processing*,
Advances in Intelligent Systems and Computing 1101,
https://doi.org/10.1007/978-981-15-1884-3_39

is focused in section three. A case study of Min-Min scheduling algorithm is presented in section four with its performance evaluation using CloudSim simulation tool. Finally, section five concludes the paper.

2 Basic Concepts and Background

This section gives an overview of brokering concept and its major components. The necessity of brokering services, techniques, and types are presented to strengthen the foundation concepts related to brokering services.

2.1 Brokering

The term cloud brokering was coined at the beginning of the appearance of cloud computing. According to the National Institute of Standards and Technologies (NIST) [18], cloud broker is "An entity that manages the use performance, and delivery of cloud services, and negotiates relationships between Cloud Providers and Cloud Consumers." In more precise terms, a cloud broker is a third-party auditor between cloud service consumers and cloud service providers. It manages multiple services such as data management, budgeting management, provisioning needs, and work processes. The cloud service broker is not a cloud service provider but a mediator that works on typical brokerage process principles [13].

2.2 Components of Brokering

Cloud brokerage is a time and cost saving technology, that enables the management of all operations between cloud service consumers and cloud service providers. The principal components of cloud broker are presented in Fig. 1.

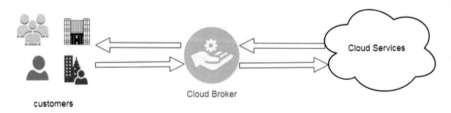

Fig. 1 Cloud broker

2.2.1 Service Consumer

The service consumer is the individual or organization who pays for the service. They use an interface to gain access and submit their requests to the broker. In addition, they are able to monitor and manage the services after deployment by use of management console [8].

2.2.2 Cloud Broker

The main objective of cloud broker is to negotiate relationships between cloud providers and cloud consumers. In addition, cloud broker delivers the cloud services and manages performance by offering attractive value-added services satisfying the users' service requirements. The components of cloud broker are illustrated in Fig. 2.

- Identity Manager: This component handles user authentication and admission while ensuring IDs and roles enforcements.
- SLA manager: Service level agreement (SLA) is an agreement between the service consumer and the service provider to ensure the maintenance of the minimum level of service [12]. SLA manager is responsible for the negotiation to create and handle the SLA provisioning.
- The Monitoring and Discovery Manager: This component of cloud broker is responsible for monitors the SLA metrics and queries resource information.
- Match Maker: It performs a matching process for selecting the best cloud providers using different matching algorithms to make a swap between the SLA characteristics and cost.
- Deployment Manager: This component is responsible for deploying the cloud service on the selected provider.
- Persistence: It is responsible for storing service status and resource information of broker specific data like resources data, monitoring, and SLA templates.

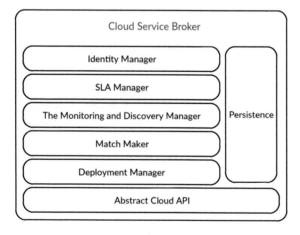

Fig. 2 Component of cloud broker

- Abstract Cloud API: This component is used to manage cloud resources and provide an API to the cloud user to interact with heterogeneous clouds and different cloud providers.

2.2.3 Cloud Provider

Cloud provider is a virtual resources pool accessed by service consumers to achieve specific tasks. Inter-cloud gateway is an interface hosted on the provider side to manage and respond to the request that came from users through the broker.

2.3 The Need for Cloud Brokering

The huge volume of data transferred in cloud of things requires access authorization, identity management, ease of storing in the right location, and many other processes that make the data safer and accessible by the owner in an easy way with good performance and low cost. Cloud broker is a business model that acts as a third party to help the customers to choose the right resources and track their activities with less time and minimal effort [1, 2]. Moreover, cloud broker facilitates the customer in choosing and contracting the services, and deploys the applications onto the cloud platform. It is responsible for event or incident reporting such as scaling request or renewal request. The cloud broker supervises the billing and pricing service, which is considered as a serious aspect from the point of view of customers and service providers. The cloud broker performs several operations in the cloud such as politics management, security, identity management, access management, scheduling data, and store data.

2.4 Cloud Brokering Techniques

Cloud brokerage is classified into multiple techniques such as pricing, multi-criteria, optimization, quality of services (QoS), and Trust.

- Pricing: The cloud of things services are based on *pay-as-you-use* principle. Several models are proposed to reduce the cost of paid services [7, 26]. One such model uses multiple brokers [7] where, mobile devices reserve services in two different strategies. In the first strategy, there is no cooperation between cloud brokers, and there is competition to reserve cloud resources between them. In the second strategy, cloud brokers cooperate to share resources in order to reduce the total cost. However, the cloud broker is committed to the price that customers will pay.

- Multi-criteria: It is a technique used for minimizing energy cost, solving multiple conflicting criteria problems, resource usage, and maximize user satisfaction.
- Optimization: It is a technique to find the most suitable services for the clients and providers. The broker sets conditions to minimize the cost, time, and energy with maximize trust and performance.
- Quality of services: Quality of service (QoS) is the measurement of the overall performance of a cloud broker. QoS helps manage service performance and gives a detailed description of quality of service.
- Trust: Resources are shared among hundreds of thousands of users [7]. Trust is required to build confidence, especially for sensitive information. The cloud service provider does all operations such as executing applications, and hosting. [7]. Cloud broker is a third party between the user and service provider which concerned with access control, availability, confidentiality, security, and integrity of user's data [22].

2.5   Challenges

- Trust-based Cloud Brokering: Trust is a major challenge in heterogeneous environment as the cloud resources are geographically distributed. The cloud brokering must be trusted by both cloud service consumer and cloud service provider for acceptance of cloud computing as a utility. Efficient techniques are required to ensure privacy and reliability in storing data and executing applications in cloud data centers [7].
- Service Allocation: Service allocation is a challenge as it has to fulfill requirements of cloud service consumers within constraints of cloud service providers [3].
- Service Provisioning: Service provisioning is an operation to reserving resources and utilizing as on-demand. Service provisioning is a challenge for cloud broker as resources are heterogeneous and distributed. Migrating a service from one cloud service provider to another is critical and needs to be achieved with good performance [7].
- Service Selection: Selection services are a challenging task as cloud service providers offer various services as per their interfaces and proprietary models [7].

2.6   Cloud Broker Services

The cloud broker provides the customer three categories of services such as service aggregation, service arbitrage, and service intermediation [10, 16].

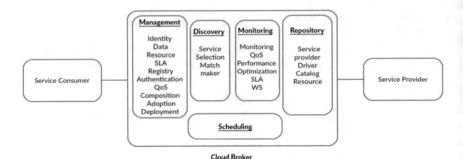

Fig. 3 Cloud broker services

- Service Aggregation: It is done by a cloud broker to create new services by integrating multiple services. The broker ensures that data traffic between the cloud consumer and the cloud provider is secure.
- Service Arbitrage: It is the broker's ability to choose services from multiple agencies. The difference between service arbitrage and service aggregation is the services that aggregated are not fixed.
- Service Intermediation: The cloud broker manages access to cloud services, enhanced security, performance reporting, identity management, and data transfer.

Moreover, the cloud broker offers many services that aid the service consumers and service providers to achieve good performance with high reliability. Figure 3 shows the common services that provided by a cloud broker [20].

The common services for most cloud brokers in cloud environments are:

- Management: The cloud broker aim to manage the functionalities and parameters such as identity, data, resource, SLA, registry, authentication, QoS, composition, adoption, and deployment.
- Monitoring: A service responsible for controlling and checking operations and functions in the cloud, in order to ensure good performance and quality of service (QoS), and to detect and alert possible errors (performance QoS optimization WS SLA).
- Discovery: This service by cloud brokers aims to target process of service selection and service ranking as per customer requirements. The broker discovers services from multiple providers to achieve good performance.
- Repository: This service helps to store all related information of services, cloud provider drivers, service providers, catalog and possible resources.
- Scheduling: Scheduler processes the data that move between the cloud provider and customer. The scheduler designed to help in automated and self-adoption processes. Provisioning, automated and self-adoption considered as main objectives of cloud brokering.

3 Scheduling as Brokering Service

Scheduling is one of the most important services provided by the broker, where the broker organizes traffic between users and cloud resources. It is a set of policies that dispatch workflow to properly node machines to improve system efficiency, reduce the total execution time and computing cost [23]. The scheduling and controlling the workflow in cloud of things environments by the broker is presented in Fig. 4.

3.1 Scheduling Operation

The cloud broker plays a responsible role for autonomous scheduling decisions while transferring data from users to cloud and vice versa. The aim of scheduling is to optimize techniques such as data integrity, performance, and execution time [14]. The role of a cloud broker in a cloud of things environments [24] is shown in Fig. 5.
   The following steps summarize the process of scheduling data in cloud broker.

- Step 1: Registration of resources in cloud information services (CIS).
- Step 2: User submits tasks to broker with complete specification.
- Step 3: The broker places all submitted tasks in a task set.
- Step 4: The broker queries resources from cloud information services (CIS).
- Step 5: Cloud information services (CIS) returns the attribute of resources such as the number of processing elements, allocation policy, and number of virtual machines.

Fig. 4 Scheduling as service

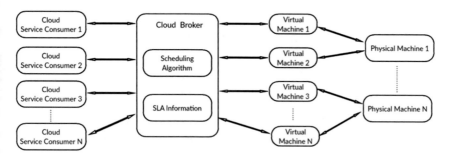

Fig. 5 Role scheduling

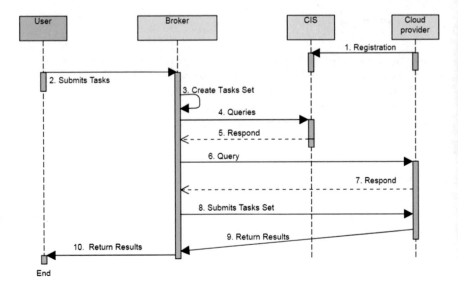

Fig. 6 Sequence diagram of the scheduling process

- Step 6: Broker ensures the registered resources are available by sending a query to resources.
- Step 7: Based on responses, the broker makes resources available to allocate process.
- Step 8: Broker submits tasks set to an available virtual machine by the aid of a specific scheduling algorithm such as first come first service scheduling algorithm (FCFS), Min-Min scheduling algorithm, round robin scheduling algorithm (RR), and Max-Min algorithm,.
- Step 9: Tasks executed and results returned to the broker.
- Step 10: Broker returns tasks execution status and results to the user.

The sequence diagram for the process of scheduling data in cloud broker is shown in Fig. 6.

3.2 Survey of Scheduling as Service in Cloud of Things

In 2019, Arunarani et al. [4] presented a comprehensive survey of task scheduling strategies suitable for cloud computing environments. They discussed on various issues related to scheduling methodologies. The survey was organized on three categories, that is, applications, methods, and parameter-based measures.

Chakravarthi and Vijayakumar [6] discussed scheduling algorithms in cloud environment with focus on Infrastructure as a Service (IaaS). The survey was conducted on the scheduling algorithms published between 2011 and 2016.

In 2018, Dewangan et al. [9] presented a survey on resource scheduling algorithms in cloud computing. This paper provides a comparative analysis in terms of different performance metrics for scheduling algorithms in cloud. They provided observation of some research gaps and discussed about various issues related to scheduling algorithms. They concluded that automatic resource scheduling algorithms may be achieved by opting self-characteristics methods such as configuring scheme, healing, protecting, and self optimization. The survey was based on a comparison of scheduling algorithms published between 2013 and 2016.

Varshney and Singh [27] described different types of resource scheduling algorithms that aims to optimize the QoS such as reliability, makespan, cost, and priority. The authors analyzed and compared the current works with their methodologies and objectives for each category, and they conducted a comprehensive survey on the scheduling algorithm published between 2009 and 2015.

In 2017, Kaur and Sidhu [15] presented a review on different approaches to cloud scheduling. They classified various scheduling algorithms based on objectives considered for optimization.

Ghomi et al. [11] presented the literature on load balancing algorithms and task scheduling. They proposed a new classification of scheduling algorithms. The aim of this survey was to identify the guidelines for future research and open issues. The survey was performed on the scheduling algorithms between 2012 and 2016.

In 2016, Rodriguez and Buyya [23] reviewed the challenges and studied the existing algorithms in cloud computing. The survey presented a taxonomy that focuses on features of clouds. The aim of this survey was to provide aid to researchers for exploring different types of algorithms in cloud. The survey was carried out on the scheduling algorithms between 2011 and 2015. Excellent surveys on scheduling algorithms are reported in [17, 25, 28].

4　Case Study (Min-Min Algorithm)

Cloud of things utilizes a suite of heterogeneous machines that are interconnected by high-speed networks. There are many sophisticated algorithms for mapping tasks to machines, including the Min-Min algorithm. The Min-Min algorithm is a simple algorithm and still is the basis of present cloud scheduling algorithm [29, 30].

4.1　Process Scheduling

The Min-Min algorithm executes by computing the minimum completion time for each task. The task with the overall minimum completion time is executed first and submitted to the corresponding machine for processing. The task with next minimum completion time is scheduled next for execution. This process repeats until all tasks mapped. The completion time $CT(j, r)$ a task is computed as:

Algorithm 1 The Min-Min Scheduling Algorithm
1: **for** all submitted tasks in the set; T_i **do**
2:       **for** all resources; R_j **do**
3:            $CT_{ij} = ET_{ij} + RT_{ij}$;
4:       **end for**
5: **end for**
6: **Do** while tasks set is not empty
7:       Find task T_k that cost minimum execution time.
8:       Assign T_k to the resource R_j which gives minimum expected complete time
9:       Remove T_k from the tasks set
10:      Update ready time RT_{ij}
11:      Update CT_{ij} for all T_i
12: End **Do**

Fig. 7 Min-Min scheduling algorithm

$$CT(j, r) = ET(j, r) + RT(j, r) \tag{1}$$

where ET(j, r) is the execution time of task $T(j, r)$, and RT(j, r) is ready time for resource j to take the job r. The Min-Min scheduling algorithm is shown in Fig. 7.

4.2 Performance Evaluations

The Min-Min algorithm performance was evaluated by using Cloudsim simulation tool. Cloudsim is a virtual tool that allows to evaluate the hypothesis before deployment in the real environment [5].

4.2.1 Simulation Setup

- Virtual machines (VM), OS—Xen for VMM, Bandwidth—10 MB/s, CPU—1000 MIPS, RAM—512 MB, Storage—1000 MB.
- Hosts, CPU—1000 MIPS, Bandwidth—1000 Kbits/s, Storage—1,000,000 MB, RAM—2048 MB.
- Cloudlets: All tasks generated dynamically and randomly. With twenty tasks (Task0, Task1, Task2, Task3, ... and Task19), each task has 64 KB lengths of instructions, 30 Mb input file size, 30 Mb output file size.
- Finally, 0.11 s of running time for each task.

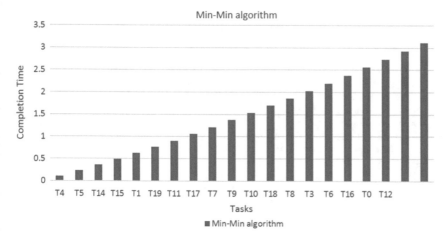

Fig. 8 Graphical representation to show of Min-Min algorithm

4.2.2 Performance Impact of Min-Min Algorithm

Assume that we have twenty tasks (Task0, Task1, Task2, Task3, ... and Task19) submitted by different users. The service broker receives tasks from customers with the "REQUEST" tag and schedules the tasks by applying the Min-Min algorithm. The task with the overall minimum completion time overall task is executed first (T4), than the task with next minimum completion time (T5) is executed and so on. The result of scheduling the twenty submitted tasks is showed in Fig. 8.

5 Conclusion

Cloud broker is considered as a major part of cloud environment. This paper focused on the concepts and requirements of cloud broker in cloud of things environment. It explores the challenges and open research issues associated with the cloud brokers focused on scheduling as a service by taking Min-Min scheduling as a case study. This paper will provide initial background knowledge to researchers on cloud brokering services and platform to explore the scheduling as a service in cloud of things.

References

1. Aazam, M., and E.N. Huh. 2017. Cloud broker service-oriented resource management model. *Transactions on Emerging Telecommunications Technologies* 28 (2): e2937.
2. Alkhanak, E.N., S.P. Lee, and S.U.R. Khan. 2015. Cost-aware challenges for workflow scheduling approaches in cloud computing environments: taxonomy and opportunities. *Future*

Generation Computer Systems 50: 3–21.

3. Anuradha, V., and D. Sumathi. 2014. A survey on resource allocation strategies in cloud computing. In *International Conference on Information Communication and Embedded Systems (ICICES2014)*. IEEE.

4. Arunarani, A., D. Manjula, and V. Sugumaran. 2019. Task scheduling techniques in cloud computing: a literature survey. *Future Generation Computer Systems* 91: 407–415.

5. Calheiros, R.N., et al. 2011. CloudSim: a toolkit for modeling and simulation of cloud computing environments and evaluation of resource provisioning algorithms. *Software: Practice and Experience* 41 (1): 23–50.

6. Chakravarthi, K.K., and V. Vijayakumar. 2018. Workflow scheduling techniques and algorithms in IaaS cloud: a survey. *International Journal of Electrical and Computer Engineering* 8 (2): 853.

7. Chauhan, S.S., et al. 2018. Brokering in interconnected cloud computing environments: a survey. *Journal of Parallel and Distributed Computing*.

8. Chen, H., et al. 2016. A cloud service broker based on dynamic game theory for bilateral SLA negotiation in cloud environment. *International Journal of Grid and Distributed Computing* 9 (9): 251–268.

9. Dewangan, B.K., et al. 2018. Resource scheduling in cloud: a comparative study. *International Journal of Computer Sciences and Engineering* 6 (8): 168–173.

10. Fowley, F., et al. 2016. A classification and comparison framework for cloud service brokerage architectures. *IEEE Transactions on Cloud Computing* 6 (2): 358–371.

11. Ghomi, E.J., A.M. Rahmani, and N.N. Qader. 2017. Load-balancing algorithms in cloud computing: a survey. *Journal of Network and Computer Applications* 88: 50–71.

12. Hani, A.F.M., I.V. Paputungan, and M.F. Hassan. 2015. Renegotiation in service level agreement management for a cloud-based system. *ACM Computing Surveys (CSUR)* 47 (3): 51.

13. Hlavac, J. 2017. Brokers, dual-role mediators and professional interpreters: a discourse-based examination of mediated speech and the roles that linguistic mediators enact. *The Translator* 23 (2): 197–216.

14. Hongyan, T., et al. 2016. Evaluating performance of rescheduling strategies in cloud system. In *2016 IEEE Trustcom/BigDataSE/ISPA*. IEEE.

15. Kaur, J., and B.K. Sidhu. 2017. Task scheduling in cloud computing using various techniques. *International Journal of Advanced Research in Computer Science* 8 (5).

16. Khurana, R., and R.K. Bawa. 2017. Quality based cloud service broker for optimal cloud service provider selection. *International Journal of Applied Engineering Research* 12 (18): 7962–7975.

17. Masdari, M., et al. 2016. Towards workflow scheduling in cloud computing: a comprehensive analysis. *Journal of Network and Computer Applications* 66: 64–82.

18. Mell, P., and T. Grance. 2010. National Institute of Standards and Technology (NIST). Definition of Cloud computing. *Communications of the ACM* 53 (6): 50.

19. Mohammed, A., et al. 2018. Towards on cloud of things: survey, challenges, open research issues, and tools. In *2018 Second International Conference on Inventive Communication and Computational Technologies (ICICCT)*. IEEE.

20. Mostajeran, E., et al. 2015. A survey on SLA-based brokering for inter-cloud computing. In *2015 Second International Conference on Computing Technology and Information Management (ICCTIM)*. IEEE.

21. Obaid, A.M.H., S.K. Pani, and P.K. Pattnaik. 2018. A priority based on Min-Min algorithm for reducing make span task scheduling in cloud computing. *International Journal of Engineering and Technology* 7 (4): 5.

22. Rizvi, S., et al. 2018. A security evaluation framework for cloud security auditing. *The Journal of Supercomputing* 74 (11): 5774–5796.

23. Rodriguez, M.A., and R. Buyya. 2017. A taxonomy and survey on scheduling algorithms for scientific workflows in IaaS cloud computing environments. *Concurrency and Computation: Practice and Experience* 29 (8): e4041.

24. Singh, L., and S. Singh. 2013. A survey of workflow scheduling algorithms and research issues. *International Journal of Computer Applications* 74 (15).
25. Smanchat, S., and K. Viriyapant. 2015. Taxonomies of workflow scheduling problem and techniques in the cloud. *Future Generation Computer Systems* 52: 1–12.
26. Tang, S., et al. 2018. Long-term multi-resource fairness for pay-as-you use computing systems. *IEEE Transactions on Parallel and Distributed Systems* 29 (5): 1147–1160.
27. Varshney, S., and S. Singh. 2018. A survey on resource scheduling algorithms in cloud computing. *International Journal of Applied Engineering Research* 13 (9): 6839–6845.
28. Wadhonkar, A., and D. Theng. 2016. A survey on different scheduling algorithms in cloud computing. In *2016 2nd International Conference on Advances in Electrical, Electronics, Information, Communication and Bio-Informatics (AEEICB)*. IEEE.
29. Yu, X., and X. Yu. 2009. A new grid computation-based Min-Min algorithm. In *2009 Sixth International Conference on Fuzzy Systems and Knowledge Discovery*. IEEE.
30. Zhang, Q., and Z. Li. 2009. Design of grid resource management system based on divided Min-Min scheduling algorithm. In *2009 First International Workshop on Education Technology and Computer Science*. IEEE.

Application of Neural Networks in Model Updation Using Frequency Response and Radial Basis Function Network

Tumbanath Samantara

Abstract The main objective of this paper is to construct a function, out of the given data which are the functional values defined at some discrete points. Many existing traditional methods like numerical as well as statistical methods are available and give good approximate result up to some extent. But if the data set is very large then the computational complexity rises. Again in case of dynamic data, the working procedure starts from the beginning at every time and a new function is constructed. In this paper, a model is developed by using neural network method, a non-traditional algorithm, which overcomes above difficulties and give results more close towards accurate. Also when the input data changes, the problem need not to be solved afresh starting from beginning but can be achieved good approximation solution by updating the parameter generated from previous data set.

Keywords Neural network · Frequency response function · Radial basis function network · Model parameter

1 Introduction

The model updation can be defined as modifying the mathematical model so that the relation between input and output data has very good agreement. Many engineering and scientific problems can be modelled and the output obtained very close to accurate one by using neural network. The common objective of engineering or scientific problem is formation of a function on the basis of knowledge obtained from some examples of input–output pairs. The neural network community calls this process as supervised learning [1].

This paper presents a technique that enables updating the model of weakly non-linear in nature. Model updating of engineering problems has been an important topic of research during the last 25 years [2]. Whenever we are going to find a good

T. Samantara (✉)
Centurion University of Technology and Management, Bhubaneswar, Odisha, India
e-mail: tnsamantara@gmail.com

© Springer Nature Singapore Pte Ltd. 2020

439

D. Swain et al. (eds.), *Machine Learning and Information Processing*,
Advances in Intelligent Systems and Computing 1101,
https://doi.org/10.1007/978-981-15-1884-3_40

approximation model to the existing accurate model, we should not minimize the difference between the predicated output and the actual output only rather the difference between existed and updated parameters also.

In most of the optimization direct problems for getting optimal solution, the problem has been solved many times. Once the problem is solved for a basic feasible point in the search space and if that point does not results the optimal solution, then the solution along with the point is discarded and then approached to another point in search space. Next time, when we again solve the problem, the previous values are not used. But in case of our proposed method, i.e. neural network method, a mapping is created between frequency response functions and model parameters by using the results of discarded searching solutions. If the parameter of the model changes in future, the earlier created mapping can be applied to update this parameter without solving the problem from beginning. Now by using these updated parameters, a model can be created which is consistent with experimental data.

Out of different type of neural network organization, radial basis function networks are used here for model updating problems.

2 Problem Statement

In this paper, we analysed the relation between load and frequency in a structure and constructed a map by using non-traditional algorithm like neural network. The map consists of parameters that can change with time. An adaptive model [3] has been developed such that for any change of model parameters [4] in future, the map constructed can be applied to update these parameters without again solving the direct problem from beginning. By using these updated parameters, an updated model [5] can be formulated which is very much consistent with the experimental data given earlier.

3 Method of Solution

3.1 Radial Basis Function

$$X = \begin{bmatrix} H_1(\omega_1) & H_2(\omega_1) & \cdots & H_N(\omega_1) \\ H_1(\omega_2) & H_2(\omega_2) & \cdots & H_N(\omega_2) \\ H_1(\omega_3) & H_2(\omega_3) & \cdots & H_N(\omega_3) \end{bmatrix}$$
$$Y = \begin{bmatrix} k_1 & k_2 & \cdots & k_N \\ c_1 & c_2 & \cdots & c_N \end{bmatrix} \tag{1}$$

Here, $H(\omega_i)$ is a frequency response function calculated at ω_i.

The number of rows of X and Y represents the number of neurons in the input layer (NI) and output layer (NO). The number of columns of X and Y represents the number N_t of sample pairs used to train the network.

In this paper, the centres of RBFNN [6] (matrix C) are chosen to be the input vectors (matrix X) used to train the network.

First, we train the network and then compare the output with the exact solution of the problem. Solving N times, a set of model parameters generated randomly.

The output corresponding to these N set of model parameters are input to the network and the corresponding results are compared with the model parameters used to generate the input data. Then, the network is measured. If performance is low, then go for more training.

The network output Y is computed as,

$$Y = WH \tag{2}$$

where W = weight matrix representing the connection of hidden layer and output layer.

Solving we get

$$W = YH' \tag{3}$$

The generalization characteristics of network can be measured in different ways. Some of them used here are

(a) Mean absolute error:

$$e_p^{\text{avg}} = \sum_{k-1}^{N_t} \left| \frac{y_{p,k}^a - y_{p,k}^d}{y_{p,k}^d} \right|, \quad k = 1, 2, \ldots, N \tag{4}$$

Here e_p^{avg}: mean absolute error for estimating pth parameter, y^a: actual output of the network, y^d: desired output. This formula shows, on average, how much deviations may occur in a network during computing of model parameter.

(b) Maximum error:

$$e_p^{\text{max}} = \max_k \left| \frac{y_{p,k}^a - y_{p,k}^d}{y_{p,k}^d} \right|, \quad k = 1, 2, \ldots, N \tag{5}$$

(c) Sum-squared-error:

$$\varepsilon_p = \frac{1}{2N} \sum_{k=1}^{N_t} \left(y_{p,k}^d - y_{p,k}^a\right)^2 \tag{6}$$

(d) Euclidean norm

$$d = \left(\sum_{m=1}^{NI} (x_m - c_m)^2\right)^{1/2} \tag{7}$$

If x and c are, for example,

$$x = \begin{Bmatrix} x_1 \pm \sigma_1 \\ x_2 \pm \sigma_2 \end{Bmatrix} \text{ and } c = \begin{Bmatrix} c_1 \\ c_2 \end{Bmatrix} \tag{8}$$

Here x_1 and x_2 are the mean of the first and second positions of x and σ_1 and σ_2 are the expected variations of those positions,
   Then distance

$$d = \left((x_1 - c_1 \pm \sigma_1)^2 + (x_2 - c_2 \pm \sigma_2)^2\right)^{1/2} \tag{9}$$

4   Normalization and choice of the spread constants

During formulation of a problem, we consider different type of variables as per our requirements. All the variables do not have data of same magnitude variance. Also do not have same range. One category variable have very magnitude data, whereas other have very small magnitude. During calculation the high magnitude data has domination over small magnitude data. But every variable has importance in formulation of a model. To overcome this situation, we normalize every data and transform such that every entry should be in the interval {0,1}. The linear transformation is given by

$$x_m^{new} = \frac{x_m^{old} - \min\left(x_m^{old}\right)}{\max\left(x_m^{old}\right) - \min\left(x_m^{old}\right)} \tag{10}$$

where x_m is a row of the input matrix.

The normalization only could not overcome all short of problems. Scaling is required to get better results in distance measure of two vectors. This is accomplished by using the spread constants vector σ that is associated with each row vector of the input matrix.

It was found that for the model updating problem an appropriate way to choose the vector σ is to make it dependent on the standard deviation [7] of the rows of the normalized set of training vectors. The vector σ is then calculated by

$$\sigma_m = \gamma \ \text{std}(x_m) \tag{11}$$

where std($\cdot$) is the standard deviation and is a scaling constant multiplying it, being the value $\gamma = 1$ a very good initial guess.

When training a neural network, one is always interested in minimizing the generalization error, i.e. the resulting output error when an input vector not used to train the network is presented to the network. Let us assume that N input vectors are presented to the network. Each input has an output error vector associated with it, defined as

$$e_i = y_i^d - y_i^a \tag{12}$$

where y_i^d is the vector of desired outputs and y_i^a is the actual output vector when the vector x_i is presented to the network. It is desirable to minimize the squared sum of all error vectors. Minimizing this scalar is synonymous to improving the generalization characteristics of the network. Therefore, it is desirable to find the constant such that the sum-squared-error

$$\varepsilon = \frac{1}{2N} \sum_{i=1}^{N} \sum_{q=1}^{NO} e_q^2(i) \tag{13}$$

is minimized. Any minimization technique can be used to find the optimal value of γ. Here, a gradient descent formulation is presented. The gradient of ε with respect to γ is given by

$$\frac{\partial \varepsilon}{\partial \gamma} = \sum \frac{\partial \varepsilon}{\partial e_j(i)} \frac{\partial e_j(i)}{\partial y_k^a(i)} \frac{\partial y_k^a(i)}{\partial \gamma} \tag{14}$$

where $y_k^a = \sum_m w_{k,m} h_m^a$ and $h_m^a = e^{-\gamma d_m}$.

Analysing term by term, Eq. (2.14) yields

$$\frac{\partial \varepsilon}{\partial e_j(i)} = \frac{1}{N} e_j(i) \tag{15}$$

$$\frac{\partial e_j(i)}{\partial y_k^a(i)} = -\delta_{jk} \tag{16}$$

$$\frac{\partial y_k^a(i)}{\partial \gamma} = \sum_{p=1}^{N_t} \frac{\partial w_{k,p}}{\partial \gamma} h_p^a(i) + \sum_{p=1}^{N_t} w_{k,p} \frac{\partial h_p^a(i)}{\partial \gamma} \tag{17}$$

The first term of the above expression can be calculated by using Eq. (2.3)

$$\frac{\partial W}{\partial \gamma} = \frac{\partial \gamma H^{-1}}{\partial \gamma} = Y \frac{\partial H^{-1}}{\partial \gamma} = -Y H^{-1} \frac{\partial H}{\partial \gamma} H^{-1} = -W \frac{\partial H}{\partial \gamma} H^{-1} \tag{18}$$

where H is the matrix output of the hidden layer when presented with the training vectors. Writing the expression above in summation form and calculating the derivative of H with respect to γ yields

$$\frac{\partial \sum_{p=1}^{N_t} w_{k,p}}{\partial \gamma} = \sum_{p=1}^{N_t} \sum_{n=1}^{N} w_{k,p}(d_{p,n} h_{p,n}) h_{n,l}^l, \qquad \begin{array}{l} k = 1, 2, \ldots, \text{NO} \\ l = 1, 2, \ldots, N_t \end{array} \tag{19}$$

where $h_{n,l}^l$ is the (n, k) element of the matrix H^{-1}. The second term of Eq. (17) is given by

$$\frac{\partial h_k^a(i)}{\partial \gamma} = -\left(d_k^a(i) h_k^a(i)\right) \tag{20}$$

Collecting the results from Eqs. (19) and (20), Eq. (17) can be written as:

$$\frac{\partial y_k^a(i)}{\partial \gamma} = \sum_{p=1}^{N_t} w_{k,p} \left(\sum_{n=1}^{N_t} \sum_{l=1}^{N} (d_{p,n} h_{p,n}) h_{n,l}^l h_k^a(i) - \left(d_p^a(i) h_p^a(i)\right) \right) \tag{21}$$

Or, in matrix form:

$$\frac{\partial y^a(i)}{\partial \gamma} = W\left(\bar{H} H^{-1} h^a(i) - \bar{h}^a(i)\right) \tag{22}$$

where $H_{p,n} = d_{p,n} h_{p,n}$, and $\bar{h}_p^a(i) = d_p^a(i) h_p^a(i)$. Now, collecting the results obtained from Eqs. (15), (16) and (21), one can write $\frac{\partial \varepsilon}{\partial \gamma}$ as

$$\frac{\partial \varepsilon}{\partial \gamma} = -\frac{1}{N} \sum_{i=1}^{N} e_j(i) \sum_{p=1}^{N_t} w_{k,p} \left(\sum_{n=1}^{N_t} \sum_{l=1}^{N} (d_{p,n} h_{p,n}) h_{n,l}^l h_k^a(i) - \left(d_p^a(i) h_p^a(i)\right) \right) \tag{23}$$

Or, in matrix form as

$$\frac{\partial \varepsilon}{\partial \gamma} = -\frac{1}{N} \sum_{i=1}^{N} e_j^T(i)\left(W\left(\bar{H} H^{-1} h^a(i) - \bar{h}^a(i)\right)\right) \tag{24}$$

The change in the γ parameter should be in the direction opposite to the gradient $\frac{\partial \varepsilon}{\partial \gamma}$, therefore, the iteration scheme for minimizing ε is

$$\gamma^{k+1} = \gamma^k + \eta \frac{\partial \varepsilon}{\partial \gamma} \tag{25}$$

where η is the step size, has value between 10^{-1} and 10^{-2}. The estimation problem has to be solved at each iteration. This can be efficiently done by recognizing that

$$e^{-\gamma d} = \left(e^{-d}\right)^{\gamma} \tag{26}$$

Therefore, it is not necessary to measure the distance matrix at each iteration. One need only to perform the following operation:

$$d\left(\gamma^{k+1}\right) = (d(\gamma = 1))^{\gamma^{k+1}} \tag{27}$$

Equations (24), (25) and (27) represent the necessary steps to efficiently adjust the constant γ in a gradient descent [8] fashion in order to minimize the sum-squared-error of the network over the validation set.

5 Sensitivity of FRFs with Respect to Parameter Changes

Frequency response functions (FRFs) contain a large amount of redundant information since there are many more points in a FRF than parameters in the model. It is, therefore, necessary to choose which points of the FRF should be used in updating the model. Intuitively, one expects that these points should be the ones that are most sensitive to changes in the model parameters. To determine which points are most sensitive [9], one should first write the expression for the frequency response matrix. In state-space notation, the system is described by a set of first-order differential equations

$$\begin{aligned} \dot{x} &= Ax(t) + Bu(t) \\ y(t) &= Cx(t) + Du(t) \end{aligned} \tag{28}$$

where A is the system's state-matrix, B is the matrix of inputs, C is the matrix of measurements, D is the feed-through matrix, $x(t)$ is the state vector, $u(t)$ is the vector of control forces, and $y(t)$ is the vector of measurements. Laplace transform [10] of the above equation yields a set of algebraic equations in the Laplace variable s. Assuming zero initial conditions, Eq. (28) becomes

$$\begin{aligned} sX(s) &= AX(s) + BU(s) \\ Y(s) &= CX(s) + DU(s) \end{aligned} \tag{29}$$

The first equation can be solved for $X(s)$ yielding

$$X(s) = (sI - A)^{-1}BU(s) \tag{30}$$

This expression for $X(s)$ can be substituted into Eq. (29) to yields

$$Y(s) = [C(sI - A)^{-1}B + D]U(s) \tag{31}$$

Or, in condensed form,

$$Y(s) = H(s)U(s) \tag{32}$$

Assuming the force vector is periodic, with frequency ω, the equation above can be rewritten as

$$Y(\omega) = H(\omega)U(\omega), \Rightarrow Y_i(\omega) = \sum_{j=1}^{N_t} H_{i,j}(\omega)U_j(\omega) \tag{33}$$

where $H(\omega)$ is the matrix of frequency response functions.

The sensitivity of $H(\omega)$ with respect to a variation of the generic parameter α is determined next. The derivative of $H(\omega)$ with respect to α is

$$\frac{\partial H(\omega)}{\partial \alpha} = \frac{\partial [C(j\omega I - A)^{-1}B + D]}{\partial \alpha} = C\frac{\partial (j\omega I - A)^{-1}}{\partial \alpha}B + C(j\omega I - A)^{-1}\frac{\partial B}{\partial \alpha} \tag{34}$$

Making the reasonable assumption that C and D are not functions of the model parameter α. The derivative of the inverse of a matrix [11] can be easily calculated using Eq. (2.45), yielding

$$\frac{\partial H(\omega)}{\partial \alpha} = -C(j\omega I - A)^{-1}\frac{\partial (j\omega I - A)}{\partial \alpha}(j\omega I - A)^{-1}B + C(j\omega I - A)^{-1}\frac{\partial B}{\partial \alpha}$$

$$= C(j\omega I - A)^{-1}\left(\frac{\partial A}{\partial \alpha}(j\omega I - A)^{-1}B + \frac{\partial B}{\partial \alpha}\right) \tag{35}$$

Equation (35) is the sensitivity of the FRF due to model parameter changes and now the issue of which points to select can be analysed. For values of ω close to a natural frequency of the system, the determinant of the matrix $(j\omega I - A)$ becomes a small number, since this expression is very similar to the eigen value problem. In fact, it becomes zero for undamped systems at the natural frequency. In the general damped case, this small number makes the sensitivity $\frac{\partial H(\omega)}{\partial \alpha}$ a large number. Therefore, the general rule for selecting points of the frequency response function to be used in the updating procedure is that the chosen points should be as close as possible to the natural frequencies. In general, however, experimental frequency response functions are not very accurate at the natural frequencies. This occurs for many reasons, among them, for lightly damped systems, the response become too large and a linear model is no longer valid. Because of this, the points selected for updating should be nearby the natural frequencies so a high sensitivity is obtained, but not exactly at the natural frequencies, so inexact values are not used in the updating.

An extra guideline is that one should use the smallest number of frequency points necessary to solve the problem. Extra points do not contain any new information, increase the computational load, and can cause numerical instabilities.

6 Conclusion

In this paper, the objective and the procedure followed are discussed in detail. By use of neural network technique, a frequency domain data set can be converted to a set of parameters which model the system accurately. Once the knowledge of conversion is gained that help us in estimating the parameter being updated and the technique become more vival for updating systems that change over time frequently.

References

1. Satish, Kumar. 2007. *Neural networks*. New Delhi: TataMcGraw-Hill Publishing Company Limited.
2. Friswell, M.I., and J.E. Mottershead. 1995. Finite element model updating in structural dynamics. Kluwer Academic Publishers https://doi.org/10.1007/978-94-015-8508-8.
3. Rao, Vittal, Rajendra Damie, Chris Tebbe et al. 1994. The adaptive control of smart structures using neural networks. *Smart Materials and Structures* 3:354–366.
4. Szewczyk, Z.P., and Prabhat Hajela. 1993. Neural network based selection of dynamic system parameters. *Transactions of the Canadian Society of Mechanical Engineers* 17 (4A): 567–584.
5. Natke, H.G. 1988. Updating computational models in the frequency domain based on measured data: a survey. *Probabilistic Engineering Mechanics* 3 (1): 28–35.
6. Park, J., and I.W. Sandberg. 1991. Universal approximation using radial-basis-function networks. *Neural Computation* 3 (2): 246–257.
7. Gupta, S.C., and V.K. Kapoor. 2005. *Fundamentals of Mathematical Statistics*.
8. Smith, S.W., and C.A. Beattie. 1991. Secant-method adjustment to structural models. *AIAA Journal* 29 (1): 119–126.
9. Sastry, Shankar, and Marc Bodson. 1989. *Adaptive Control—Stability, Convergence, and Robustness*, Prentice Hall. https://doi.org/10.1121/1.399905.
10. Nayfeh, A.H. 1998. *Introduction to Perturbation Techniques*, Wiley.
11. Simonian, S.S. 1981. Inverse problems in structural dynamics ii-applications. *International Journal of Numerical Methods in Engineering* 17 (3): 367–386.

A Review on SLA-Based Resource Provisioning in Cloud

Prashant Pranav, Naela Rizvi, Naghma Khatoon and Sharmistha Roy

Abstract Traditional IT infrastructures are gradually becoming obsolete and an alternative way to store, manipulate, and retrieve data. Namely, cloud computing is gaining momentum to replace the traditional computing environment. Sharing of resources over a distributed network is the main motive of cloud computing in order to provide consistency and reliability of the shared resources while keeping a check on the monetary factor involved. The resources available in cloud can not only be shared by multiple users, but are also be facilitated to reallocate with every demand. So, there has been always a focus on best techniques to provision the available resources in the cloud. Cloud resource provisioning mechanisms must follow some service-level agreements (SLAs) in order to abide by customers demand properly. This paper focuses on various research works undertaken on cloud computing resource provisioning techniques by taking SLA into account.

Keywords Cloud computing · Virtual machines (VM) · SLA · Resource provisioning

1 Introduction

As a new computing technique, cloud computing gained momentum in late 2007. Delivery of resources over the Internet is the main methodology behind the cloud environment. Computational services can be accessed by the users whenever they

P. Pranav · N. Khatoon · S. Roy (✉)
Faculty of Computing and Information Technology, Usha Martin University, Ranchi, India
e-mail: sharmistharoy11@gmail.com

P. Pranav
e-mail: prashantpranav19@gmail.com

N. Khatoon
e-mail: naghma.bit@gmail.com

N. Rizvi
Department of Computer Science and Engineering, Indian Institute of Technology, Dhanbad, India
e-mail: naelarizvi92@gmail.com

© Springer Nature Singapore Pte Ltd. 2020
D. Swain et al. (eds.), *Machine Learning and Information Processing*,
Advances in Intelligent Systems and Computing 1101,
https://doi.org/10.1007/978-981-15-1884-3_41

449

actually need them. This can be realized because of the utility-based nature of cloud. Users use the resources or computing infrastructure of the cloud and pay only for those things which they use and not for the whole infrastructure. The resources of the cloud are distributed across the globe so that individual or companies can access and utilize the resources and services from anywhere. Due to many exciting features such as reduced computational cost, flexibility, and very high degree of reliability, cloud has become one of the technologies to look for.

The resource requests from the users are handled by cloud service providers by creating and deploying enough number of virtual machines (VMs) where the requests are actually tackled. This allocation of resources by cloud service providers is achieved by utilizing some resource allocation or provisioning technique. Two main provisioning techniques used in cloud environment are static provisioning and dynamic provisioning. Some parameters such as cost minimization, maximization of resources, and response time for each request are also to be considered while allocating resources. Resources are to be allocated in a way to follow the service level agreements (SLAs) as prescribed by the cloud service provider to its users. SLAs typically are defined in terms of mean time to failure (MTTF) and mean time to recover (MTTR). The MTTR metric measures the availability of the system, and MTTF metric computes the reliability of the system. These two together are used to establish a contract between the service provider and the user.

Cloud computing provides many services such as infrastructure to be used by the users, platform to develop and deploy software, and software created by providers to be used by the users. Based on these different services, different service models are available in a cloud environment which is discussed below:

Infrastructure-as-a-Service (IaaS):

- In an IaaS environment, cloud resources such as storage, computing power, bandwidth, and databases are provided on demand to the users.
- Many users can work simultaneously on a single hardware. Examples include Amazon Web Service (AWS) and GoGrid.

Platform-as-a-Service (PaaS):

- In a PaaS environment, cloud service providers provide platforms for users to build and deploy different Web-based applications.
- The constraint of software download and installation for developers is removed while providing facilities which are required during a complete life cycle of building and developing applications.
- Different UI scenarios are created, modified, tested, and deployed using Web-based tools which also handles billing and subscriptions.
- Examples include Microsoft Azure and Salesforce's Force.com.

Software-as-a-Service (SaaS):

- SaaS model distributes ready software, where applications are hosted by service provider which is made available to customer through the Internet.
- SaaS applications are designed for users and delivered over the Internet.
- Software is managed from a central location, so there is no need of users to handle and control infrastructures like network, operating system, servers, storage, etc. Companies that offer SaaS are Google, Microsoft, Zoho, etc.

Some of the benefits attributed to the use of cloud computing are:

- **Reduced Cost**: Cloud computing provides the facility of pay as per usage, thus reducing the initial and recurring expenses.
- **Increased Storage**: As cloud provides huge infrastructure, storage and maintenance of data in quite a large volume becomes easy.
- **Increased Flexibility**: Cloud computing facilitates employees as there is no restriction on resources and locations.

The rest of the chapter is organized as follows: In Sect. 2, we give an overview of resource provisioning policies in cloud computing. Section 3 describes the SLA-oriented system architecture for resource provisioning in cloud computing. Section 4 presents comparison study of various resource provisioning techniques highlighting its methodology, result analysis, advantages, and disadvantages. The comparison of the architectures used in the papers is mentioned in Sect. 5. Section 6 concludes the paper by mentioning the future scope of work.

2   Provisioning of Resources in Cloud Computing

The process of assigning resources over the network as needed for the application requiring different types of computational resources is called resource provisioning in cloud computing. The term resource provisioning includes a vast area such as selection, allocation, and dynamic management of both the hardware and software resources. The provisioning and allocation of resources must be done in way to abide by the prescribed SLA of the service provider to their users and also to meet Quality of Services (QoS) parameters such as availability of resources, response time of each job, throughput and security of data stored by the users.

Two policies for resource provisioning are:

- **Static Provisioning**: All resources required for the completion of a job are submitted before the submission of any job by application in static provisioning. The resources are released gradually as the jobs finish their executions. So, static provisioning technique is best suited for applications which have predictable and static demands. A cloud provider in this type of provisioning provides the users with maximum of resources in order to avoid SLA violation. Due to the over allocation of resources, a lot of resources are wasted, and as such, both users and providers suffer loss.

- **Dynamic Provisioning**: Resource allocation to a job by the provider during run-time of a job is done in dynamic provisioning. The available resources are stored in a resource pool which grows and shrinks according to the varying needs of the user-based applications. Dynamic provisioning is more suitable for those applications which have varying demand of resources. The cloud service provider allocates resources to the job when needed and removes the resources when not being used by that specific job to allocate to some other waiting applications. Customers are charged on pay per use basis. Maximization of profit for both the user and the provider is guaranteed in a dynamic provisioning technique.

Various parameters which define resource provisioning are:

- **Minimize Cost**: The user of a cloud service should pay minimum possible money on behalf of using the resources of the cloud.
- **Resource Maximization**: The cloud service provider should be allowed to have maximum possible resources.
- **Response Time**: Such scheduling algorithms must be employed which takes minimum amount of time for job completion.
- **Reduced SLA Violation**: Violation of SLA must be minimum possible by the algorithms designed to allocate resources.
- **Reduced Power Consumption**: Consumption of power should be least possible by VM placement and migration techniques.
- **Fault Tolerant**: The algorithms should work without stopping in case of failures of nodes.

Although dynamic resource provisioning has many advantages over static resource provisioning, it has many issues and challenges to be resolved. The provisioning should be done in way to remove the wastage of both money and resources. This must be done in an efficient manner and on time. The users should be satisfied as far as the meeting of their QoS requirement is concerned. Lastly, the provisioning must be done by taking into account the SLA as prescribed by the provider.

Some of the well-known scheduling techniques used in cloud computing as of now are FCFS scheduling, priority scheduling, round-robin scheduling, simulated annealing technique, genetic algorithm-based scheduling, etc.

3 System Architecture Supporting SLA-Oriented Resource Provisioning in Cloud

The four components of a SLA-oriented architecture are shown in Fig. 1 and described henceforth.

- **User/Broker**: The interaction between the user and the cloud management is done through the brokers. The requests of the users' are submitted to the cloud server by the brokers.

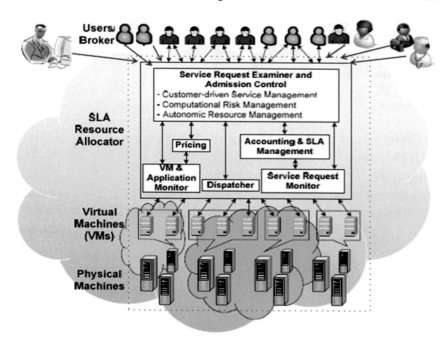

Fig. 1 SLA-oriented resource provisioning in cloud [1]

- **SLA Resource Allocator**: It acts as a medium between the cloud environment and users/brokers. A SLA resource allocator provides the following services:
- **Service Request Examiner and Admission Control**: Over- and under-provisioning resources are tackled by this unit in order to avoid SLA violation. It renders latest updates regarding availability of resources so as to formulate better decision making.
- **Pricing**: It manages service demand and maximizes profit of cloud providers.
- **Accounting and SLA Management**: It keeps track of SLA of users with cloud providers. Accounting mechanisms keep details of actual usage of resources to compute the final cost.
- **VM and Application Monitor**: It monitors and keeps track of the VMs availability.
- **Dispatcher**: The deployment of applications on appropriate VMs is done by this unit. It also creates the image of VMs.
- **Service Request Monitor**: The progress of execution of service request is tracked by this unit.

- **Virtual Machines**: Multiple virtual machines are created to meet service request on the same physical machines.
- **Physical Machines**: Multiple servers are hosted on data centers to provide resources to meet demand.

4 Comparison of Various Resource Provisioning Techniques

There are many resource provisioning techniques examined so far. These techniques are detailed in Table 1 describing various factors, viz methodologies used, results obtained, and their advantages and disadvantages.

5 Architectures Used

Various architectures used in the above techniques are discussed in Table 2.

Table 3 shows the limitations and future scope of the above-mentioned algorithms.

6 Conclusion and Future Work

Apart from the various resource provisioning techniques available, none of them are fully devoted to SLA monitoring. Some techniques require the requirements of users' resources to be known in advance, whereas in other techniques, time and cost optimizations are not possible. Moreover, multilayer implementation of cloud services is under process because of migration cost of VMs.

Out of the discussed techniques, the Mixed Workload-Aware Policy (MWAP) using artificial neural network (ANN) is the most suitable one as it focuses on many factors while allocating resources, and thus, the chances of SLA violation are very less. Also, the utilization of cloud data centers is the most in this technique. So, in the future, flexible algorithms can be designed by taking into consideration several aspects of SLA.

Table 1 Comparison of various resource provisioning techniques

Description	Methodology	Result	Advantages	Disadvantages
Paper [1] describes a SLA-based resource provisioning methodology using Aneka platform. SLA-oriented dynamic provisioning algorithm in Aneka is described here where users' SLA is handled by Aneka	SLA-oriented dynamic provisioning algorithm is used which works by getting executed on the entry or exit of a job. Completion time is calculated and is checked against the deadline. Request for extra resources is submitted only if completion time is greater than the deadline. If this is not the case, then the resources are released	QoS requirements of applications and hence the prescribed SLA can be effectively met by Aneka by allocating the resources dynamically	Cost optimization	More cost requirement in case of time-based optimization
Authors in paper [2] present a deadline-constraint provisioning policy of resources in cloud using Aneka cloud platform. Resources obtained from a variety of sources such as cluster grid, desktop grid, and private grid can be effectively provisioned by Aneka	The scheduler first determines the deadline of a task, number of tasks in the queue, and run time estimation for each task in order to predecide if the deadline can be met or not. The scheduler also determines the number of extra resources required by each process and submits a resource request to the provisioner service which contains the number of resource request and the request	Execution time for each job was calculated without any deadline first and then after fixing many different deadlines. The result shows that execution time of each job decreases significantly when the deadlines are made stricter	Resource allocation from different sources can be done effectively while keeping a check on application and execution time	Further improvements in the dynamic resource provisioning algorithms of Aneka are under developed

(continued)

Table 1 (continued)

Description	Methodology	Result	Advantages	Disadvantages
In [3], the authors have optimized the VMs placement over the host for minimizing power consumption by the use of simulated annealing optimizing technique	At first, the total non-optimized power consumed by data centers is measured by fixing the temperature and cooling rate Possible VM migration is carried out further in order to reduce the total power consumption	The optimization of VMs in accordance with the current CPU utilization decreases the energy consumed but increases the violation of SLA	Reduced energy consumption and hence profit maximization	The constraint of knowing the requirements in advance for initial allocation makes this methodology not much feasible for practical scenario
In [4], authors have allocated resources in cloud computing using multi-parameter bin-packing problem. This technique uses n number of computing resources, each having m number of parameters such as RAM, CPU, and storage. Resource requests come in the form of certain amount of these parameters	The use of simulated annealing technique is done to design algorithm to achieve optimal solution. By considering random requests, at first, initial solution is developed. Then swapping is carried out the costliest function with a bin parameter which allows violation of soft constraints using the cost function. The process of swapping is carried out till we reach a threshold	Simulated annealing algorithms show better results as compared to FCFS. In the case of variant cooling rate and high temperature, SA shows better results	Simulated annealing gives efficient result	On the top of cloud environment, implementation of this algorithm in a multilayer system is not done

(continued)

Table 1 (continued)

Description	Methodology	Result	Advantages	Disadvantages
Authors in [5], discussed an efficient resource provisioning approach in cloud computing which serves the SaaS users request for fulfilling the QoS requirements and maximizing profit utilization	ORCP includes both resource-level and application-level optimization algorithms and is economic-based algorithm. The price of SaaS provider is taken as input in the optimization algorithm. By using this price, the optimal payment for SaaS provider is calculated by SaaS user. The SaaS provider and then optimal SaaS compute new SaaS price and update it to all users	Performance of ORCP is measured under metrics like budget and load factors. The system burden increases when load factor is increased	QoS is optimized by allocating efficiently cloud resources for SaaS users within budget and deadline	Applicable only for SaaS users and SaaS providers
In paper [6], the authors have mentioned resource allocation problem within a data centers that run different types of application like non-interactive and transactional without SLA violation. For this, Mixed Workload-Aware Policy (MWAP) is used using artificial neural network (ANN)	A forecasting model based on ANN is used, and CPU utilization is predicted through the standard back-propagation algorithm. The use of ANN can forecast the utilization of CPU with minimum RMSE network	The MWAP of resource provisioning was compared against existing techniques used. On an average, the MWAP utilized 60% less number of servers. Migration overhead also reduces significantly by using this policy	Less SLA violation and more data center utilization	Considers only heterogeneous workloads

(continued)

Table 1 (continued)

Description	Methodology	Result	Advantages	Disadvantages
In paper [7], demand is predicted by using elastic resource framework technique. An efficient cost model is also described where resource allocation in access results in over allocation cost and under allocation as penalty cost. Minimization of excess cost is done through a confidence interval for predicted workload	Workload for different applications is predicted and optimal resource is found by a forecasting engine. Confidence interval is used by the forecasting engine to optimize the resource allocation. The accuracy of forecasting is calculated by SMAPE technique	After certain iterations, the response time starts decreasing, and finally, comes under the predefined SLA. The violation of SLA is observed only in 2.5% of workloads coming to the data center	Access cost and SLA violation decrease considerably	Waiting time of each workload increases
Paper [8] describes Aneka, a framework which is used to develop, manage, and deploy applications. Aneka helps in resource provisioning from different sources and supports different application models	Anekas' spot-instance-aware provisioning algorithm is used. This algorithm checks if currently available resources are enough for completing jobs within their deadlines	Default resource provisioning and spot instance algorithms are compared with the proposed algorithm and it has been found that Aneka provides resource provisioning with less VM	Saving in investment by taking the advantage of this algorithm, Aneka is able to meet even more strict application deadline with minimum budget	Not applicable when the external resources required are not known Not applicable when information about the current and predicted utilization of local infrastructure is not known

(continued)

Table 1 (continued)

Description	Methodology	Result	Advantages	Disadvantages
A two-level resource management is shown in paper [9]. The technique shown enables automatic as well as SLA-based adaptive resource provisioning. Dynamic trade-offs of service, quality, and cost are specified	A control system using fuzzy logic is used A fuzzy logic-based control system is used to signify application workloads and resource–demand relationship	Experiments carried out on local controller and global controller, and then, it was concluded that system can significantly reduce the cost of the resource and maximize resource utilization	Maximize resource utilization profit and minimize cost	Does not consider the cost of migrating virtual machines
Paper [10] proposes a RAS-M strategy where a utility function of QoS reflection is created according to different resource requirement of cloud clients. Equilibrium theory is introduced, and price adjustment algorithm based on GA is also mentioned	RAS-M determines the number and fraction workloads which can be used by one VM at a time. This can be dynamically adjusted according to changing resource demand and hence the workload. RAS-M architecture consists of three parts, namely resource agent (RA), consumer agent (CA), and market economy mechanism	Price adjustment policy based on GA and employing RSA-M can achieve equilibrium state. The demand of resources and supply by the providers is balanced in this policy	Maximize the profit of all CA, and both customer and provider gain maximum profit	Used to allocate resources only in lower level of cloud computing Manages CPU resource only

(continued)

Table 1 (continued)

Description	Methodology	Result	Advantages	Disadvantages
Genetic algorithm (GA) is used in [11] to solve the problem of resource allocation. The paper also discusses a new model to enhance the decision making result	Genetic algorithm is used to build SLA proposal. Negotiation model is built which is evaluated by genetic approach. SLA is modeled as chromosome and each resource type as a gene. Two-level evaluations are carried out. At first level, resource fitness is calculated, and in second level SLA proposal fitness is calculated	Performance evaluation of genetic algorithm and branch and bound algorithm is done. Both algorithms show identical results in case of SLA proposal fitness, but the genetic approach offers a response time which is very reasonable in the absence of any rule	The response time was very reasonable in the absence of any rule. The performance was very good when several rules were specified	Not taken for software service provisioning
In [12], dynamic resource provisioning and resource allocation are done through an auction. It also considers the SLA between the client and the service provider	The proposed architecture considers the client's SLA requirement. Resources are leased from cloud vendors in auction-based algorithm for maximizing profit utilization of the cloud provider. At first, a request is made with predefined SLA, and all of the available resources are monitored. If resources are available, it goes to job scheduling, but if not available, then it goes to a coordinator which does the auction of resources with cloud vendors and the jobs are migrated to cloud vendors	The proposed algorithm is compared with RBMA. The proposed auction-based mechanism takes a short fixed time for bidding using JAS. If JAS fails in the bidding process, resources are directly purchased from cloud vendors according to the minimum price	Final profit is greater if we use the proposed auction-based system than RBMA. SLA is followed strictly. Improved resource utilization in multiple cloud environments	The job's profit has to be greater than the rent cost, otherwise it will incur loss

Table 2 Architectures used in various algorithmic models

Architecture	Entity involved	Component involved
Elastic resource framework for IaaS	Tools, core, and drivers	Administrator tool, service manager, scheduler, cloud interfaces, information manager, accounting and auditing, authorization and authentication, image manager, federation manager, VM manager, network manager, storage manager, forecasting engine based on cost model, physical infrastructure drivers, and cloud drivers
RAS-M	Physical resources, workload, demand, supply, balance, and autonomic VMM	Virtual machines, consumer agent, resource agent, market mechanism, CPU, memory, disk, and network
MOSAIC's cloud agency	Negotiating resources, information repository, provisioning resources, operation facilitator, and monitoring data	Broker agent, achiever agent, vendor agent, mediator agent, reconfigurator agent, and monitoring agent
Auction-based system architecture	Physical machines, virtual machines, and requests with SLA	Billing system, monitor, clients' SLA decision, local VM scheduling, and coordinator components

Table 3 Algorithms' limitations and future work

Algorithm	Disadvantages	Future work
Resource allocation optimization algorithm using simulated annealing	The entire resource requirement should be known in advance for allocation to begin	Designing algorithm that offers flexibility of not knowing the requirements in advance
Optical resource cloud provisioning algorithm	Both, the SaaS provider's profit and the number of VMs employed is not reasonable enough	More profit of SaaS provider
Deadline-driven resource provisioning algorithm	Improvement in this algorithm is still underdeveloped	Some HPC applications are data intensive, viz data location-aware provisioning in resources of hybrid types

(continued)

Table 3 (continued)

Algorithm	Disadvantages	Future work
Spot-instance-aware provisioning algorithm	Not used where external resources required is not known Not suitable when no information about the current and predicted utilization of local infrastructure	Information about when external resources are required and knowledge about the current and predicted utilization of local infrastructure
SLA-oriented dynamic provisioning algorithm	Time optimization not possible	Effective policies for time-based optimization
Multi-parameter bin-packing problem using algorithm Priority_Fit_SA	Multilayer implementation on the top of cloud architecture is not done	On the top of cloud environment implementation of this algorithm in a multilayer system
MWAP-based resource provisioning algorithm	Not suitable for workflows and parallel applications	Optimizing the resource provisioning technique for other types of workload like workflows and parallel application
Use of fuzzy logic modeling	Does not consider the cost of migrating virtual machines	Consider the cost of VM migration in the proposed profit model
Job allocation with SLA (JAS) algorithm and auction model	Only FCFS scheduling policy is used	Use of job scheduling approaches like SJF and RR to maximize available resources Adoption of prediction model to maximize profit
RAS-M market-based strategy through GA-based automatic price adjusted algorithm	Only manages CPU resources	Implementing RAS-M in upper-level resource management module Manages different types of resources
Resource allocation optimization using BFO	Requirements are to be known initially; cost estimation for migration is not proposed; violation of SLA	Hybrid optimization technique that takes the benefits of BFO and genetic algorithm or greedy-knapsack to give more efficient result

References

1. Buyya, R., S.K. Garg, and R.N. Calheiros. 2011. SLA-oriented resource provisioning for cloud computing: challenges, architecture, and solutions. In *International Conference on Cloud and Service Computing*, 1–10.
2. Vecchiola, C., R. Calheiros, D. Karunamoorthy, and R. Buyya. 2012. Deadline driven provisioning of resources for scientific applications in hybrid cloud with Aneka. *Future Generation Systems* 28: 58–65.
3. Dhingra, A., and S. Paul. 2014. Green cloud: smart resource allocation and optimization using simulated annealing technique. *International Journal of Computer Science & Engineering* 5 (2): 41–49.
4. Pandit, D., S. Chattopadhya, M. Chattopadhya, and N. Chaki. 2014. Resource allocation in cloud using simulated annealing. In *Applications and Innovations in Mobile Computing*, 21–27.
5. Li, C., and L. Li. 2012. Optimal resource provisioning for cloud computing. *Journal of Supercomputing* 62: 989–1022.
6. Garg, S., A. Toosi, S. Gopalaiyengar, and R. Buyya. 2014. SLA based resource provisioning for heterogeneous workload in virtualized cloud data centres. *Journal of Network and Computer Applications* 45: 108–120.
7. Dhingra, M., J. Lakshmi, S. Nandy, C. Bhattacharyya, and K. Gopinath. 2013. Elastic resource framework in IaaS, preserving performance SLA. In *IEEE 6th International Conference on Cloud Computing*, 430–437.
8. Vecchiola, C., R. Calheiros, D. Karunamoorthy, and R. Buyya. 2012. The Aneka platform and QoS driven resource provisioning for elastic applications on hybrid cloud. *Future Generation System* 28: 861–870.
9. Xu, J., M. Zhao, R. Carpenter, and M. Yousif. 2007. On the use of fuzzy modelling in virtualized data center management. In *IEEE, ICAS, July 16, 2007*. https://doi.org/10.1109/icac.2007.28.
10. You, X., X. Xu, J. Wan, and D. Yu. 2009. RAS-M: resource allocation strategy based on market mechanism in cloud computing. In *Fourth China Grid International Conference*, 253–263.
11. Munteanu, V., T. Fortis, and V. Negru. 2013. Evolutionary approach for SLA based cloud resource provisioning. In *IEEE 27th International Conference on Advanced Information Networking and Applications*, 506–513.
12. Chang, C., K. Lai, and C. Yang. 2013. Auction based resource provisioning with SLA consideration on multi-cloud systems. In *IEEE 37th Annual Computer Software and Applications Conference Workshops*, 445–450.

An Extensive Study on Medical Image Security with ROI Preservation: Techniques, Evaluations, and Future Directions

Bijay Ku. Paikaray, Prachee Dewangan, Debabala Swain and Sujata Chakravarty

Abstract In the current era of digitization, the worldwide healthcare service is a common practice; it enables the remote healthcare service with proper digits: diagnosis and medication. The main thread for the about service is the security confidentiality and integrity of personal information of a patient. Many techniques are followed for privacy preservation in image information with descriptive aspects; however, more such strategies need to analyze and devolve for better accuracy and performance. This paper has given a broad overview of deferment techniques, their performance, and some future directions for ROI preservation in medical images.

Keywords Medical image · Data hiding · ROI · RONI · Privacy preservation

1 Introduction

In the present-day scenario, digital communication has a vital role. Internet technology along with cloud computing boosts the use of data communication exponentially. In this regard, the security issue on data communication is a sensitive matter of concern. Although many efficient and secure mythologies are proposed, still more secure robust techniques in terms of performance need to be proposed. The information over the communicating channel can be sent in two different forms that are (1) in encrypted format and (2) in hidden format [1]. In a cryptography system, the encryption technique is used to convert the original information into arbitrary encrypted form of transmission. The decryption system is used to regenerate the arbitrary information to the original form; the cryptography system is of two types, i.e., symmetric and asymmetric. Information hiding techniques involve steganography, watermarking, reversible data hiding, etc., where the digital image video, audio,

B. Ku. Paikaray · S. Chakravarty
Department of CSE, Centurion University of Technology and Management, Bhubaneswar, Odisha, India

P. Dewangan (✉) · D. Swain
Department of Computer Science, Rama Devi Women's University, Bhubaneswar, India
e-mail: dewanganprachee7@gmail.com

© Springer Nature Singapore Pte Ltd. 2020
D. Swain et al. (eds.), *Machine Learning and Information Processing*,
Advances in Intelligent Systems and Computing 1101,
https://doi.org/10.1007/978-981-15-1884-3_42

and textual contents can be hidden in the original file. Also, several techniques use a combination of both cryptography and steganography for better security and privacy preservation of sensitive data. In the data transmission process, medical image transmission plays a significant role in telemedicine applications. There are various sensitive contents present in the medical image along with the EPR which needs to be shared between the patient and the authenticated receiver. During the transmission of medical images, security issues are quite questionable in terms of confidentiality, reliability, and integrity [2]. To preserve the patient's private data, that images need to be encrypted; also, certain information needs to be embedded in the encrypted sensitive contents. After transmission, the decrypted system also plays an essential role in data recovery and integration. This paper describes the above discussion in a precise way.

The rest of the paper is organized as follows: Sect. 2 describes major security issues on medical data image with significant ROI preservation. Section 3 represents recent related works on ROI preservation for medical images. Sections 4 and 5 represent a comparative view on the performance of different reviewed techniques using various performance parameters. Section 6 summarizes the overall contribution of the paper and future directions.

2 Security Issues in Medical Images

A medical image can be divided into two regions based on content sensitivity: (1) region of interest (ROI) and (2) region of non-interest (RONI), as demonstrated in Fig. 1. The ROI contains the sensitive information required for medical diagnosis, so it needs to be preserved over the transmission media. Any kind of tampering or noise addition to the ROI should be detected, located, and recovered at the receiver end using various data hiding techniques. In this way, the veracity of ROI can be well protected even if it is attacked and tampered by any third party, whereas RONI recovery cannot be achieved losslessly. Further, segmentation of the medical image

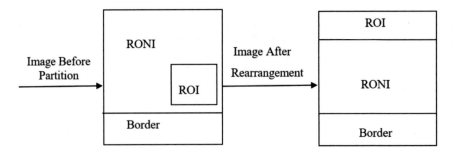

Fig. 1 Illustration of image partition and rearrangement

includes ROI and RONI in the spatial domain using watermark embedding which needs extra security as it is easy to destroy the embedded information in RONI.

In this paper, various data hiding techniques for protecting and integrating medical images are reviewed. Also, the compression and evaluation of their performance are discussed.

3 Related Works

3.1 A Novel Robust Reversible Watermarking Scheme for Protecting the Authenticity and Integrity of Medical Images [3]

This technique was proposed by Liu et al. [3] for verifying the integrity and authenticity of medical images. In this scheme, from a hospital logo, the use of authenticity is to generate data from the hash values. The entire medical image used hash for the function of tamper detection. Before finding the tampered reasons for ROI, it is divided into 16×16 non-overlapping blocks. Then, for tampered localization in each ROI block, CRC was adopted for generating tamper localization information of ROI.

For tamper recovery in ROI regions, the IWT coefficients were used [3]. Further to minimize the tamper recovery in ROI, BTC techniques were used.

There are four basic phases in all reversible watermarking schemes: watermark generation, watermark embedding, watermark extraction, and security verification as illustrated in Fig. 2.

3.1.1 Watermark Generation Phase

This phase includes four sub-phases.

(i) In the first sub-phase, the authenticity of the data is generated. As there is a possibility that different messages may have the same hash function close to 0, so a new hash function $A = f(L)$ is applied using SHA-1, where A is the 160-bit authenticity data and L is a hospital logo.

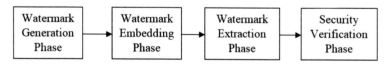

Fig. 2 Different phases of the reversible watermarking scheme

(ii)   In the second phase, the tampered detection information is generated using another $D = f(M)$ using SHA-1, where D is a 160-bit tampered detection information and M is a medical image.

(iii) In the third phase, tamper localization information is generated using CRC-16 of ROI. It includes ROI normalization, ROI division into non-overlapping blocks, conversion of generator function to binary digits, non-version of each pixel of 8-bit to a vector, appending 0's to the vector, and dividing the vector by polynomial generator to 16-bit reminder for tamper localization.

(iv)   In the fourth phase, the information regarding tamper recovery is generated. This phase is also responsible for the quality recovery of ROI. This process executes in four steps using the coefficient approximation matrix of IWT, division of the CA into non-overlapping block, reconstruction of them, and further conversion of the reconstructed matrix to binary and re-arranging them into a vector to generate the tamper recovery information.

3.1.2   Watermark Embedding Phase

In this phase, watermark embedding used the technique of SLT. S is the singular matrix using SVD and the most significant value. Further RDM-based function is applied to embed watermarks for restoring lossless medical images. During watermarking, the embedding into ROI and RONI is done without dividing the medical images.

3.1.3   Watermark Extraction Phase

Watermark extraction technique is the inverse process of the watermarked medical image of the embedding process. First, the ROI shifted pixels are located and recovered. Then, the watermark is divided into non-overlapping blocks and the embedding sequence; the coefficient matrix can be obtained using SLT. Further, the SVD is applied to produce the singular matrix and watermark bits can be extracted to restore the original medical images.

3.1.4   Security Verification Phase

This phase ensures the authenticity and integrity of the quality and source of the medical images. This phase includes the authenticity of the hospital logo using a hash function and the authenticity of a medical image using integrity verification. If tamper detected using hash function then a series of operations using normalization, bit mapping, reconstruction, and inverse IWT can be applied to restore the medical images (Fig. 3).

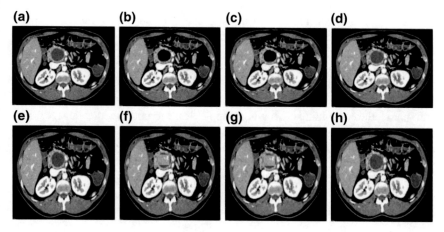

Fig. 3 Test images for ROI tamper detection and recovery: **a** image watermark, **b** tampered erased image, **c** tampered localization, **d** tampered erased recovery image, **e** watermarked image, **f** copy–paste tampered image, **g** localization of copy–paste tampered blocks, and **h** recovery of copy–paste tampered image

3.2 A ROI-Based Reversible Data Hiding Scheme in Encrypted Medical Images [4]

This method was proposed by Lie et al. for data hiding on medical images in encrypted form. It comprises three phases: (1) image encryption, (2) data embedding, and (3) data extraction and recovery, as illustrated in Fig. 4.

3.2.1 Image Encryption

In this phase, the image partitioned into non-overlapping blocks then the encryption is done. The original image is divided into three parts: ROI, RONI, and border area. The ROI is selected using any polygon function, and the bottom line of the image can be selected as a border. Then, the hash value of the ROI can be calculated using MD5. Then, the ROI can be concatenated by the RONI border area. After this rearrangement, the encryption is done using stream cipher with pseudorandom key

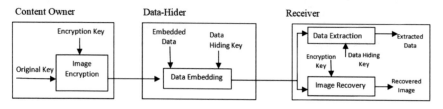

Fig. 4 RDH framework

bits. Finally, the ROI bit and the hash values are embedded into the LSBs of the border area to generate the encrypted image.

3.2.2 Data Embedding

In this phase, the embedding processing is applied on encrypted medical image. First, the ROI portion is identified from the encrypted image; then, the LSBs of the ROI pixels are manipulated according to the embedded data. The EPR gets concatenated on the LSBs to generate the embedded data.

3.2.3 Data Extraction and Recovery

The extraction and recovery from the embedded image can be done in three ways, using the key for data hiding or key for encryption or both. If the data hiding key is only used, the receiver can de-embed the hidden data from the received image but cannot recover the original image. If the receiver has only the encryption key, then the received image can be decrypted but the hidden data cannot be decrypted. So the receiver must have both keys for encryption and data hiding for extraction and recovery of the lossless original image. The execution of the above-mentioned technique is illustrated in Fig. 5, and its performance is analyzed in Table 1.

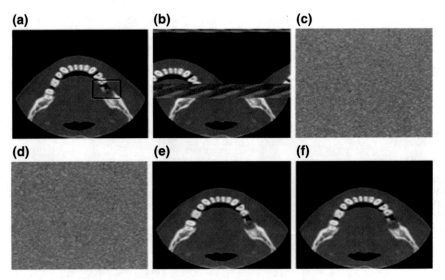

Fig. 5 Test image analysis: **a** original image, **b** rearranged image, **c** encrypted image, **d** embedded image, **e** decrypted image, and **f** recovered image

Table 1 PSNR and SSIM analysis with various sizes of ROI

ROI size (%)	0.05	0.1	0.2	0.3	0.4
D_{lsb} (bits)	13,108	26,215	52,429	78,644	104,858
EPR (bits)	1308	2616	13,080	26,161	52,322
Hiding payload (bits)	14,416	28,831	65,509	104,805	157,180
PSNR (dB) (recovered image)	131.16	120.97	107.22	103.94	100.19
SSIM (recovered image)	0.9999	0.9999	0.9999	0.9999	0.9999
PSNR (dB) (directly decrypted image)	113.55	110.46	106.24	103.87	100.05
SSIM (directly decrypted image)	0.9999	0.9999	0.9999	0.9999	0.9999

3.3 Secure Telemedicine Using RONI Halftoned Visual Cryptography Without Pixel Expansion [5]

The technique uses visual cryptography for secure telemedicine data for proper diagnosis. It involves five algorithms. The details of encryption and decryption process are explained in Fig. 6. The first algorithm identifies the RONI for data embedding. It analyzes the min threshold value of the pixel and then divides the image into rows and

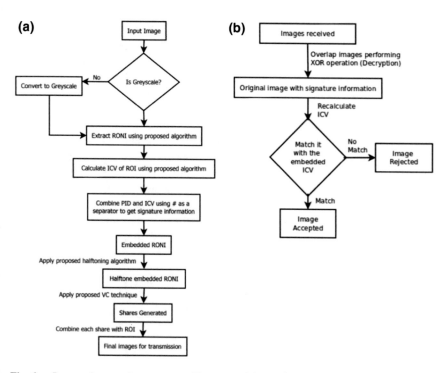

Fig. 6 **a** Proposed encryption process and **b** proposed decryption process

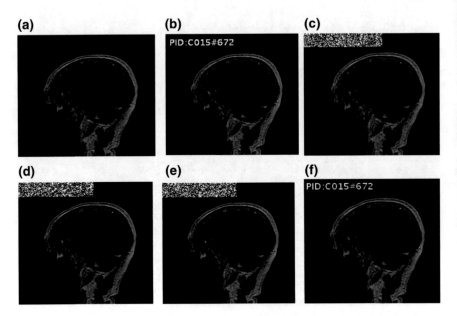

Fig. 7 **a** Original image [6], **b** embedded image, **c** share 1, **d** share 2, **e** share 3, **f** overlapped share 1, share 2, and share 3

columns. Then, it analyzes the pixels whose threshold value is greater than the min threshold and marks them as RONI. In this way, ten RONI regions are located. The second algorithm operates on a portion of the ROI which gets embedded into RONI for integrity verification. For integrity verification in ROI, the algorithm divides the image into two portions. The first portion is taken from the top left corner to the bottom right corner of the image. The second portion is taken from the top right corner to the bottom left corner of the image. So, it obtains three ICV s1, s2, and s3 to be used for future integrity verification. The algorithm three converts the grayscale image format to binary and uses pushing error forward-based error diffusion technique for halftoning. The algorithm four analyzes the original image by comparing the original pixel value greater than 255 and less than 255 and replaces the values in three shares. The fifth algorithm implements all above-mentioned algorithms and other algorithms as required to generate the embedding information in RONI. The performance is analyzed in Fig. 7.

4 Performance Evaluation Parameters

The assessment of different techniques can be expressed using the following parameters [7–10].

4.1 Peak Signal-to-Noise Ratio (PSNR)

The PSNR can be expressed in terms of dB as follows:

$$\text{PSNR (in dB)} = 10 \log_{10} \left(\frac{255^2}{\text{MSE}} \right) \tag{1}$$

The mean square error (MSE) between the original and recovered images can be calculated as follows:

$$\text{MSE} = \frac{\sum_{i=1}^{N} (C_i - C_i')^2}{N} \tag{2}$$

4.2 Structural Similarity Index Measure (SSIM)

This parameter is used to compare the similarity between the recovered and the original images. It can be calculated as follows:

$$\text{SSIM}(x, y) = \frac{(2\mu_x \mu_y + c_1)(2\sigma_{xy} + c_2)}{(\mu_x^2 + \mu_y^2 + c_1)(\sigma_x^2 + \sigma_y^2 + c_2)} \tag{3}$$

where μ_x, μ_y are the averages of x and y, σ_x^2, σ_y^2 are the variances, and σ_{xy} are covariances for x and y, respectively. c_1 and c_2 are balancing constants.

4.3 Payload Capacity

It represents the hidden information present in the transmitted image, expressed in terms of bits per pixel (BPP). The more the payload capacity represented, the more the privacy in the sent image.

$$\text{BPP} = \frac{\text{Number of secret bits embedded}}{\text{Total pixels in the cover image}} \tag{4}$$

5 Comparative Review on ROI Preservation Technique

The techniques discussed in Sect. 3 are summarized in Table 2.

Table 2 Comparative review of different ROI preservation techniques

References	Basic parameters	Methodology description	Advantages	Disadvantages	Results
Liu et al. [3]	ROI and RONI for watermarking generation	Reversible watermarking is based on recursive dither modulation (RDM). RDM is combined with SLT-SVD hybrid transform and singular value decomposition for authenticity	– Spatial image segmentation can be avoided during embedding as ROI and RONI are not getting divided – Reliable authentication tampered detection localization and recovery – The RDM-based function enables lossless restoration of ROI and RONI. – Using SLT-SVD hybrid transform, the algorithm became robust against attack – Using IWT with BTC, the tamper ROI can be easily recovered	– The poor robustness against rotational attachment – Low payload capacity	PSNR = 41.92995 SSIM = 0.9607

(continued)

Table 2 (continued)

References	Basic parameters	Methodology description	Advantages	Disadvantages	Results
Liu et al.[7]	LSB in ROI and EPR	The LSB of the encrypted ROI and the EPR are concatenated. Then, they are embedded into the encrypted image by LSB substitution method. The embedded data can be instructed using a data hiding key, and the original image can be recovered using the encryption key. The image can be authenticated by the ROI hash message	– Using the data hiding key and encryption, their image can be extracted without any error losslessly – The ROI embedding is completely reversible – Decrypted image quality was improved – Gain in PSNR and embedding rate	– Error was detected in RONI recovery – The ROI and RONI partitioning was not spontaneously done and poor in terms of time complexity – The LSB embedding in RONI was less robust	$D_{lsb} =$ 104,858 EPR $=$ 52,322 Hiding payload $=$ 157,180 PSNR $=$ 100.19 SSIM $=$ 0.9999 (with embedding rate $= 0.4$)
Bakshi and Patel [5]	Visual cryptography, ROI and RONI integration	The RONI is extracted from a grayscale image. ICV of ROI is calculated and combines with PID to get signature information; then using a halftoning algorithm, the RONI is embedded. The VC techniques are used to generate the shares. Then, ROI is combined to get the final transmitted image. The decryption and recovery process is completely reversible	– Secure image transmission – Better confidentiality of patient records – Decryption and recovery with constant time complexity	– Weak RONI recovery – ICV algorithm is less secured	PSNR $=$ 22.9452 SSIM $=$ 0.9701 Accuracy $= 99.874$

6 Conclusion

This paper explores different techniques proposed for security issues on ROI-based medical images and the solution for ROI-based privacy preservation. Different techniques are analyzed with basic parameters, methodology advantages, and drawbacks with performance parameters. Although the region-based analysis is done in all reviewer papers, still the embedding technique differs. Based on this review, it can be concluded to adobe heterogeneous parameters for ROI manipulation to achieve robust security and better privacy and preservation. The comparison table illustrates different techniques with a qualitative and quantitative way that source future direction for the development of new algorithms of medical images.

References

1. Cheddad, A., J. Condell, K. Curran, and P.M. Kevitt. 2010. Digital image steganography survey and analysis of current methods. *Signal Processing* 90: 727–752. https://doi.org/10.1016/j.sigpro.2009.08.010.
2. Coatrieux, G., H. Maitre, and B. Sankur, et al. 2000. Relevance of watermarking in medical imaging. In *Proceedings of IEEE EMBS International Conference on Information Technology Applications in Biomedicine*, 250–255.
3. Xiyao, Liu, Jieting Lou, Hui Fang, Yan Chen, Pingbo Ouyang, Yifen Wang, Beiji Zou, Lei Wang. 2019. *A Novel Robust Reversible Watermarking Scheme for Protecting Authenticity and Integrity of Medical Images.* IEEE. https://doi.org/10.1109/access.2019.2921894.
4. Liu Yuling, Qu, and Xin Guojiang Xinxin. 2016. A ROI-based reversible data hiding scheme in encrypted medical images. *Journal of Visual Communication and Image Representation* 39: 51–57. https://doi.org/10.1016/j.jvcir.2016.05.008.
5. Arvind, Bakshi, and A.K. Patel. 2019. Secure telemedicine using RONI halftoned visual cryptography without pixel expansion. *Journal of Information Security and Applications* 46: 281–295. https://doi.org/10.1016/j.jisa.2019.03.004.
6. http://imaging.cancer.gov/.
7. Liu, Y., X. Qu, G. Xin, et al. 2015. ROI-based reversible data hiding scheme for medical images with tamper detection. *IEICE Transactions on Information and Systems* E98-D (4): 769–774.
8. Lizhi, Xiong, and Dong Danping. 2019. Reversible data hiding in encrypted images with somewhat homomorphic encryption based on sorting block-level prediction-error expansion. *Journal of Information Security and Applications* 47: 78–85. https://doi.org/10.1016/j.jisa.2019.04.005.
9. Koley, Subhadeep. 2019. A feature adaptive image watermarking framework based on Phase Congruency and Symmetric Key Cryptography. *Journal of King Saud University—CIS.* https://doi.org/10.1016/j.jksuci.2019.03.002.
10. Yan, X., L. Liu, Y. Lu, et al. 2019. Security analysis and classification of image secret sharing. *Journal of Information Security and Applications* 47: 208–216. https://doi.org/10.1016/j.jisa.2019.05.008.

A Competitive Analysis on Digital Image Tamper Detection and Its Secure Recovery Techniques Using Watermarking

Monalisa Swain, Debabala Swain and Bijay Ku. Paikaray

Abstract Digital images play a vital role in human life. Hence, its protection from unauthorized access is a serious matter of concern. Even if the contents are modified then its detection and recovery must be defined. Nowadays, a number of methods are proposed to protect digital images based on digital watermarking. But all are not with similar capability in terms of security, authenticity, recovery. This paper represents the basics of digital watermarking techniques along with their competency and weakness for the detection of tampered images and their recovery process. A series of watermarking techniques with simulated results show their working efficiency with quantitative result analysis.

Keywords Digital image · Tamper detection · Image recovery · Watermarking · Digital security

1 Introduction

With the rapid improvement of the information sharing over the internet, digital images are hugely transmitted, whereas digital images are vulnerable to modifications and manipulation with widely available image manipulation tools. The wholeness and genuineness of digital images can be assured by using a digital watermarking process to which embed special information (text or image), known as a watermark, to digital source [1]. To maintain the full integrity of digitalized images research into watermarking-based image, authentication has been evolved. The first watermarking-based image authentication technique was proposed by Van Schyndel et al. [2] in 1994. The major concern for watermarking the digital image was that there should be

M. Swain · D. Swain
Department of Computer Science, Rama Devi Women's University, Bhubaneswar, India

B. Ku. Paikaray (✉)
Department of CSE, Centurion University of Technology and Management, Bhubaneswar, Odisha, India
e-mail: bijaypaikaray87@gmail.com

© Springer Nature Singapore Pte Ltd. 2020
D. Swain et al. (eds.), *Machine Learning and Information Processing*,
Advances in Intelligent Systems and Computing 1101,
https://doi.org/10.1007/978-981-15-1884-3_43

477

the lowest modification, which can be recovered at the receiver end. We are reviewing previous fragile watermarking techniques in this paper.

The rest of this paper is arranged as follows. In Sect. 2, basics of watermarking are discussed. In Sect. 3, some of the existing tamper detection and recovery methods are discussed along with the comparative analysis. At last, the paper is concluded in Sect. 5.

2   Digital Watermarking Essentials

The watermarking approach can be categorized based on some attributes [3]. Watermarking concept categorized into spatial domain and frequency domain when watermark data is embedded in the form of spatial and frequency domain representation of the original image. In the spatial domain watermarking method, the image pixel value is directly manipulated by changing some bits of pixel with watermarked bits. In frequency domain watermarking method watermark embedding, the first original image is transformed then watermark data is embedded to the changed coefficients.

According to the robustness of the algorithm, watermarking is classified into three types, robust, fragile and semi-fragile. The robust watermarking method can accept to eliminate the watermarks from images in different kinds of attacks and hard. Thus, this technique can be used to insert copyright information. The fragile watermarking technique can detect a tampered image by any attack with removed watermark. The semi-fragile watermarking is similar to fragile watermarking with additional features but it can bear some unintentional attack (Figs. 1 and 2).

During watermark image transmission, the image can face various attacks. The attacks are broadly categorized as unintentional attack (UA) and intentional attack

Fig. 1 Classification of watermarking approach

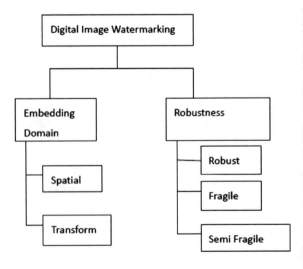

Fig. 2 Watermarking classifications

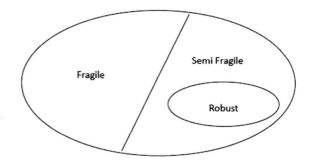

(IA). During the unintentional attack, noise may be inserted due to signal fluctuations and toggling of bits due to various reasons. An intentional attack, the image is tampered intentionally at the time of its storage and transmission. It is also known as man in the middle attack.

There are various intentional attacks that can be more classified. A removal attack is an IA, where the attacker removes the original content of the image. It may be possibly copyright information. Addition attack is an IA where some additional content is added on the watermark image. Cryptographic attack is an IA, where the watermarking scheme is cracked to embed misleading watermark by removing the original watermark. Copy-paste attack is an IA where attacker duplicates the watermarked data illegally so that the second copy of the watermarked image is generated. Geometric attack is an IA which involves all distortion on the image generated due to transformations like translation, rotation, scaling, shearing, cropping, line or column removal, etc.

3 Performance Measuring Parameters of Watermarking Algorithms

Any watermarking techniques performance can be measured against the following parameters.

3.1 Peak Signal to Noise Ratio (PSNR)

This parameter is used to determine reconstructed or restored image quality concerning its original image. A high PSNR value signifies an image with good quality.

The PSNR can be expressed in terms of dB as follows.

$$\text{PSNR (in dB)} = 10 \log_{10}\left(\frac{255^2}{\text{MSE}}\right) \quad (1)$$

3.2 The Mean Square Error (MSE)

It is used in the reconstructed image and original image to find out the cumulative squared error. The error value of MSE is lowered off the error.

It can be calculated as follows.

$$\text{MSE} = \frac{\sum_{i=1}^{N}(C_i - C_i')^2}{N} \quad (2)$$

3.3 Structural Similarity Index Measure (SSIM)

This method is used in between two images for measuring the similarity and comparing the watermarked image with the original. It can be calculated as follows.

$$\text{SSIM}(x, y) = \frac{(2\mu_x\mu_y + c_1)(2\sigma_{xy} + c_2)}{(\mu_x^2 + \mu_y^2 + c_1)(\sigma_x^2 + \sigma_y^2 + c_2)} \quad (3)$$

3.4 Payload Capacity

It represents the hidden information present inside the transmitted image, which is expressed by bits per pixel (BPP). More about payload capacity represented more privacy in the sent image.

$$\text{BPP} = \frac{\text{Number of secret bits embedded}}{\text{Total pixels in the cover image}} \quad (4)$$

4 Literature Review

In 2005, Hsieh et al. [4] proposed a hierarchical fragile watermarking scheme for image tamper detection and recovery by using a parity check and comparison between the pixels average value of each block. To improve tamper detection accuracy, three detection levels were used. In this approach, when two correlative blocks are tampered with, recovered image quality reduced.

In 2008, Lee and Lin [5] proposed a dual watermark scheme for tamper detection and recovery. In this approach, the watermarked data for each non-overlapping block were maintained in two copies. If one copy is destroyed, then the second chance was provided for block recovery. For tamper detection, parity check was used and a public chaotic algorithm was used and for recovery a secret key used. The use of dual watermark ensured a higher image quality of the recovered image even a watermarked image contains large tampered regions.

In 2010, Yang and Shen [6] proposed a watermarking scheme by creating an index table of the original image by vector quantization (VQ) technique and embedded into an original image for recovery. Wong's watermarking scheme [7] was used for tampered detection. In this scheme, higher probability of image recovery because of the VQ index table can be embedded into the cover image several times.

In 2011, Li et al. [8] proposed a watermarking scheme based on a dual-redundant-ring structure (DRRS). For each block, two copies of the watermark data were embedded in two different blocks. Tamper was detected through its mapping block in the block-neighborhood by comparing it with inconsistent blocks. In this process, it provides two copies of watermark; first chance was to recover the block and second chance was to destroy the block.

In 2012, He et al. [9] proposed a watermarking scheme based on self-recovery fragile using block-neighborhood tamper detection. This technique used to generate the nonlinear block mapping with a pseudorandom sequence and find out the tamper detection an optimized neighborhood characterization method.

In 2012, Qin et al. [10] proposed a fragile watermarking scheme with content restoration capability by using non-sub sampled contourlet transform (NSCT) coefficients. NSCT coefficients are utilized the concept of adaptive bit allocation mechanism to encoding in restoration bits efficiently. Here, all the blocks are divided according to their degree of smoothness, such as smooth block and complex block, where smooth blocks are allocating fewer bits than complex blocks.

In 2013, Adiwijaya et al. [11] proposed a watermarking scheme based on the separation of region of interest (ROI) and region of non-interest (RONI). Watermark data was embedded in ROI using a block-based watermarking method and Huffman compression was used in RONI. It represents a reversible approach to ROI. The approach was reversible in the ROI region. It was used to recover the attacks of tamper detection.

In 2013, Tong et al. [12] proposed a watermarking approach for providing tampering localization and self-recovery. By applying two authentication bits MSB and LSB of combination its output is better results in the tampered detection and defense attacks. Their recovery tampering was improved by using the sister block embedding and optimization method, which was used to find the average value of the valid blocks from their eight neighboring blocks and recover those unreconstructed blocks.

In 2014, Shivananda and Chetan [13] proposed a scheme based on a fragile watermarking for digital image tamper detection and recovery. In this method, images were divided into uniform-sized blocks and least significant bit of each block was used to create authentication data. Tamper detection was performed using a dual-option parity check method. The scheme has achieved a higher quality of recovery in tampered regions.

In 2015, Dhole and Patil [14] proposed a scheme based on fragile watermarking for tamper detection and recovery using self-recovery blocks. In this method by using block chaining first watermark image obtained and shuffled image of the original image was merged with the original image to get the final shuffled image. Then it has the final watermarked image after merging the first watermark image with the final shuffled image.

In 2016, Sreenivas and Prasad [15] proposed a scheme based on a self-embedding watermarking for tamper detection and self- recovery. This proposed method takes the average intensity of blocks with variable length and comparing the block with the watermarking bits.

In 2016, Molina-Garcia et al. [16] proposed a watermarking algorithm with authentication and self-recovery of tampered images using DWT. In this proposed method first, digital image was protected by halftoning, DWT and QIM methods and then through IDWT, where it inverses the halftoning then applies the median filtering authentication. In the end, the tamper detection of watermarked image and self-recovery was performed.

In 2018, Shehab et al. [17] proposed a scheme based on fragile watermarking for image authentication and self-recovery in medical applications. In this approach, images are divided into block and singular value decomposition using block division. Then singular matrices in each block were used as block authentication. The first five MSB were used as the self-recovery information. It could survive against the vector quantization attack using two bits called block authentication and self-recovery bits.

In 2019, Liu et al. [18] proposed a reversible watermarking scheme, where the images are divided into ROI and RONI. The tamper detection information was generated using the hash function then the tamper localization can be done by CRC calculation in each block of ROI. Further, the information regarding tampered recovery was generated using integer wavelet transform. All required data for authentication, tamper detection, in the whole image the tamper localization and recovery are embedded.

In 2019, Tamal et al. [19] proposed a watermarking method for tamper detection. This method used IWT for watermarking. Here, the images were divided into ROI and RONI. Logistic mapping, ROI hash values and the average value of the last bit in a 5×5 mask of ROI were used to generate the watermark information. Here, tamper detection was performed by two levels using hash value and last bits (Table 1).

Table 1 Comparative analysis of different watermarking schemes in spatial and frequency domains

Methods	Basic domain	Tamper detection and recovery method	Watermarked image PSNR	Recovered image PSNR
[4]	Spatial	Yes	>40 dB	>30 dB
[5]	Spatial	Yes	40.68 dB	20 dB
[6]	Spatial	Yes	30–40 dB	>30 dB
[8]	Spatial	Yes	44.26 dB	44–50 dB
[9]	Spatial	Yes	30–44.15 dB	32.04–62 dB
[10]	Spatial	Yes	51 dB	41–48 dB
[11]	Spatial	Yes	47–48.6	Not specified
[12]	Spatial	Yes	30–40	20–30
[13]	Spatial	Yes	30–44.15 dB	32.04–62 dB
[14]	Spatial	Yes	34–38 dB	36–44 dB
[15]	Spatial	Yes	30–44.15 dB	32.04–62 dB
[16]	Frequency	Yes	>32 dB	>32 dB
[17]	Spatial	Yes	Not specified	30.25–38.96 dB
[18]	Spatial	Yes	Not specified	41.2995 dB
[19]	Frequency	No	51.90 dB	Not specified

5 Conclusion

This paper represents various techniques of watermarking which were discussed for tamper detection, localization and recovery. All techniques classified in two domains i.e., spatial and frequency. Each technique outperforms others in terms of PSNR value of watermarked and recovered image. The basic techniques for tampered detections and recovery can also be analyzed in the context of processing time and security. Also, new experimentation can be done for minimizing error detection and correction in the tampered regions. The above discussions reconstruct the idea for developing new algorithms of the discussed issues in digital images.

References

1. Bhargava, N., M.M. Sharma, A.S. Garhwal, and M. Mathuria. 2012. Digital image authentication system based on digital watermarking. In *IEEE International Conference on Radar, Communication, and Computing (ICRCC)*. https://doi.org/10.1109/icrcc.2012.6450573.
2. Van Schyndel, R.G., A.Z. Tirkel, and C.F. Osborne. 1994. A digital watermark. In *Proceedings of the IEEE International Conference on Image Processing*, vol. 2, 86-90.

3. Tamal, T.A., C. Saha, M.D. Foisal Hossain, and S. Rahman. 2019. Integer wavelet transform based medical image watermarking for tamper detection. In *International Conference on Electrical, Computer and Communication Engineering (ECCE)*, 7–9. https://doi.org/10.1109/ecace.2019.8679152.
4. Lin, P.L., C.K. Hsieh, and P.W. Huang. 2005. A hierarchical digital watermarking method for image tamper detection and recovery. *Pattern Recognition* 38: 2519–2529.
5. Lee, T.Y., and S.D. Lin. 2008. Dual watermarking for image tamper detection and recovery. *Pattern Recognition* 41: 3497–3506.
6. Yang, C.W., and J.J. Shen. 2010. Recover the tampered image based on VQ indexing. *Signal Processing* 90: 331–343.
7. Wong, P.W., and N. Memon. 2001. Secret and public key image watermarking schemes for image authentication and ownership verification. *IEEE Transactions on Image Processing* 10: 1593–1601.
8. Li, Chunlei, Y. Wang, B. Ma, and Z. Zhang. 2011. A novel self-recovery fragile watermarking scheme based on dual-redundant-ring structure. *Computers and Electrical Engineering* 37: 927–940.
9. He, H., F. Chen, H.M. Tai, T. Kalker, and J. Zhang. 2012. Performance analysis of a block-neighborhood-based self-recovery fragile watermarking scheme. *IEEE Transactions on Information Forensics and Security* 7: 185–196.
10. Qin, C., C.C. Chang, and P.Y. Chen. 2012. Self-embedding fragile watermarking with restoration capability based on adaptive bit allocation mechanism. *Signal Processing* 92: 1137–1150.
11. Adiwijaya, P.N. Faoziyah, F.P. Permana, T.A.B. Wirayuda, and U.N. Wisesty. 2013. Tamper detection and recovery of medical image watermarking using modified LSB and Huffman compression. In *Second International Conference on Informatics & Applications (ICIA)*, 129–132.
12. Tong, X.J., Y. Liu, M. Zhang, and Y. Chen. 2013. A novel chaos-based fragile watermarking for image tampering detection and self-recovery. *Signal Processing: Image Communication* 28: 301–308.
13. Shivananda, N., and K.R. Chetan. 2014. A new fragile watermarking approach for tamper detection and recovery of document images. In: *IEEE International Conference on Advances in Computing, Communications and Informatics (ICACCI)*, 1494–1498.
14. Dhole, V.S., and N.N. Patil. 2016. Self embedding fragile watermarking for image tampering detection and image recovery using self recovery blocks. In *International Conference on Computing Communication Control and Automation*, 752–757.
15. Sreenivas, K., and V.K. Prasad. 2016. Improved block encoding method for an image self-recovery approach. In *International Conference on Information Communication and Embedded Systems (ICICES)*, 3–7.
16. Molina-Garcia, J., R.R. Reyes, V. Pronomaryov, and C.C. Ramos. 2016. Watermarking algorithm for authentication and self-recovery of tampered images using DWT. In *9th International Kharkiv Symposium on Physics and Engineering of Microwaves, Millimeter and Submillimeter Waves (MSMW)*, 1–4.
17. Shehab, A., M. Elhoseny, K. Muhammad, and A.K. Sangaiah. 2018. Secure and robust fragile watermarking scheme for medical images. *IEEE Access* 6: 10269–10278.
18. Liu, X., J. Lou, H. Fan, Y. Chen, P. Ouyang, Y. Wang, B. Zou, and L. Wang. 2019. A novel robust reversible watermarking scheme for protecting authenticity and integrity of medical images. *IEEE Access* 7: 76580–76598.
19. Tamal, T.A., C. Saha, and S. Rahman. 2019. Integer wavelet transform based medical image watermarking for tamper detection. In *International Conference on Electrical, Computer and Communication Engineering (ECCE)*. https://doi.org/10.1109/ecace.2019.8679152.

Elliptic Curve Cryptography-Based Signcryption Scheme with a Strong Designated Verifier for the Internet of Things

Biswojit Nayak

Abstract The Internet of Things (IoT) is an emerging technology that grows across the World Wide Web. It has scenarios in which the real-world object is transferring data over an insecure wireless network. The security in IoT became more challenging because of the low computational and communication capacity of the object. The proposed signcryption scheme is a combination of a digital signature and symmetric key encryption in a single logical unit, which reduces the computational complexity as compared to the traditional signature, then encryption process along with the digital signature of the sender can only verify by the designated verifier. The computational and communication overhead of the elliptic curve cryptography (ECC) scheme have less because of short key length with the same security level as compared to other public key cryptosystem. The security hardness of the scheme is based elliptic curve discrete logarithm (ECDLP) and also provided various security requirements. The proposed scheme has low computational and communication overhead with low-power efficiency for IoT scenarios.

Keywords Signcryption · Digital signature · Encryption · Elliptic curve cryptography · IoT

1 Introduction

Internet of Things (IoT) comprises of scenarios where all entities are connected and configured with unique identifiers for transferring data via an unsafe wireless network. Through their identifiers, the host machines identify senders and handle and manipulate these items. The IoT system is not only accountable for sensing, but can also take action and computations based activities. The security and efficiency [1, 12]

B. Nayak (✉)
Utkal University, Bhubaneswar, Odisha, India
e-mail: biswojit22@gmail.com

© Springer Nature Singapore Pte Ltd. 2020
D. Swain et al. (eds.), *Machine Learning and Information Processing*,
Advances in Intelligent Systems and Computing 1101,
https://doi.org/10.1007/978-981-15-1884-3_44

are two significant difficulties in IoT environments. Typical low-power IoT devices have low computing ability and restricted storage. As a consequence, it becomes an significant problem for IoT to set up safe and effective communication connections. The public key cryptosystem is commonly accepted in the IoT to guarantee secure communications for sensor node.

Computation and communication overhead are two important criteria of computer network communication. To reduce the computational and communication overhead, Zheng [13] in 1999 proposed a concept of signcryption. In order to attain both authentication features and confidentiality, signcryption provides both the notion of digital signature and symmetric key encryption in a single logical step. Zheng also proves that the computational and communication cost of signcryption has less than the traditional signature and encryption process. There is a part of research which has been done in the range of signcryption since 1997 [2–5, 11, 14].

The one-time symmetric key is used to encrypt messages in the signcryption scheme. The sender and receiver's private and public key, respectively, are used to produce the one-time symmetric key. At the receiver end, the receiver and sender's public and private key are used to derive the same session key. In most of the real-time application communication, a specific receiver can only verify the signature that is called designated verifier. Jakobsson [6] was first proposed the concept of designated verifier signature, where only designated verifier can verify the signature of the signer. The third party cannot recognize in the Jakobsson scheme, whether the signature was given by the signer or designated verifier.

In 2004, Saeednia et al. [10] proposed a scheme called strong designated verifiable scheme to overcome the Jakobsson scheme. But later in 2008, Lee–Chang [7] identify that this scheme not only verify by designated verifier's private key but also by the signer's private key. So, they proposed a strong designated verifiable signature scheme.

To overcome the computational and communication overhead in 2012, Mohanty and Majhi [9] consider the mechanism of the Lee–Chang [7] scheme and proposed a signcryption scheme based on the hardness of the discrete logarithm problem. But the scheme has one disadvantage that it takes large size of one-time symmetric key. So, the scheme has more computational and communication overhead. The proposed scheme can provide a smaller key with the same level of security by using elliptic curve cryptography [8].

This paper suggested a novel strongly designated verifier signcryption scheme, where only the signature can be verified by the designated verifier. The suggested scheme's hardness is based on the elliptic curve discrete logarithm problem (ECDLP) with smaller key length. The suggested scheme can therefore be used in IoT technologies.

2 Preliminaries

The proposed scheme is based on the below computational hard problem [3, 4].

Elliptic Curve Discrete Logarithm Problem (ECDLP): If X and Y are two points of an elliptic curve and $X = k.Y$, where k is a scalar value. Given X and Y, it is difficult to derive k.

3 Formal Model of Signcryption

A formal model of a signcryption scheme goes through the following three modules [15].

Setup: In Setup phase, the signcryptor takes contribution as security parameter k and produces the system's public parameters. At that point, signcryptor randomly picks his/her private key and generates his/her public key. The receiver also picks his/her private key and generates his/her public key.

Signcrypt: In this phase, if the sender wants to send a message M to receiver, then he/she has to run Signcrypt(M, SK_A, PK_B) to generates the ciphertext δ, where SK_A and PK_B is private and public key of the sender and receiver, respectively.

Unsigncrypt: After receiving ciphertext δ, receiver computes Unsigncrypt(δ, PK_A, SK_B) and obtains the plain text M, if δ is an valid ciphertext, where PK_A and SK_B is public and private key of the sender and receiver, respectively.

4 Proposed Scheme

This scheme arrangement involves two social affairs explicitly, Signcryptor and Recipient. There are three phases in the arrangement: Setup, Signcrypt and Unsigncrypt. In the fundamental stage, signcryptor makes and circulates all open parameters of an elliptic curve and each user creates his own specific private key and the related public key.

Setup: The signcryptor selects an elliptic curve that defines set of points which satisfy the equation $y^2 = x^3 + ax + b \bmod q$. The parameters of elliptic curve are as follows:

q: a large prime number

(a, b): coefficients of elliptic curves, whose satisfy the equation $4a^3 + 27b^2 \neq 0 \bmod q$

G: generator point of elliptic curve

0: point of elliptic curve at infinity

n: order of generator point

$H: \{0, 1\}^* \to Z_q$ (Hash function)

open parameters (q, a, b, G, n, H).

Signcryptor picks his private key $SK_S \in [1, 2, \ldots, (n-1)]$ and registers his public key as

$$PK_S = SK_S.G = (PK_{S1}, PK_{S2})$$

Recipient likewise picks his private key $SK_R \in [1, 2, \ldots, (n-1)]$ and registers his public key as

$$PK_R = SK_R.G = (PK_{R1}, PK_{R2})$$

Signcrypt: Signcryptor picks an arbitrary number $x \in [1, \ldots, (n-1)]$ and computes

$$K = x.PK_R = (K_1, K_2)$$
$$c = E_{K_1}(M)$$
$$h = H(M \parallel K_2)$$
$$s = (x - h.SK_S) \bmod n$$

Then, the signcryptor sends encrypted message $\delta = (c, h, s)$ to receiver
Unsigncrypt: The receiver gets encrypted message $\delta = (c, h, s)$ and processes

$$K' = s.PK_R + h.SK_R.PK_S = (K'_1, K'_2)$$

The original message is recovered from the encrypted message $\delta = (c, h, s)$ as seeks after.

$$M' = D_{K'_1}(c)$$

At that point, receiver confirms the authenticity by checking the accompanying condition.

$$h' = H(M' \parallel K'_2)$$

5 Security Analysis of the Proposed Scheme

The correctness of the suggested signcryption scheme is assessed in this section. Then, the security aspect of the suggested scheme is briefly discussed. We define the security goals used to demonstrate the various types of attacks in the security proof.

5.1 Correctness

The $\delta = (c, h, s)$ signcrypted text is a valid one; its correctness is shown below.

$$
\begin{aligned}
K &= s.PK_R + h.SK_R.PK_S \\
&= (x - h.SK_S).PK_R + h.SK_R.PK_S \\
&= x.PK_R - h.SK_S.PK_R + h.SK_R.SK_S.G \\
&= x.PK_R - h.SK_S.PK_R + h.SK_S.PK_S \\
&= x.PK_R = (K_1, K_2)
\end{aligned}
$$

5.2 Security Proof

Definition 1 In the following game, a signcryption system is said to be indistinguishable from adaptive chosen ciphertext attack property if no polynomial bounded adversary has non-negligible advantages [9].

(i) The challenger C obtains the common parameter by running the setup algorithm with security parameter K, and the public parameter is sent to the adversary A.

(ii) The adversary A asks the following queries adaptively to the algorithms signcryption and unsigncryption, respectively, with a polynomially bounded number of times.

Queries to signcryption algorithm: The challenger performs the signcryption algorithm called signcrypt(M, SK_S, PK_R) with the private key of signcryptor SK_S after getting the message M and an arbitrary recipient public key PK_R chosen by adversary A. Then, the result is sent to the adversary.

Queries to unsigncryption algorithm: The challenger runs the unsigncryption algorithm called unsigncrypt(δ, SK_R, PK_S) after getting the ciphertext δ from the adversary A. The result obtained here is a signed plaintext M. The challenger returns the plaintext M to adversary A if it is valid for the recovered signcryptor's public key PK_S, otherwise return the symbol.

(iii) There are two plaintexts, M_0 and M_1, and A will select an arbitrary private key, SK_S, and he will challenge it. Then, the challenger chooses a bit b randomly, which depends on M_0 and M_1 and computes $\delta = Signcrypt(M_b, SK_S, PK_R)$. Final result sends to A.

(iv) Again, adversary A adaptively requests the predefined number of queries as in the second stage, but it is not permitted to query the unsigncryption algorithm corresponding to δ.

(v) Finally, if $b' = b$, A will generate a bit of b' and win the game. A is provided as an advantage

$$
Adv(A) = Pr[b' = b] - 1/2
$$

Proof From the above game, the suggested system is demonstrated to be more secure against the adaptive chosen ciphertext attack.

Here, the following assumptions are taken:

The adversary selects the private key of a signcryptor, and two messages of the same length M_0 and M_1 are selected by an adversary A and sent to challenge. Then, the challenger randomly selects a bit b and calculates the δ ciphertext with the specified recipient public key PK_R using the following formula.

$$\delta = (c, h, s)$$

After the challenger receives the δ ciphertext, the adversary guesses a bit b' and produces a new ciphertext by randomly selecting a message M'. Then, he selects a private key $SK_S \rightarrow Z_q*$ randomly and computes the following necessary parameters for δ.

$$PK'_S = SK'_S.G$$
$$K' = x'.PK_R = (K'_1, K'_2)$$
$$c' = E_{K1'}(M')$$
$$h' = H(M' \parallel K'_2)$$
$$s' = (x' - h'.SK'_S) \bmod n$$

Then, adversary A sends back δ' to the challenger for unsigncryption.

Then, the challenger uses $\delta' = (c', h', s')$ for computing the one-time symmetric key. $K' = s'.PK_R + h'.SK_R.PK_S = (K'_1, K'_2)$ and $M' = D_{K'_1}(c)$.

In case $h' = H(M' \parallel K'_2)$, then the challenger returns the message M, otherwise rejects the message. The response is automatically dismissed because it is impossible to calculate $K' \neq K$ as SK'_R from PK_R, whose complexity is a elliptic curve discrete logarithm problem. It is therefore found that the suggested scheme is safe against the adaptive chosen ciphertext attack.

Definition 2 The suggested signcryption scheme against chosen message attack is unforgeable [9].

Proof In the aforementioned scheme, the attacker cannot generate a valid (c, h, s) ciphertext without the private key of the sender because the computational complexity of the private key of the sender falls under the hardness of the discrete logarithm elliptic curve issue.

The attacker is unable to produce if the value of K is leaked or compromised as it needs two parameters of x and SK_S. An attacker needs to solve ECDLP before deriving x from $K = x.PK_R$. Again, the solution for ECDLP is computationally infeasible. Hence, the proposed scheme is unforgeable.

Definition 3 No other person will verify the signcrypted message in the suggested signcryption scheme except for the designated receiver [9].

Proof The sender generates a one-time symmetric key using the public key obtained from the designated receiver. At the receiver end, the receiver generates the one-time symmetric key by its own private key. This work cannot be done by another receiver except the designated receiver because this step is computationally hard under ECDLP.

6 Conclusion

Our proposed scheme based on elliptic curve discrete logarithmic problem simultaneously provides unforgeability and secure against adaptive chosen ciphertext attack with shorter key length. The proposed system achieves the security properties with computational cost savings compared to the traditional signature and then the encryption scheme, thus making the new scheme more suitable for IoT environments with restricted energy. At long last, the proposed scheme is based on elliptic curve cryptography, and it has great advantages to apply in the IoT environment more efficiently because of the low computational and communication cost.

References

1. Alsaadi, Ebraheim, and Abdallah Tubaishat. 2015. Internet of things: features, challenges, and vulnerabilities. *International Journal of Advanced Computer Science and Information Technology* 4 (1): 1–13.
2. Baek, Joonsang, Ron Steinfeld, and Yuliang Zheng. 2007. Formal proofs for the security of signcryption. *Journal of Cryptology* 20 (2): 203–235.
3. Elkamchouchi, Hassan M., Eman F. Abu Elkhair, and Yasmine Abouelseoud. 2013. An efficient proxy signcryption scheme based on the discrete logarithm problem. *International Journal of Information Technology*.
4. Hwang, Ren-Junn, Chih-Hua Lai, and Feng-Fu Su. 2005. An efficient signcryption scheme with forward secrecy based on elliptic curve. *Applied Mathematics and Computation* 167 (2): 870–881.
5. Hyun, Suhng-Ill, Eun-Jun Yoon, and Kee-Young Yoo. 2008. Forgery attacks on Lee–Chang's strong designated verifier signature scheme. In *Second International Conference on Future Generation Communication and Networking Symposia, 2008. FGCNS'08*, vol. 2. IEEE.
6. Jakobsson, Markus, Kazue Sako, and Russell Impagliazzo. 1996. Designated verifier proofs and their applications. In *Advances in Cryptology—EUROCRYPT96*. Berlin, Heidelberg: Springer.
7. Lee, Ji-Seon, and Jik Hyun Chang. 2009. Comment on Saeednia et al.'s strong designated verifier signature scheme. *Computer Standards & Interfaces* 31 (1): 258–260.
8. Lopez, Julio, and Ricardo Dahab. 2000. *An Overview of Elliptic Curve Cryptography*.
9. Mohanty, Sujata, and Banshidhar Majhi. 2012. A strong designated verifiable DL based signcryption scheme. *JIPS* 8 (4): 567–574.
10. Saeednia, Shahrokh, Steve Kremer, and Olivier Markowitch. 2004. An efficient strong designated verifier signature scheme. In *Information Security and Cryptology—ICISC, 2003*, 40–54. Berlin, Heidelberg: Springer.
11. Steinfeld, Ron, and Yuliang Zheng. 2000. A signcryption scheme based on integer factorization. *Information Security*, 308–322. Berlin, Heidelberg: Springer.

12. Ting, Pei-Yih, Jia-Lun Tsai, and Tzong-Sun Wu. 2017. Signcryption method suitable for low-power IoT devices in a wireless sensor network. *IEEE Systems Journal* 12 (3): 2385–2394.
13. Zheng, Yuliang. 1997. Digital signcryption or how to achieve cost (signature & encryption) ≪ cost (signature) + cost (encryption). In *Advances in Cryptology—CRYPTO'97*, 165–179. Berlin, Heidelberg: Springer.
14. Zheng, Yuliang, and Hideki Imai. 1998. How to construct efficient signcryption schemes on elliptic curves. *Information Processing Letters* 68 (5): 227–233.
15. Yu, Yong, et al. 2009. Identity based signcryption scheme without random oracles. *Computer Standards & Interfaces* 31 (1): 56–62.

Performance of C++ Language Code in Quantum Computing Environment

Rajat Rajesh Narsapur, Rishabh Arora, Meet K. Shah and Prasant Kumar Pattnaik

Abstract This paper considers the execution of different C++ language program codes based on simple program, functions, class inheritance, recursion and file handling under Quantum Computing Environment using the Quantum++ simulator and the execution of similar C++ language program codes on the traditional computer. The comparison of the results has been presented, analyzed and concluded with. The need, advantages and disadvantages of a Quantum Computing Environment have also been discussed briefly in this paper.

Keywords Quantum++ · C++ · Quantum Computing Environment · Sequential Computing Environment · Classical Computing Environment

1 Introduction

The world became familiar with Quantum Computing with the creation of the first Quantum Turing Machine (QTM) in 1985. With this, Deutsch also introduced a new variety of Quantum Computing Environment which supported a feature known as Quantum Parallelism, which was absent in Classical Computing Environment. If replicated in a Classical Computing Environment, it would result into a significant slowdown in the performance. Climbing the steps in 1996, researchers proved that a Universal Quantum Simulator could be developed [1]. The Quantum Computing

R. R. Narsapur (✉) · R. Arora · M. K. Shah · P. K. Pattnaik
School of Computer Engineering, Kalinga Institute of Industrial Technology (KIIT-DU),
Bhubaneswar, Odisha 751024, India
e-mail: rjtnarsapur575@gmail.com

R. Arora
e-mail: arorarishabh607@gmail.com

M. K. Shah
e-mail: meet.shah1598@gmail.com

P. K. Pattnaik
e-mail: patnaikprasantfcs@kiit.ac.in

© Springer Nature Singapore Pte Ltd. 2020
D. Swain et al. (eds.), *Machine Learning and Information Processing*,
Advances in Intelligent Systems and Computing 1101,
https://doi.org/10.1007/978-981-15-1884-3_46

Environment relies on an idea of superposition of logical pathways that could interfere with one another [2]. In a classical non-deterministic Turing Machine, the number of steps involved in computations and simulations for probabilistically weighted quantum processes grows exponentially which makes it very difficult to perform the computations efficiently on a typical probabilistic or non-deterministic automation. But in 1986, Richard Feynman proved that computations can be performed on a Quantum Computing Environment which could act as a simulator for the probabilistically weighted quantum processes. Reference [3] having gone through the basic introduction, basic information about a Quantum Computing Environment has been given in Sect. 2. Further, the need of a Quantum Computing Environment in today's world has been briefly described in Sect. 3. The advantages and disadvantages of a Quantum Computing Environment are listed in Sect. 4. The results of the practical implementation of the program codes are presented in Sect. 5, which have been further analyzed in Sect. 6. The conclusion has been presented in Sect. 7 and Sect. 8 lists the references and citations.

2 Quantum Computing Environment

A Quantum Computing Environment (QCE) is an edge above the Classical Computing in terms of many aspects. A brief explanation on the internal terminologies of the QCE is given below.

In Classical Computing Environment, a bit is always represented by a 0 or a 1, both of which are binary values, whereas in a Quantum Computing Environment, the smallest unit is represented by a two-dimensional vector known as a qubit. The two basic states of a qubit are $|0\rangle$ and $|1\rangle$, pronounced as ket 0 and ket 1, respectively [4–8]. While any calculation is done using this 1-bit, the result obtained is a superposition combining the answers of the computation having been applied to a 0 and a 1. Any calculation is performed simultaneously for both 0 and 1 [1, 9]. After having gone through this, the application of such an environment is today a requirement, the reasons for which follow in the next section.

3 Need of a Quantum Computing Environment

With a person being able to cross seven seas in a finite amount of time through accelerated speed of transport mechanism in today's world, the world needs computations to be accelerated too. A computer has a major amount of contribution in every field of study and also, the daily life of individuals. Nowadays, people need their computers to be as much fast as they can be in terms of CPU Processing time. In an era of transforming and evolving technology, it is a necessity to introduce a faster environment for a better human–computer interaction. This need can be fulfilled by the use of Quantum Computing Environments. All the tricky algorithms that require higher execution time can be executed more than 100 times faster using a Quantum Computing Environment. The comparison of execution times between a Traditional and a Quantum Computing Environment is shown in later sections. While searching a database, a Quantum Computing Environment provides up to four times the speed in the number of queries needed. Using the principle of quantum superposition, a database with k number of items can be searched using $k^{1/2}$ queries. This square root speed can have a significant impact on the time required for a search [10]. The Quantum Computing Environment offers a healthy environment of an efficient, time-saving and a faster computing capability which shall basically accelerate the computing speed. When it comes to transforming an idea into reality, it is very important to consider the practical implications. The advantages and disadvantages of Quantum Computing have been listed in the next section.

4 Advantages and Disadvantages of a Quantum Computing

4.1 Advantages

The main advantage of quantum computing is that it can execute any task very fast when compared to the Classical Computing Environment. A Quantum Computing Environment is able to perform a class of computational problems, like prime factorization. Diving further into the topic, we come to know that the atoms exchange their positions and transfer signals faster in the case of a Quantum Computing Environment than in case of a Classical Computing Environment. In case of any classical algorithm (non-randomized or randomized), that is capable of solving the problem with bounded error, needs $O(n^3)$ queries to the multiplication constants of the algebra. In a Quantum Computing Environment, it requires $O(n^{3/2})$ queries to the multiplication constants of the algebra [11]. Quantum Computing Environments have the potential to satisfy the need for extra computing power needed by the next generation to transform modern society. With this era, lives could be saved by more methodical and effective drug designs, discoveries in material science could be revolutionized by simulations of quantum processes, and Internet encryption could be replaced by much more reliable methods [12].

4.2 Disadvantages

The major disadvantage of Quantum Computing lies in the technology required to implement it. The technology that is required is not available abundantly and requires a lot of research and innovation. The reason that has been observed is that an electron, which is essential for the functioning of a Quantum Computing Environment, is damaged as soon as it comes in contact with the quantum environment. So far there has not been any positive progress on this research.

5 Performance of Quantum Computing Over Sequential Computing

This section contains a practical implementation of a Quantum Computing environment through the means of a simulator and observes the difference in relative performances when the programs are executed in the two different environments, a Quantum Computing Environment and a Sequential (or Classical) Computing Environment. The execution of programs has been done using a Quantum Computing Simulator: Quantum++. It is a multithreaded modern Quantum Computing library that solely consists of header files and has been written in C++ 11 standard. The library is capable of simulating arbitrary quantum processes and classical reversible operations on billions of bits. It also supports the traditional reversible logic. Quantum++ can successfully simulate the evolution of 25 qubits in a pure state or of 12 qubits in a mixed state reasonably fast, if operated on a typical machine (Intel i5 8 GB RAM) [13]. The execution of the programs has been carried out through a GCC-GNU Compiler. The observations of the performance of a Classical Computing Environment versus the performance of a Quantum Computing Environment (through a simulator) while executing a C++ program are shown in Table 1 that includes serial number, the type of program, execution time in Quantum Computing Environment and execution time in Classical (or Sequential) Computing Environment.

6 Result Analysis

During the experiment, it has been observed that the execution times of different types of programs in two different environments, namely Quantum Computing Environment and Classical (or Sequential) Computing Environment has been recorded in Table 1. The graphical representation of the execution time of different types of programs (listed in Table 1 with serial no. 1, 2, 3, 4, 5, 6 and 7) in a Quantum Computing Environment (abbreviated as QCE in the graph) and the execution time of different types of programs (listed in Table 1 with serial no. 1, 2, 3, 4, 5, 6 and 7) in a Classical Computing Environment are shown in Fig. 1.

Figure 1 indicates that for every test case that was performed in both the computing environments, Quantum Computing Environment has executed the code faster. For example, in the case of a simple program (listed at Serial No. 1 in Table 1), the execution time taken by a Quantum Computing Environment is 0.0001 s whereas the execution time taken by a Classical Computing Environment is 0.126 s. It is evident from this test case that the QCE executes the code faster.

The ratio of the performance of Quantum Computing Environment over Classical Computing Environment is calculated as per Equation no. (1).

$$P.R. = \frac{\text{Execution Time taken by a Classical Computer}}{\text{Execution Time taken by a Quantum Computing Environment}} \quad (1)$$

Table 1 Execution times (in seconds) obtained after executing C++ program codes of different types of programs in a Quantum Computing Environment (using quantum++) and Classical (or Sequential) computer, respectively

S. No.	Program type (in C++)	Execution time in Quantum Computing Environment (in seconds)	Execution time in Classical Computing Environment (in seconds)
1	Simple program (Hello World)	0.0001	0.126
2	Program containing a Predefined function (factorial using random function)	0.002	0.125
3	Class Object (operations on a data member associated with the Class Object)	0.008	0.151
4	Inheritance of classes (operations on data members of inherited classes)	0.007	0.133
5	Subroutine execution (bubble sorting of an array using a static array)	0.011	0.143
6	File handling (784 characters)	0.013	0.15
7	Factorial program using recursion	0.0009	0.144

where the performance ratio has been denoted by P.R.

The calculated ratio portrays the enhanced performance of Quantum Computing Environment over a Classical Computing Environment. The following points may be noted after the analysis:

1. On an average, the Quantum Computing Environment is approximately 220 times faster than a Classical Computing Environment in executing basic C++ programs.
2. The Quantum Computing Environment performs extremely well in the case of a recursive algorithm. According to the test case no. 7, it has been recorded that the Quantum Computing Environment is 160 times faster than the Classical Computing Environment.
3. In programs involving simple output statements, the performance of a Quantum Computing Environment is very high and appreciable. According to the test case no. 1, it has been recorded that the Quantum Computing Environment is 1260 times faster than the Classical Computing Environment.
4. However, in programs involving file handling and subroutine execution, the performance of a Quantum Computing Environment over the Classical Computing Environment has been observed to be significantly lower than the other test cases.

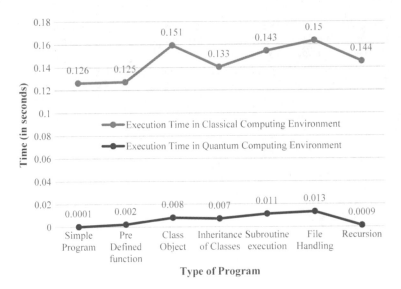

Fig. 1 Graphical representation of the execution times (in seconds) obtained after executing C++ program codes of different types of programs in a Quantum Computing Environment (using Quantum++) and Classical (or Sequential) computer, respectively

The graphical representation of the performance ratio of a Quantum Computing Environment over the Classical Computing Environment for different types of programs (listed in Table 2 with serial no. 1, 2, 3, 4, 5, 6 and 7) is given in Fig. 2.

After analyzing the observations and the obtained data, we arrived upon the following conclusion.

7   Conclusion

Classical (or Sequential) Computing Environment has contributed enormously in the progress of science and technology while spreading its reach to different professional fields and areas of the society, whereas Quantum Computing Environment, an emerging technology may escalate each process and reduce the execution time taken. In this work attempt, we implemented the C++ program codes with the use of a Quantum++ Simulator. The experimentation of the genetic algorithm in both the prime environments will be a future scope of work.

Table 2 Performance ratio of the Quantum Computing Environment over the Classical Computing Environment for different types of C++ program language codes

S. No.	Program type	Performance ratio of Quantum Computing Environment over Classical Computing Environment
1	Simple program (Hello World)	1260
2	Program containing a Predefined function (factorial using random function)	62.5
3	Class Object (operations on a data member associated with the Class Object)	18.875
4	Inheritance of classes (operations on data members of inherited classes)	19
5	Subroutine execution (bubble sorting of an array using a static array)	13
6	File handling (784 characters)	11.54
7	Factorial program using recursion	160
Average ratio		220.702

This table also consists the average of all the performance ratios obtained after executing different C++ program codes

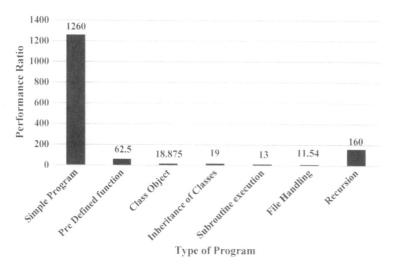

Fig. 2 Graphical representation of the performance ratio between the Classical Computing Environment and the Quantum Computing Environment for different types of C++ program language codes

References

1. Rylander, Bart, Terence Soule, James Foster, and Jim Alves-Foss. 2000. Quantum genetic algorithms, 373.
2. DiVincenzo, D.P. 1995. Principles of quantum computing. In *Proceedings ISSCC '95—International Solid-State Circuits Conference*, San Francisco, CA, USA, 312–313. https://doi.org/10.1109/isscc.1995.535569.
3. Narayanan, A. 1999. Quantum computing for beginners. In *Proceedings of the 1999 Congress on Evolutionary Computation-CEC99 (Cat. No. 99TH8406)*, vol. 3. Washington, DC, USA, 1999, 2231–2238. https://doi.org/10.1109/cec.1999.785552.
4. Jain, S. 2015. Quantum computer architectures: a survey. In *2015 2nd International Conference on Computing for Sustainable Global Development (INDIACom)*, New Delhi, 2165–2169.
5. Kasivajhula, Siddhartha. 2006. Quantum computing: a survey, 249–253. https://doi.org/10.1145/1185448.1185504.
6. Nielsen, Michael A., and Isaac L. Chuang. 2010. *Quantum computation and quantum information*, 13. Cambridge University Press. ISBN 978-1-107-00217-3.
7. Ying, Mingsheng. 2010. Quantum computation, quantum theory and AI. *Artificial Intelligence* 174: 162–176. https://doi.org/10.1016/j.artint.2009.11.009.
8. Altenkirch, T., and Grattage J. 2005. A functional quantum programming language. In *20th Annual IEEE Symposium on Logic in Computer Science (LICS' 05)*, Chicago, IL, USA, 249–258.
9. Ankenbrandt, C.A. 1990. An extension to the theory of convergence and a proof of the time complexity of genetic algorithms. In *FOGA*.
10. Humble, T. 2018. Consumer applications of quantum computing: a promising approach for secure computation, trusted data storage, and efficient applications. *IEEE Consumer Electronics Magazine* 7 (6): 8–14.
11. Combarro, E.F., J. Ranilla, and I.F. Rúa. 2019. A quantum algorithm for the commutativity of finite dimensional algebras. *IEEE Access* 7: 45554–45562.
12. McGeoch, C.C., R. Harris, S.P. Reinhardt, and P.I. Bunyk. 2019. Practical annealing-based quantum computing. *Computer* 52 (6): 38–46.
13. Gheorghiu, V. 2018. Quantum++: a modern C++ quantum computing library. *PLoS One* 13 (12): e0208073.

Retrieval of Ontological Knowledge from Unstructured Text

Dipak Pawar and Suresh Mali

Abstract In this article, we examined the issue of automatic ontology formation process from unstructured text data. To understand the ontology of the domain, ontology should be expressed in terms of information tables and ontology graphs. Ontology graph consists of taxonomic and non-taxonomic relations. Non-taxonomic relations are easier to understand to non-expert users. Extracting non-taxonomic relations from ontology is a challenge. In order to improve ontology of the domain, appropriate machine learning classifier needs to be investigated for feature classification.

Keywords Ontology · Taxonomic relations · Non-taxonomic relations · Information tables

1 Introduction

Use of Internet is rapidly increasing day by day. A survey suggests that nearly 50% of world population is using Internet. Volume of data generation is increasing tremendously. All human beings have 340 times digital data in his/her account. This digital data cannot be processed by orthodox digital data processing instruments within real time. The inconvenience integrates the areas of trapping, repository, find, sharing, interchange, investigation, and representation of this digital data. Data generated can be heavy text files or multimedia. Large amount of data generated at data warehouses, which can be in structured, semi-structured, and unstructured format. Structured data is in organized format. It can be represented into relational databases. Meaningful information can be extracted from relational database by performing upfront search queries and search algorithms. Semi-structured data cannot be represented into relational databases. Semi-structured data has some organizational attributes which make it somewhat easier to analyze. Unstructured data is complex and textual. It does not have any organization properties. Unstructured data cannot be represented into relational databases. Unstructured data can be human generated or machine generated.

D. Pawar (✉) · S. Mali
SKNCOE, Pune, India
e-mail: deepak.pawar@vit.edu

© Springer Nature Singapore Pte Ltd. 2020
D. Swain et al. (eds.), *Machine Learning and Information Processing*,
Advances in Intelligent Systems and Computing 1101,
https://doi.org/10.1007/978-981-15-1884-3_47

501

Study states that the volume of unstructured data is 90% and structured data is 10%. Unstructured data can be in the form of text or multimedia (image/audio/video). Analysis, organization, integration, and search for unstructured data are very important to infer knowledge which will help practically in various domains. Unstructured data is present in various domains such as healthcare applications, social media text analysis, industry applications, automatic personality detection, education, transportation, and many more. To infer knowledge from unstructured data, one needs to extract ontology of the domain. Ontology is a data model that can be defined as the formal representation of knowledge of domain which is expressed in terms of concepts and attributes by considering semantic relation between classes. Attributes and concepts have association between them. This association can be expressed in terms of taxonomic and non-taxonomic relations. Taxonomy relation is type_of or is_a (instance_of) and part_of relation between ontology classes whereas non-taxonomic relation is any relation between concepts except taxonomy relation. Ontology of the domain needs to be expressed in terms of information tables and ontology graph which make it easier to understand to non-expert users. Identifying non-taxonomic relation between concepts and attributes is challenge. Manuscript is divided into Sect. 2 Ontology Formation Process, Sect. 3 Mathematical modeling of ontology formation process, Sect. 4 Literature Review, and Sect. 5 Conclusion.

2   Ontology Formation Process

Ontology represents knowledge of the selected domain. Input to ontology formation process is unstructured text file. Words are extracted from unstructured text using WordNet library. Words are classified into classes and concepts using machine learning by considering semantics and sentiments of the selected domain. Output of ontology formation process is ontology graph and information tables. Graphs and information tables can be understood by non-expert users. Ontology formation process is depicted in Fig. 1.

Ontology formation process is illustrated with example as shown in Fig. 2.

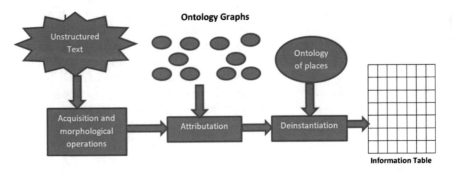

Fig. 1 Ontology formation process

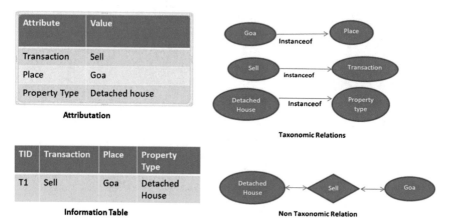

Fig. 2 Ontology graphs and information table

3 Mathematical Modeling of Ontology Formation Process

Knowledge system (KS) is fourfold.

$$KS = (X, T, Y_a, f_{inf})$$

where $a \in T$

X = Nonempty finite set of objects. Ex. Sentences from Unstructured Text.

T = Nonempty finite set of attributes/characteristics.

$\{Y_a\}\ a \in T$ = Family of sets of attribute/characteristic values.

$f_{inf} = T \times X \to \sum_{a \in T} Y_a \to$ Information function such that $f_{inf}(a, x) \in Y_a$ for each $a \in T$ and $u \in X$.

Let O_n be a given ontology. Formally, an ontological graph is a fourfold.

$$O_n G = (N, L, Z, \rho),$$

where:

- N is the nonempty, finite set of concepts/nodes representing concepts in the ontology O_n
- $L \in N \times N$ is the finite set of edges symbolizing semantic relations between concepts from N
- Z is series of type of relation between edges and concepts
- $\rho: L \to Z$ is the function attributing semantic relation to every edge.

Let $O_n G = (N, L, Z, \rho)$ is an ontological graph.

A regional ontological part of graph of $O_n G$ is a graph in a form

$$LOG = (N_L, L_L, Z_L, \rho_L),$$

where $N_L \in L_L \in Z_L \in$ and ρ_L is a function ρ restricted to L_L.

Formally, a simple knowledge system SIS^{OG} over ontological graphs is a twofold.

$$\text{SIS}^{OG} = (X, T\{OG_a\}_a \in_T f_{\text{inf}}),$$

where:

- X is the nonblank, finite set of entities,
- T is the nonblank, finite set of characteristics,
- $\{OG_a\}_a \in_T$ is the family of ontological graphs associated with characteristics,
- $f_{\text{inf}} : T \times X \rightarrow \sum_{a \in T} Y_a \rightarrow$ Knowledge function such that $f_{\text{inf}}(a,u) \in V_a$ for each $a \in T$ and $u \in X$, N_a is the set of concepts from the graph OG_a.

4 Literature Review

Literature survey consisting of seven different methodologies to retrieve ontology from unstructured data as follows:

4.1 Classification of Unstructured Data

Eduardo Castro et al. [1], proposed Associative Classifier for Entity Resolution (AssocER), which is used for the classification of unstructured short texts. Results are not optimized. Not able to identify novel classes. It is not able to perform self-training during prediction phase. Pratibha [2] emphasized on characteristic-based learning which is used to find out features and also to estimate attributes of unstructured text. It is a content-based classification technique. This method is not finding non-taxonomic relations. Bartoli [3] proposes entity extractor technique with less human intervention. Improvements are needed in this technique in order to increase accuracy. Hassan [4] proposes joint convolutional neural network (CNN) and recurrent neural network (RNN) approaches for sentence categorization. This research has tremendous part in decreasing the number of parameters and designing convolutional layer after recurrent layer as replacement for pooling layer. This technique can be used in information retrieval. Tekli [5] proposes a review paper which supplements summarized and exhaustive review of the methods related to semi-structured data processing. Leng et al. [6] proposed the extraction of relationships from high level of noisy and unrelated information unstructured text. It is based on semi-supervised learning approach. This method requires annotation from user. Ritter et al. [7] performed survey on techniques to fetch semantic information from unstructured data. Harleen et al. [8] analyzed unstructured data on Amazon Elastic Cloud and analyzed unstructured data for real-time data analytics. Author is not focusing on investigating the ontology behind domain. Gabriele et al. [9] presented review paper that shows the types of unstructured data available to researchers supporting basic data mining techniques to investigate them. This is a survey paper. Chiange [10] proposes a graph model and an agglomerative algorithm for text document clustering. Proposed algorithm is heavily better than orthodox clustering algorithms, such as k-means clustering, division partitioning, auto class, and hierarchical clustering algorithm. Between documents, there can be overlap which can be resolved by this method which supplies more summarized and detailed clustering results which allow concept overlap. Bafna [11] proposed a feature learning framework which automatically detects features or abstractions. This research investigated the importance of unsupervised learning and non-Euclidian distance techniques. This research also explored dimensionality reduction. Main drawback of this research is that it is only identifying taxonomic relations. Non-taxonomic relations need to be explored for correct ontology framework formulations. Reyes Ortiz [12] performed survey based on natural language processing (NLP) approached for unstructured data. Fang et al. [13] presented the Unified Automata Processor (UAP), a novel construction that supplies general and efficient support for Finite Automata (FA) required for NLP. Proposed architecture is implemented on CPU and GPU. The main drawback of this system is that it requires external hardware to run proposed algorithm. Islam et al. [14] applied standard encryption techniques on unstructured data. Researchers show that unstructured text categorization with respect to delicacy levels improves the attainment of the system.

4.2 Attribute Association Modeling Using Taxonomic Relations

Shen et al. [15] proposed a general architecture for connecting named attributes in Internet-free text with a hybrid knowledge network; a probabilistic linking model, which consolidates an attribute popularity model with an attribute object model; and also a knowledge population algorithm to rigorously augment the network. This paper does not address about non-taxonomic relation mapping. Sriraghav et al. [16] proposed attribute-based opinion mining algorithm which focuses on user centric attributes. Algorithm forces user to take well-informed decisions focusing on the attributes user is relying most. Author proposes a tool that can be applied in any domain which can analyze unstructured data by accepting domain and user-specific attributes. Human intervention is needed to input domain-specific attributes. Tarasconi [17] proposes emergency management tool using tweeter tweets. Author has performed sentiment analysis on twitter dataset. However, this research is not able to accurately find the notion of in-formativeness within hazard-related streams. Ahmad et al. [18] proposed framework for text information extraction in terms of context vectors. This architecture is dependent on mappers and reducers developed on Apache Hadoop. Large dimensionality is handled by clustering. Extracted context vectors have large dimensions. Domain-specific attributes entered manually. Fikry et al. [19]

proposed business analytics framework for unstructured data extraction. However, author is not handling non-taxonomic relations. Istephan et al. [20] proposed framework for unstructured medical data. This framework has in-built module. User can add his/her own unstructured data operation module. Query can be fired to framework. This framework is not able to identify non-taxonomic relations. It also needs to manually enter domain-specific attributes. Lee et al. [21] proposed multidimensional model to extract information. This paper presents a text cube model which is based on multidimensional text database. This model is not able to identify non-taxonomic relations. Domain-specific attributes need to be manually entered. Saini et al. [22] proposed domain-independent emotion mining framework for unstructured data. This method implements self-learning dictionary. This method is less accurate because of semantic ambiguity in statements.

4.3 Attribute Association Modeling Using Non-taxonomic Relations

Ali et al. [23] proposed process of learning non-taxonomic relationships of ontology (LNTRO) from unstructured data. This paper focuses on non-taxonomic relations. Little human intervention is required to perform attribution. If there is a semantic ambiguity, accuracy degrades.

4.4 Generation of Information Tables

Rajpathak [24] proposes D-Matrix generation technique from unstructured data for fault diagnosis in automobile domain. While constructing D-matrix from unstructured data, non-taxonomic relations are not considered. Also, it requires annotation for entity extraction. Krzysztof et al. [25] proposed the generation of information tables from unstructured data from real estate domain. It is not generalized solution, and it requires human intervention. Sadoddin et al. [26] investigated the issue of extracting not known relation between 'concepts.' Author has examined the attainment of various correlation methods in finding both 'direct' and 'indirect' relations between concepts. Concept list treated as dictionary given as input to algorithm. Every entity from dictionary is considered as independent concept. The significance of this method is automatic investigation of concepts.

4.5 Graphical Visualization of Ontology

Gianis et al. [27] proposed framework for graphically representing knowledge from structured and unstructured data. It can be understood by non-expert users. It forms query based on graphs. Logical operators like 'AND,' 'OR,' and 'NOT' are not considered in framework. Categorization of concepts requires human intervention. It is time-consuming for complicated query. Mallek et al. [28] graphically represented unstructured data using hidden statistics. Graphical representation is dynamic means changing according to live data. Ontology creation is manual which requires human intervention. Work is limited to single domain; it is not generalized solution.

4.6 File System-Based Search

Alexandru et al. [29] presented indexing solution, named FusionDex, which gives beneficial model for questioning over distributed file systems. It outperforms Hadoop, Grep, and Cloudera. But graph-based user-oriented query retrieval is not addressed in this paper. Zhu et al. [30] proposed mixed architecture which is the mixture of structured and unstructured text data. The combined index is also called joint index. It is semantic index which represents semantic association between attributes and their various resources. The main concern here is how to calculate combined joint index? It is totally complex task and difficult to implement in real time as query processing will be too high.

4.7 Graph-Based Search

Sheokand et al. [31] proposed query-based answering model. It stores relationships between keywords extracted from the user queries, improves query-answering services, and adds additional services like re-ranking of results and exploration recommendations. This method is based on Web search engine. However, search latency degrades for complicated query.

   Figures 3 and 4 show the summary of reviewed papers based on ontology retrieval methods and evaluation parameters, respectively. Figure 3 show that less attention is given on non-taxonomic relation identification from unstructured data, conversion of unstructured data into information tables, representation of relational databases and information tables graphically, and graph-based search on unstructured data. Figure 4 shows the summary of evaluation parameters in current state of art.

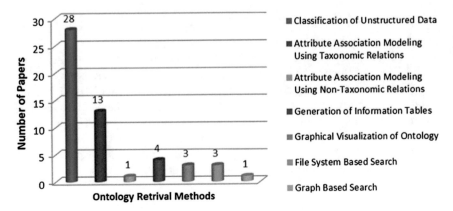

Fig. 3 Summary of reviewed papers based on ontology retrieval methods

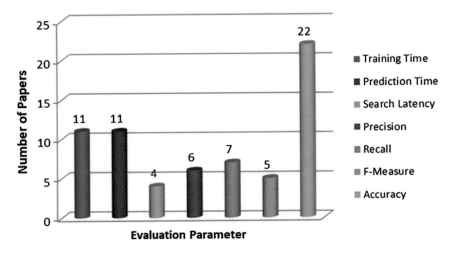

Fig. 4 Summary of reviewed papers based on evaluation parameters

5 Conclusion

In this paper, we have reviewed work done by previous researchers in the ontology formation process. We have illustrated ontology formation process with appropriate example. To accurately represent ontology of the domain in terms of information tables and ontology graph, focus needs to be given on non-taxonomic relations. More work needs to be done on attribute association modeling using non-taxonomic relations. For feature selection, appropriate machine learning algorithm needs to be explored. Best effort query answering tool need to be designed to apply business intelligence on unstructured text data.

References

1. Castro, P.S., et al. 2017. Classifying short unstructured data using Apache Spark Platforms Eduardo. In *ACM/IEEE Joint Conference.*
2. Pratibha, P. 2014. Attribute based classification and annotation of unstructured data in social networks. In *IEEE International Conference on Advanced Computing.*
3. Bartoli, Alberto. 2017. Active learning of regular expressions for entity extraction. In *IEEE transaction on Cybernetics.*
4. Hassan, Abdalraouf. 2017. Convolutional recurrent deep learning model for sentence classification. *Journal of IEEE Access.*
5. Tekli, Joe. 2015. An overview on XML semantic disambiguation from unstructured text to semi-structured data: background, applications, and ongoing challenges. *IIEEE Transaction on Knowledge and Data Engineering.*
6. Leng, Jiewu, et al. 2016 Mining and matching relationships from interaction contexts in a social manufacturing paradigm. *IEEE Transactions on Systems, Man, and Cybernetics.*
7. Ritter et al. 2017. Toward application integration with multimedia data-Daniel. In *IEEE International Enterprise Distributed Object Computing Conference.*
8. Harleen. 2016. Analysis of hadoop performance and unstructured data using zeppelin. In *IEEE International Conference on Research Advances in Integrated Navigation Systems.*
9. Gabriele, et al. 2016. Mining unstructured data in software repositories: current and future trends. In *IEEE International Conference on Software Analysis, Evolution and Reengineering.*
10. Chiange, I-Jen. 2015. Agglomerative algorithm to discover semantics from unstructured big data. In *IEEE International Conference on Big Data.*
11. Bafna, Abhishek. 2015. Automated feature learning: mining unstructured data for useful abstractions. In *IEEE International Conference on Data Mining.*
12. Reyes Ortiz, Jose A., et al. 2015. Clinical decision support systems: a survey of NLP-based approaches from unstructured data. In *IEEE International workshop on Database and Expert Systems Applications.*
13. Fang, Yanwei, et al. 2015. Fast support for unstructured data processing: the unified automata processor. In *ACM Proceedings of International Symposium on Microarchitecture.*
14. Islam, Md. Rafiqul. 2014. An approach to provide security to unstructured big data. In *IEEE International Conference on Software Knowledge, Information Management and Applications.*
15. Shen, Wei, et al. 2018. SHINE+: a general framework for domain-specific entity linking with heterogeneous information networks. *IEEE Transactions on Knowledge and Data Engineering.*
16. Sriraghav, K. et al. 2017. ScrAnViz-A tool to scrap, analyze and visualize unstructured data using attribute based opinion mining algorithm. In *IEEE International Conference on Innovations in Power and Advanced Computing Technologies.*
17. Tarasconi, Frensacsco. 2017. The role of unstructured data in real time disaster related social media monitoring. In *IEEE Inernational Conference on Big Data.*
18. Ahmad, Tanvir, et al. 2016. Framework to extract context vectors from unstructured data using big data analytics. In *IEEE Conference on Contemporary Computing.*
19. Mohammad Fikry Abdullah et al. 2015. Business intelligence model for unstructured data management. In *IEEE Conference on Electrical Engineering and Informatics.*
20. Istephan, Sarmad, et al. 2015. Extensible query framework for unstructured medical data—a big data approach. In *IEEE International Conference on Data Mining Workshops.*
21. Lee, Saun, et al. 2014. A multi-dimensional analysis and data cube for unstructured text and social media. In *IEEE International Conference on Big Data and Cloud Computing.*
22. Saini, Akriti, et al. 2014. EmoXract: domain independent emotion mining model for unstructured data. In *IEEE Conference on Contemprory Computing.*
23. Ali, Mohamaed, et al. 2017. The problem learning non taxonomic relationships of ontology from unstructured data sources. In *IEEE International Conference on Automation & Computing.*
24. Rajpathak, Dnyanesh. 2014. An ontology-based text mining method to develop D-matrix from unstructured text. In *IEEE Transactions on Systems, Man, and Cybernetics: Systems.*

25. Krzysztof, et al. 2017. From unstructured data included in real estate listings to information systems over ontological graphs. In *IEEE Conference on Information and Digital Technologies*.
26. Sadoddin, Reza, et al. 2016. Mining and visualizing associations of concepts on a large-scale unstructured data. In *IEEE Conference on Big Data Computing Service & Application*.
27. Gianis, et al. 2017. Graph Based Information Exploration Over Structured and Unstructured Data. In *IEEE Conference on Big Data*.
28. Mallek, Maha, et al. 2017. Graphical representation of statistics hidden in unstructured data: a software application. In *IEEE International Conference on Systems, Man & Cybernetics*.
29. Alexandru, et al. 2017. Toward scalable indexing and search on distributed and unstructured data. In *IEEE International Congress on Big Data*.
30. Zhu, Chunying, et al. 2015. A combined index for mixed structured and unstructured data. In *IEEE International Conference Web Information Systems*.
31. Sheokand, Vishal, et al. 2016. Best effort query answering in data spaces on unstructured data. In *IEEE International Conference on Computing, Communications and Automation*.

Critical Non-functional Requirements Modeling Profiler with Meta-Model Semantics

Mahesh R. Dube

Abstract The system is defined in multiple ways by various engineering and business domains which is an organized set of elements with capabilities that are integrated and interoperable. Every system has a scope and mission-criticality which is determined by user-stakeholder agreement based on both functional and non-functional requirements. The profiling and modeling of requirements are possible at structural levels by making use of Unified Modeling Language (UML). The proposed framework is organized by a meta-model and notation base for modeling critical non-functional requirements. There is a possibility of capabilities overlap and redundancy if the components to be developed are not unique causing an irregular and useless repository of components fulfilling requirements. The model transformations can be performed using a model transformation language that generates target model from source model with the help of transformation rules. The proposed framework focuses on theory of model generation with incorporation of notations to represent a set of critical non-functional requirements.

Keywords Unified Modeling Language · Model-driven engineering · Meta-model · Non-functional requirement

1 Introduction

The system is defined using well-formed goals and objectives requiring management support to deal with the strategic and tactical planning. System has operational as well as functional performance which can be assured by performing variety of tests with the aid from analysis design specification. Quality assurance indicates that the system processes are verified and the system outcomes are validated. The capabilities are integrated using hierarchical structure and architectural baselines. The system classification can be seen in many multiple contexts and domains. The external

M. R. Dube (✉)
Department of Computer Engineering, Vishwakarma Institute of Technology,
Pune, Maharashtra, India
e-mail: mahesh.dube@vit.edu

© Springer Nature Singapore Pte Ltd. 2020
D. Swain et al. (eds.), *Machine Learning and Information Processing*,
Advances in Intelligent Systems and Computing 1101,
https://doi.org/10.1007/978-981-15-1884-3_48

stimuli change the behavior of open system responding to boundary-based stimuli whereas a closed system is self-performing system [1]. The proposed framework focuses on creating mathematical meta-model on the basis of UML meta-model.

The problem of having correct and stable notations for requirement diagrams, especially addressing non-functional requirements is addressed by the proposed the framework. Majority of the systems belong to open systems. In static systems, components or characteristics are stable over time whereas dynamic systems components or characteristics are changing. Logical and conceptual systems comprise of ideas, often represented as models. System Engineering (SE) consists of devising solutions from available alternatives with the help of scientific, mathematical, and analytical principles which satisfy the user expectations along with low risk and high efficiency. A capability is equivalent to a feature that responds to the stimulus generated by external environment with adequate performance. Resource non-availability or completion within stipulated time is the reasons for termination of capability execution. Fitness for purpose is an associated term with the capabilities which is set of characteristics that need to be preserved by making use of configuration management technique. A function can be an internal activity or operation or task that has unique purpose and objective with performance prescription.

The problem-solving process involves multistep activities beginning with problem identification or proposing statement of work and terminating with the best fit solution. Requirements are the statements indicating the features, functions, operations, tasks, or attributes that the software has to possess. Changes in requirements affect the analysis and design phase requiring repeated cycles of implementation and testing [2, 3].

The requirements engineer plays central role in initial phases of software development lifecycle and has greater responsibility when it comes to construction release phases. It is therefore that requirement engineering is just not limited to documentation of needs or features but to the collaborative and coordination skills. The capabilities of a requirement engineer are as follows:

- The requirements engineer must have process knowledge along with ability to interpret and synthesize the knowledge base or empirical evidences.
- The requirements engineer must have analytic thinking capability to thoroughly investigate familiar and unfamiliar domains under study or observations. The requirement engineering must be able to classify the problem domain into simple, complicated, and complex.
- The elicitation of stakeholder requirements and its interpretation is partly based on communication skills of requirement engineer. Best practices for communication like ability to listen, frequency of meaningful questionnaire to be asked, picking useful contextual information leads to better outcomes.

A traceability matrix provides visibility into completeness of the quantitative definition and testability of each requirement. The critical non-functional requirements considered are availability, efficiency, flexibility, and performance. As requirement inconsistency is an indicator of different views by different users the tool attempts to note the comments and indicates such requirements [4, 5]. Section 2 focuses on the

UML context, features of models. Section 3 describes the knowledge dimension in meta-model generation and semantics of requirement profiler. Section 4 represents the critical non-functional requirements notation catalog using which the instance level models can be created by the modeler. It is followed by the conclusion and references.

2 The Unified Modeling Language Context

The approach for indicating the requirement evolution management is through Version Control visualized in terms of a Version-Graph. The input for this graph will be through the configuration information of requirement. The main argument for visualization is that it can represent information in ways that are easier to understand, therefore, enhancing communication. When applied to requirements engineering, it is expected that the requirements specification can be transformed into a form, which the customer and the developer can both comprehend more easily, thus bridging the communications gap between the two [6].

Object Management Group (OMG) came up with Unified Modeling Language (UML) facilitating the system modelers to model the system based on use case-based approach that can be initiated in project analysis phase to deployment structure that can be observed in the product release phase. Over a period of two decades, UML became standard in modeling arena incorporating various methods as well methodologies for software development. The concepts and diagrams represented in UML are based on object-oriented paradigm that considers abstraction, hierarchy, modularity, and encapsulation to be the prime properties [7–9].

Since UML is flexible and customizable, the multiple models can be created indicating organizational business processes, application design, database architecture design, and workflow structures. UML foundation was laid by adopting three methodologies namely object-oriented analysis and design (OOAD, Booch Method), object-oriented software engineering (OOSE, Jacobson) and object-modeling technique (OMT, Rumbaugh). A model can represent minimal as well as maximal representations of objects. This means that the object design can be extended as a shallow or deep structures indicating levels of abstractions within the scope and limits of the objects. Each object carries features, properties, aspects or characteristics that can be static or dynamic. The models are thought as a repository of knowledge about object characteristics and the cycles the object is expected to undergo during its construction a, and deployment [10–12].

Features of Models: A model needs to possess the following three features:

- Mapping feature: A model is based on an abstraction that is observable.
- Reduction feature: A model cause reflection of a selection of an abstraction properties leading to a set of homogenous abstractions in a view.
- Pragmatic feature: A model replaces an abstraction with respect to defined intention, theme, scope and some purpose.

An engineering model must possess the following five characteristics:

- Abstraction: It highlights essential aspects while suppressing the others.
- Understandability: A model must rise to the modeler's intuition and interpretable by the project stakeholders.
- Accuracy: A model must be as close as possible to the reality which indicates that the model exactly represents features as system would have performed.
- Predictiveness: Predictability of models is an asset to the designers in terms of estimating the success of modeled system.
- Inexpensiveness: The analysis, design, and construction activities carried out using modeling languages must not involve recurring costs including the repair and maintenance costs.

3   Knowledge Engineering Dimensions

A subject area understanding is represented as knowledge. Subject area covers concepts and facts along with relations and mechanisms for problem-solving in the area. A notation is required for developing knowledge representation technique. First-order predicate calculus having formal notations is used for representing knowledge. To represent knowledge formally, the description logics notations are required to be expressive. Knowledge structures are commonly represented by using XML syntax.

The semantics of formal language under consideration consists of the graphical or textual syntax of the UML, a well-formed mathematical theory representing the semantic domain, and a semantic mapping indicating the relationship between syntactic elements and the semantic domain elements [13–15]. The syntactic construct are mapped onto semantic constructs individually generating the denotational semantics. The grammar on the meta-model can be called as semantics. UML abstract syntax is used to generate context-free grammar. The semantic domain is capable of defining structural and behavioral properties for a class of systems [16–18].

- Let there be two models m and n *where* $m, n \in UML$. *These models can be* joined into a composition $m \oplus n$ where or $\oplus$ indicates the syntactic domain composition operator. The $SemDomain(m \oplus n) = SemDomain(m) \cap SemDomain(n)$ defines semantics of $m \oplus n$. The semantics for the UML documents consists of multiple views with multiple UML documents is given as $SemDomain(\{document_1, ..., document_n\}) = SemDomain(document) \cap \cdots \cap SemDomain(document_n)$. If $SemDomain(documents) \neq \emptyset$ is satisfied, the UML models' documents are consistently bearing the correct properties. View integration and model consistency verification are supported by examining the semantic domain consistency. Refinement of $m \in UML$ *belonging to* $n \in UML$ is exactly the same if $SemDomain(n) \subseteq SemDomain(m)$.
- The buffer of the events to be processed is maintained by event store. The universe of types is defined by UETYPE and universe of values is defined by UEVAL. A

universe of class names ids defined by UECLASS with universe of object identifiers UEOBJECT-ID with UECAR as a set of values used for associating the values assignment to each type. A series of semantic properties are described below, each describing a semantic domain in an environment. *UETYPE* indicates the universe of all type names. Members of all type of names are gathered in the universe *UEUVAL* of values. A $T \in UETYPE$ denotes type name.

3.1 Types and Values

UETYPE UEUVAL UECAR: UETYPE → P(UVAL)
$\forall u \in UETYPE: UECAR(u) \neq \emptyset$
UETYPE is the type names universe. *UEUVAL* is the values universe.
UECAR maps nonempty carrier sets to type names to; meaning that carrier sets are not disjoint.

3.2 Basic Types

Bool, Int ∈ UETYPE; true, false ∈ UVAL; UECAR (Bool) = {true,false}
true ≠ false
UECAR (Int) = Z ⊆ UVAL; UETYPE consists of type names Bool and Int.
UEUVAL consists of Boolean and integer values.
Vobject-id; Vobject-id ∈ UETYPE; vobject-id ∈ UVAL
UECAR (Vobject-id) = {vobject-id}
vobject-id returns control without an actual return value.

3.3 Variables, Attributes, Parameters

Variable UEVAR; vtype: UEVAR → UETYPE;
vsort: UEVAR → P(UEVAL); VarAssigned = (v: UEVAR ~ vsort(v))
a: T indicates "*a* is a variable of type *T*" [i.e., *vtype(a) = T*].
$\forall v \in UEVAR: vsort(v) = UECAR (vtype(v)) \land \forall val \in VarAssigned: val(v) \in vsort(v)$
UEVAR is the universe of all variable names.
VarAssigned is the set of all assignments of values.

3.4 Classes and Instances

Class UECLASS; UEOBJECT-ID INSTANCES
attr: UECLASS → Pf (UEVAR)
object-ids: UECLASS → P(UEOBJECT-ID)
objects: UECLASS → P(INSTANCES)
objects: UEOBJECT-ID → P(INSTANCE)
classOf: INSTANCES → UECLASS; classOf: UEOBJECT-ID ~ UECLASS
∀object-id ∈ object-ids(C): classOf (object-id) = C∧ objects(object-id) = {(object-id, r) | r ∈ VarAssigned ∧ dom(r) = attr(C)}
∀o ∈ objects(C): classOf (o) = C
UECLASS is the universe of class names. *attr* assigns each class name with a finite set of attributes. UEOBJECT-ID is the universe of object identifiers.
INSTANCES is the universe of objects and *r* is a variable assignment.

3.5 Attribute Access

Attribute *this: INSTANCES → UOBJECT-ID*
getAttr: INSTANCES × UEVAR ~ UEUVAL
attr: INSTANCES → Pf (UEVAR); attr: UEOBJECT-ID → Pf (UEVAR)
o.this is used instead of *this(o); o.a* is used instead of *getAttr(o, a)*
∀o, (object-id, r) ∈ INSTANCES: this((object-id, r)) = object-id getAttr((object-id, r), a) = r.a attr(object-id) = attr(classOf (object-id)) attr(o) = attr(classOf (o));o.this
is a dummy attribute.

4 Non-functional Requirements Notation Catalog

4.1 Availability

It is the degree to which the system can maintain its resource log and make them available to running program segments. The notation catalog for availability is indicated in Table 1.

Table 1 Notation catalog for availability

Parameter	Definition	Notation
Admissibility	It refers to considerable or desirable properties of the domain model	
Adaptability	It refers to the rate at which the software can demonstrate its behavior in heterogeneous environments. This property is not equivalent to extensibility	
Analyticity	It refers to a basic or essential attribute shared by all the members of the domain model or components under construction	
Attainability	It refers to ability of the software components to succeed in a directed process or progression so as to achieve its objective	
Automaticity	It refers to the ability of software components to act or operate in manner independent of external influence adhering to constraints	
Comity	It indicates referential integrity of components operating in harmony exercising expected behavior	
Commensurability	It refers to measurable attributes of the components by a common agree upon standard	
Decidability	It indicates that software components have properties which can be tested with axioms of a given logical system	
Detectability	It refers to the discovery of identifiable facts that are used to evaluate the performance of components by activity monitoring	
Liveness	It demonstrates those properties of the system components which cannot be checked without loop detection	
Localizability	It indicates the scope of application fir the software component within a spectrum of operation	
Mobility	It indicates the logical movement of component attributes cross reference and processed through an agreed upon protocol	
Monotonousness	It indicates non variant properties of the components which are designed to perform repeated tasks	
Resentfulness	It demonstrates non-agreeable properties of the system that can possibly bring unfair or displeasuring results to the stakeholders	
Resolvability	It refers to the stated facts or properties of the system that can be brought to a successful conclusion which is visible and distinct	
Suspendibility	It indicates the active components kept on hold or undetermined due to successful, definite and concluding components	

4.2 Efficiency

The efficiency requirement describes "the degree to which software makes optimal use of system resources as indicated by the attributes: time behavior and resource behavior". Table 2 indicates the notation profile for efficiency.

Table 2 Notation catalog for efficiency

Parameter	Definition	Notation
Abruptness	It indicates the sudden or unexpected series of behavioral pattern generated by the system execution	
Adequacy	It is the ability of the component to perform in order to fulfill a need or without being abundant or outstanding	
Adherence	It is the ability of the component to follow closely or exactly to the specifications during execution	
Antecedence	It refers to the priorities or precedence order of the tasks that a component can perform	
Delectability	This is a set of pleasing feature or outcomes of the system to the users	
Deliverability	It refers to the control transfers or hand—over mechanisms that the system carries successfully	
Insanity	It indicates a permanent disorder of the program of the system which is irreparable	
Mutuality	It indicates the attributes or operations which can be shared by two or more collaborating components	
Passivity	It is breadth of components which are inactive or not participating in any ongoing execution activity	
Precedency	It refers to the observable priorities of the components in time	
Promptness	It is extent to which the software generates a quick response without delay to the external environment operational requests	
Proximity	It suggests the functional closeness of two system belonging to two separate functional entities	
Quickness	This property is an indicator of the activities performed or occurred during a comparatively shorter time	
Quittance	It is a receipt that is generate as a token of successful processing from the collaborating components	
Rapidness	It is the notion of system characteristics by high speed	
Uniformity	It is the state or condition of the component in which everything is regular, homogenous or unvarying	
Utility	It is measure of gains achieved or losses suffered due to examination of sets of solution alternatives	

4.3 Flexibility

The flexibility requirement describes the "effort required to modify an operational program" or system. A software system is flexible if the changes in the operating environment do not impact its performance. Table 3 indicates the notation profile for Flexibility.

Table 3 Notation catalog for flexibility

Parameter	Definition	Notation
Alterability	It is the property of the usable components to change its behavior in some respect	
Ambivalence	It indicates the simultaneous existence of two opposed and conflicting attributes of components	
Amenability	This property indicates the component characteristics open or susceptible to suggestions. It also deals with testing and justification of component	
Compactness	It indicates well-constructed, closely .packed structure of system elements	
Concentricity	It is an indicator of centralized control applied to exercise the expected behavior from an organization of components	
Compressibility	It is the degree to which the system volume can be put together requiring comparably less memory space	
Resource Usage	It is the degree to which the components make resource consumption to initiate and terminate their operation	
Diversity	It is the relation that holds between two system elements when and only when they are not identical and existing with distinct profiles	
Duplicity	It is an indicator of dualism of interpretation derived from component semantics	
Extremity	It is the degree of acuteness or bounds of component performance resulting in the successful termination	
Omni-format	It deals with the all forms or appearances representations and specifications of the components	
Omni potency	It deals with the unlimited or great functional power of system elements which cannot be the case	
Orthogonality	It is the ability of the design elements to accommodate platforms and programming languages providing correct results	
Resilience	It is the extensive capacity of the system to address the changes and continuous improvements over a significant period of time	

4.4 Performance

The performance requirement is a "measure of process speed, response time, resource consumption, throughput, and efficiency". Performance is rated the highest in requirement hierarchy. Table 4 indicates the notation profile for performance.

5 Conclusion

Requirement profiler was identified to address the functional and non-functional requirements which were critical for developing the complex systems. The requirements were required to be categorized in a systemic manner. A notational basis for

Table 4 Notation catalog for performance

Parameter	Definition	Notation
Aperiodicity	It is the ability of the system to respond to randomly arriving events. In this case, there is no event information that can be known in advance	
Computability	It refers to a solvable problem with the help of mathematical logic or recursion theory or suitable algorithm	
Concurrency	It is the ability of the system to perform more than one tasks simultaneously and with potential interactions with each other	
Controllability	The controllability requirement describes the degree of control a software component adhering to cohesion and coupling	
Enormity	It is the extent to which the performance can go wrong as perceived by the users of the system	
Fidelity	It refers to the degree to which a system can perform without noise or distortion and generate accurate results	
Fragility	It indicates the quality of the system can be easily broken or damaged by application of rapid dynamic changes	
Latency	It the time when the system receives stimuli and generates response to it with the observable state change that includes time delays, the cause and the effect	
Memorability	It indicates characteristics of the system or ability to trace the original state information recorded periodically	
Rarity	It indicates the uncommonness or singularity of the system properties that can be observable	
Response Time	It represents the periodic or periodic responses of the system based on user request processing for a function or service	
Sensitivity	It concerns with testing the data in certain situations in algorithms, data range contained within certain extremes can lead to error-prone performance degradation	
Throughput	The throughput requirement describes the "number of processes the CPU (central processing unit) completes per time unit"	
Zestfulness	It represents lively quality of the system that increase enjoyment and bring exciting results to users	

accommodating the non-functional requirements was required to be created. The proposed framework implements the Requirement profiler with the help of requirement database which consists of permissible set of requirements.

References

1. Jarke, M., and K. Pohl. 1994. Requirements engineering in 2001: (virtually) managing a changing reality. *Software Engineering Journal* 257–266.

2. Easterbrook, S., R. Lutz, R. Covington, J. Kelly, Y. Ampo, and D. Hamilton. 1998. Experiences using lightweight formal methods for requirements modeling. *IEEE Transactions on Software Engineering* 24 (1): 04–14.
3. Kotonya, G., and I. Sommerville. 1996. Requirements engineering with viewpoints. *Software Engineering Journal* 05–18.
4. Lamsweerde, A., D. Robert, and L. Emmanuel. 1998. Managing conflicts in goal-driven requirements engineering. *IEEE Transactions on Software Engineering* 24 (11): 908–925.
5. Lamsweerde, A., and E. Letier. 2000. Handling obstacles in goal-oriented requirements engineering. *IEEE Transactions on Software Engineering* 26 (10): 978–1005.
6. Mylopoulos, J., L. Chung, and B. Nixon. 1992. Representing and using nonfunctional requirements: a process-oriented approach. *IEEE Transactions on Software Engineering* 18 (6): 483–497.
7. Cysneiros, L., and J. Leite. 2004. Nonfunctional requirements: from elicitation to conceptual models. *IEEE Transactions on Software Engineering* 30 (5): 328–350.
8. Gregoriades, A., and A. Sutcliffe. 2005. Scenario-based assessment of nonfunctional requirements. *IEEE Transactions on Software Engineering* 31 (5): 392–409.
9. Balasubramaniam, R., and M. Jarke. 2001. Toward reference models for requirements traceability. *IEEE Transactions on Software Engineering* 27 (1): 58–93.
10. Jackson, E., and J. Sztipanovits. 2009. Formalizing the structural semantics of domain-specific modeling languages. *Journal of Software and System Model* 8: 451–478. Springer-Verlag.
11. Thalheim, B., K.D. Schewe, and H. Ma. 2009. Conceptual application domain modeling. In *Proc. Sixth Asia-Pacific Conference on Conceptual Modelling (APCCM 2009)*, Wellington, New Zealand. CRPIT, vol. 96, 49–57.
12. OMG. 2010. Unified Modeling Language (OMG UML) Superstructure. Version 2.3.
13. OMG. 2010. Unified Modeling Language (OMG UML) Infrastructure. Version 2.3.
14. Meta Object Facility (MOF) Core Specification OMG Specification. 2006. Version 2.0.
15. Object Constraint Language. 2010. OMG Specification. Version 2.2.
16. Nugroho, A. 2009. Level of detail in UML models and its impact on model comprehension: a controlled experiment. *Journal of Information and Software Technology* 51: 1670–1685.
17. Aburub, F., M. Odeh, and I. Beeson. 2007. Modelling non-functional requirements of business processes. *Information and Software Technology* 49: 1162–1171.
18. Mens, T., and P. Gorp. 2006. A taxonomy of model transformation. *Electronic Notes in Theoretical Computer Science* 152: 125–142.

Some Routing Schemes and Mobility Models for Real Terrain MANET

Banoj Kumar Panda, Urmila Bhanja and Prasant Kumar Pattnaik

Abstract The primary challenges in mobile ad hoc network (MANET) are presence of obstacles, mobility, energy efficiency and network in dynamic topology environment. Efficient routing with obstacles avoidance in dynamic topology is a critical issue in MANET. Many mobility patterns have been recommended for the movement of nodes in presence of obstacles in MANET terrain. Some obstacles avoiding routing techniques are also proposed by some popular researchers. In this paper, many related articles have been reviewed and briefly discussed. The paper outlines advantages and drawbacks of each approach to get possible research scope in route planning in dynamic MANET topology in presence of obstacles.

Keywords MANET · Terrain · Routing techniques

1 Introduction

Disaster recovery during natural calamities like volcanic eruption, earthquake, tsunami, hurricanes and tornados or man-made calamity like explosions, fires and military operations are hampered when the MANET performance diminishes due to presence of obstacle in the terrain area. In addition to that performance of MANET also reduces due to node mobility, network congestion and insufficient node energy. A number of protocols are developed to improve the performance [1–4]. To develop

B. K. Panda
Department of Electronics & Telecommunication Engineering,
Utkal University, Bhubaneswar, India
e-mail: bk_panda2001@yahoo.com

U. Bhanja
Departments of Electronics & Telecommunication Engineering, IGIT, Sarang, India
e-mail: urmilabhanja@gmail.com

P. K. Pattnaik (✉)
School of Computer Engineering, Kalinga Institute of Industrial Technology (KIIT-DU),
Bhubaneswar, Odisha 751024, India
e-mail: patnaikprasant@gmail.com

© Springer Nature Singapore Pte Ltd. 2020
D. Swain et al. (eds.), *Machine Learning and Information Processing*,
Advances in Intelligent Systems and Computing 1101,
https://doi.org/10.1007/978-981-15-1884-3_49

ad hoc network, testing lab requires very high cost and so simulator tools are used for performance analysis. All simulation tools are designed considering plain terrain and do not taken into account real terrain features and obstacles in the simulation area. Presence of obstacles affects the received signal strength at the receiver and hence performance of MANET is reduced. Previously many authors have considered the geographic features and obstacles, and analysed its impact on performance on MANET but they have not developed an efficient obstacle, mobility and congestion aware optimal energy efficient routing protocol to overcome routing problems in terrain in the presence of convex and concave obstacles. Hence, in the current work, authors have reviewed the outcomes of all existing routing protocols, mobility schemes and outlined advantages and drawbacks of each approach and their scope to mitigate the challenges of mobility, network congestion and energy efficiency.

2   MANET Mobility Models

The mobility models used in MANET can be segregated into different as per their dependencies and restrictions given below.

Random based node movement-These mobility models are based on random movements of nodes. Here, any types of internodes dependencies or restrictions are not considered.
Temporal dependencies in movement-Here, present movement of a MANET node is decided based on its previous movement.
Group dependencies movement-Here moving pattern of any MANET node determined by the motion of other neighbouring nodes. *Geographic restrictions in terrain*-Some areas of the terrain are restricted from MANET node movement.

2.1   Random-Based Node Movement

Many random-based mobility models given by different authors are discussed below.
    The random way point (RWP) model for node mobility presumes a fixed number of nodes in a fixed sized rectangle terrain area. At start of simulation, the nodes are placed in a rectangular terrain uniformly. Here, each mobile node sets a destination arbitrarily and selects a random speed that distributed uniformly between $[v_{min}, v_{max}]$. When it reaches near the destination, node halts for an arbitrary time which is evenly distributed between $[P_{min}, P_{max}]$, again node selects a next direction and speed. Then node continues the same mobility pattern [5]. In case of random walk (RWM) mobility model, every nodes set an arbitrary direction which is evenly spread between $[0, 2\pi]$ (and with an arbitrary speed which is evenly spread in between $[v_{min}, v_{max}]$) and it proceeds for a set distance with that speed. After reaching at new point, same

process is repeated. The model is also known as Brownian motion, which is similar to the motion of particles present in a fluid [6].

Random direction model (RDM) allows a MANET node to set a direction and proceeds in the set direction with an arbitrary speed till an edge comes. Afterwards it selects a next direction and same process is repeated. The benefit of the process is that we get a even distribution of nodes in rectangular terrain area. Here, drawback of RDM is same as RWP and RWM that it is unrealistic [6].

RWP, RWM and RDM mobility models are unrealistic in nature and also suffer edge effects. So to avoid the above limitations, smooth mobility model (SMM) is proposed. Here, mobile nodes change the speed and direction gradually [6]. Random movement based mobility patterns of nodes covers the total terrain area. These mobility mechanisms are very easy to implement. Here, heterogeneous velocity may be combined very smoothly.

2.2 Time-Related Dependencies in Movement

Mobility mechanisms discussed in preceding section have assumed the variations of direction and speed suddenly, which is not naturalistic because of non-consideration of acceleration and deceleration. Hence, this section discusses some mobility models by using time-related dependencies.

The *Gauss-Markov* movement model [7] proposes the new value of node's velocity and direction in next instant (time slot $t + 1$) which depend on the present time slot (slot t). At the start of simulation, all nodes are assumed distributed uniformly in terrain area. Similarly, velocity and direction are assumed to be distributed in uniform pattern. Here, movement of every node is changed after a time span λt. New values of velocity and direction are selected depending on a first-order autoregressive process. Here, the mobility model does not consider presence of obstacles. The mobility model i.e. *smooth-random* model explains node movement in detailed manner [8, 9]. Here, nodes are segregated depending their highest velocity, maximum acceleration and preferred velocity. Here, next new values of velocities as well as directions are decided depending on these available parameters and the present values. During selecting velocity the direction may also selected with correlation to each other. Hence, we will record the realistic moment of the node before changing the direction. However, this model explained plain terrain.

In these above approaches, the temporal dependencies on motion of mobile nodes become smoother with respect to direction and velocity. Moreover, these approaches do not realise characteristics of tactical situations.

2.3 Group Dependencies

Apart from temporal dependencies, there is a possibility that nodes may move in groups. Hence, there is chance that a motion of a node may influence it neighbouring node or node around it.

Realising spatial dependence can be done using reference points. The paper [10] proposes a reference point to model the group node movement. Here, group movement is modelled as per to a random mobility model. Here, reference points positions changes as per random mobility model but within a group the reference point's position does not changes. Moreover, this model is silent about presence of obstacles. In the work [11], structured-group-mobility model is discussed. Here, a non-random movement vector is used. The nodes in group move depending on a fixed non varying formation. These formations are similar as the information exchanges among fire-fighters, conference participants and tanks. However, this model considers plain terrain area. In [12], social-network-inspired model uses interaction indicators, which are associated with all source-destination nodes. Here it is assumed higher value of interaction indicator to higher chance of social relational bonding, hence geographic gap is less. Here, nodes are clumped together with in a cloud depending on corresponding interaction parameter indicator. Movement of all the nodes follows random-waypoint model. Here, the waypoints are selected as per their interaction indicators. This model is described for plain terrain. In work [13], community-oriented model is proposed. Here, algorithms are utilised for the separation of terrain nodes into different group depending on interaction parameter. Here, the interaction indicators are modified over time. This does not consider real terrain situations.

In group mobility scenarios, the RPGM model performs better than others. In RPGM model, apart from the feature of batch movement, additional required features can be obtained by using a suitable model for given reference coordinate points in plain terrain.

2.4 Geographic and Obstacles Restrictions

Assumption of free node movement across entire area is also considered as unrealistic although we consider temporal and group dependencies. Many different approaches are proposed by different authors to stop the nodes motion to some parts of the terrain area where mountain, river or buildings present. The section given below describes some approaches to handle node movement avoiding restricted area.

2.4.1 Graph-Based Mobility

Graph dependant model [14] utilises a trace where vertex points are the probable destination nodes and its edges referred as routes to destinations. Depending on the trace, random-waypoint method is utilised. Here, nodes from an arbitrary position set graph initially and selects a vertex point, proceed towards that with an arbitrary velocity. Then again, it selects the next destination randomly and moves with random velocity. In weight based-waypoint model [15], the graph traces are particular areas like lecture room, meeting room, tea stall, etc. Here, every node selects sink node within terrain areas. These graph edges contain probabilities of selecting a destination.

2.4.2 Voronoi-Graph-Based Models

In real terrain, many kinds of obstacles are present like mountain, river, building and wall, etc. Hence, while doing simulation obstacles must be considered in the simulation area. In this model [16], the edges of a ground are utilised as a feed into create Voronoi plot. Here, motion graph includes the Voronoi plot and extra vertices. Extra apex points are formed due to junction of edges of Voronoi plot and building edges. Here, with the help of Voronoi diagrams, the paths are emulated having equal distance from all obstacles. These paths are not optimal paths. In this network, all lanes are assumed to be almost equal distance from entire houses and all nodes proceeds at middle of the lane. Paper [17] extends more realistically the presence of buildings and streets. Here, lanes and houses are realised more thoroughly and movement more realistic, but these movement not on optimal paths.

2.4.3 Sub Area-Based Approaches

In this approach, the simulation is divided into many sub-areas and then these used mobility modelling.

Author in paper [18] has divided the area into low density and high-density sub-area. Here, clusters are assumed as apex points of the plot and the routes are assumed as its extremities (edges). Here, each edge is associated with a probability ratio. A MANET node is assumed to proceed in the sub-area for an arbitrary time stretch as per RWP model. Then it selects one path depending on probabilities of the extremities. Then, the node proceeds forward using the selected path to the upcoming sub-area.

In *paper* [19], the entire terrain is divided into non-intersecting sub-areas. Within respective area, nodes proceed as per a random mobility pattern. Inter area transition is done in a similar process as it is done in graph-based mobility model by help of transit probabilities. When a mobile node is selected to alter area, it proceeds to a handed over area and changes to some other mobility pattern.

2.4.4 Map-Based Approaches

In the *paper* [20], the terrain is segmented to few squared pieces. Mobile nodes are modelled as walkers, which are proceeding on the apex points of the lanes. At initial time, the MANET nodes are arbitrarily scattered throughout the square area. Here, every node determines a particular direction along with a specified velocity. When a mobile node arrives a junction, it diverts its direction associated to a particular value probability. Velocity of the node is altered due course of time.

Work [21] describes urban environments vehicular traffic pattern. Here, each node selects a destination same way as RWM and selects a path with smallest moving time. Here, crossroads delays are emulated depending on number of nodes available. Here, a regular placement of the nodes along the terrain is perceived. In [22], the author evaluates routing protocols performance for different mobility like random walk (RW), random way point (RWP) and random direction (RD) considering obstacles such as mountain in the terrain which limits movement of nodes and blocks data transmission between nodes depending on the value of a new parameter i.e. probability of reachability (POR). The POR is obtained by dividing number of approachable routes with all probable routes present among all source-sink pairs. It is simulated using specially designed MATLAB simulator. It gives a noticeable difference in value of POR in presence of obstacles. Also it changes when number of obstacles varies. The paper in [23] outlines physical layer's impacts on network output efficiency of MANET. Here simulation is done considering propagation loss models that are ITU line of sight (ITU-LoS) and nonline of Sight (NLoS). First same scenario simulated using RWP mobility model for different stumbling block count in terrain area to estimate amount of data drops in different propagation path models. Here, outcomes of AODV routing protocol is analysed with 20 nodes.

3 Routing Protocol Generation Used in MANET

3.1 MANET Protocols

Some of the routing protocols mainly MANET protocols may be segregated into three main categories [24–26], they are as *Proactive l* [24], reactive [25, 26] and hybrid [27].

3.2 First-Generation MANET Protocols

3.2.1 Destination Sequenced Distance Vector (DSDV)

DSDV protocol which is discussed in [24] uses Bellman-Ford based algorithm to obtain the least count of jumps to reach at the sinking node. Here mobile node keeps a route-guiding table to store receiving node address, sequence numbers, next jump addresses, number of jumps and route-guiding table. These parameters are updated on regular time basis. DSDV uses sequence numbers and route-guiding table updates to compensate mobility. Here when a high-value sequence number route update is obtained, the available routing information will be replaced which reduces the possibility of route looping. If a high topology alteration occurs, then whole routing table information is changed that adds very high overhead to the network in high mobility scenarios.

3.2.2 Dynamic Source Routing (DSR)

DSR protocol was proposed in paper [25] and as per the author DSR protocol search a path as a source device tries to transmit data packet to a receiving node. There are mainly two activities present in DSR i.e. route discovery phase and route maintenance phase. In DSR, source routing technique is used. Here, the complete route address is stored in the route cache of a node. And this complete address guides data to the destination. Route error occurs when route breaks and that initiate the route maintenance process. So that expired routes are removed from route cache and a new route discovery is started. The work [28] compares a popular proactive protocol i.e. DSDV with a popular reactive protocol DSR and outlines individual's advantages and disadvantages.

3.3 Second-Generation MANET Routing Protocol

We have observed that from first-generation MANET mechanisms discussed above i.e. from DSR and DSDV, that their performance must enhanced in dynamic MANET scenario. Ad hoc on-demand distance vector (AODV) protocol that was suggested in [27] gives significant performance improvement on the DSDV routing protocol, the work has discussed the testing of AODV functionality in different scenario.

3.3.1 AODV Routing Mechanism

AODV do route discovery when required using route requests (RREQ) packets. Unlike as DSR, AODV uses distance vector technique; i.e. each node present along the path to keep a interim route-guiding table during communication. AODV has better technique over the DSR RRQ packet preventing excessive RREQ flooding using time-to-live (TTL) concept. Nodes within a connected route keep the source/destination IP address, senders address and sequence numbers in their routing tables. These information are used to construct reverse paths by route reply packets (RREP) [28]. AODV uses sequence numbers to track and eliminate outdated routes. Route error packets (RERR) are generated and sent to source when links break are identified to eliminate the stale links and initiate fresh route discovery process [28].

In [29], core principles of AODV protocol are discussed but it was silent about possible insight of future directions. During simulation the data of metrics like dropped packets, throughput and data delay may be considered [30, 31]. These parameters are useful indicator for grade of service and network efficiency, but testing is done taking AODV only without comparing with other protocols using random waypoint mobility model.

3.3.2   Variants of AODV

AODV protocol is the popular MANET protocol among all available protocols; so many improvements of AODV are proposed by different authors to combat some problems of MANET. Some variant AODV protocols are suggested below.

Multicast ad hoc on-demand distance vector (MAODV):

The major issue of MANET routing protocols like AODV, DSR and DSDV is lack of multicast support, which enables for communicating with multiple number of nodes and reduces control traffic overheads [32]. To tackle this issue MAODV routing technique is proposed. MAODV adds multicast operation support to existing AODV protocol. MAODV protocol uses the architecture of modified AODV and it adds multicast feature triggers (MACT) and the group level hello (GRPH) messages; here every node keeps two different routing tables, one for unicast and other for multicast function [33]. During route discovery process, MAODV broadcasts route request packets towards destination, which assists several end node IP addresses. From every IP addresses, route reply packets are returned, after receiving RREP the transmitting node will forward a MACT message towards the sink node triggering a multicast path. Multicast paths are cached in source, which keeps entire multicast end nodes in its memory and it enables the get unicast destinations from the source-tree without initiating route discovery. Leader of the group is that node which joined the group first. Leader is liable for batch management and maintenance, which is obtained by broadcasting GRPH messages that accommodate contain the group head IP address. Authors in [34] propose a modified AODV routing protocol which is aware of node energy and node mobility (MEAODV). The performance MEAODV protocol has 4–5% better in packet delivery ratio, 20–24% less in convergence time and network lifetime is increased 10–15% as compared to AODV. In [35], authors present a new route selection technique based on some points like speed of the node, direction of movement and halt time. Depending on these parameter values, a new parameter called *mobility_factor is taken* for choosing nodes to get a pathway between transmitting node and receiving node. Here, simulations of routing algorithms like DSR and AODV are done with RWP and levy walk models. It is clearly observed from simulation that using proposed method these protocols show superior performance contrast to normal AODV and DSR. However, this particular protocol is worthy for plain terrain.

3.4 Obstacle Aware Routing

Paper [36] proposes a new routing protocol known as cartography enhanced OLSR (CE-OLSR) for multi hop mobile ad hoc networks. The protocol uses an effective cartography method with a robust routing scheme. This cartography method uses the exact OLSR reduced traffic and improves activeness to the network dynamics. The proposed method is better and accurate than the normal network topology. In stability routing method, a reduced view is used from collected cartography, which only considers links not exceeding the threshold distance without crossing obstacles. In urban area, radio signal suffers very high shadowing and fading effects and many times fully obstructed by concrete structure. However, these protocols are not mobility, congestion or energy aware. The work [37] is the extended work with a distributed autonomic technique to locate obstacles (ALOE) with CE-OLSR protocol. This integration of ALOE and CE-OLSR gives very good improvement in comparison with CE-OLSR with or without knowledge of obstacle map. Here, two parameters like covering range and precision ratios are used to assess accurately the impact of this new method. Simulation outcomes indicate that suggested CE-OLSR obstacle aware technique perfectly locates the borderline obstacles present in terrain. Here, it is observed that ALOE-CE-OLSR obtains similar route viability, throughput and end-to-end delay as CE-OLSR having previous awareness of obstacles map. Moreover, these protocols are not designed mitigate mobility, congestion or energy consumption issues.

4 Conclusion

It is been observed from the above review that MANET performance depends mostly on uses of mobility models and routing protocol. The initial stage protocols were designed for ideal or plain terrain where presence of obstacles like river, mountain or building were not taken care of when designing moment of nodes in specified terrain. Later, a lot of work has been done considering realistic terrain i.e. in presence of obstacles. First-generation protocols are normally proactive protocols where control overheads are major problem which is taken care of in second-generation protocol design. In second-generation protocol, AODV is most popular and efficient one but it is not mobility, congestion or energy aware. In modified AODV protocols, few of these are taken care of but moreover they are designed for plain terrain. Hence, more research should be done to make the present protocols aware of mobility, congestion or energy consumption in presence of obstacles.

References

1. Johnson, D.B., and D.A. Maltz. 1996. Dynamic source routing in Ad Hoc wireless networks. In *Mobile Computing*, vol. 353, ed. T. Imielinski and H.F. Korth, 153–181. The Kluwer International Series in Engineering and Computer Science, US, Springer.
2. Royer, E.M., and C.K. Toh. 1999. A review of current routing protocols for ad hoc mobile wireless networks. *IEEE Personal Communications* 6 (2): 46–55.
3. Royer, E.M., P.M. Melliar-Smith, and L.E. Moser. 2001. An analysis of the optimum node density for ad hoc mobile networks. In *Proceedings of IEEE International Conference on Communications*, vol. 3, 857–861, Helsinki, Finland.
4. Chlamtac, I., M. Conti, and J.J.N. Liu. 2003. Mobile ad hoc networking: imperatives and challenges. *Ad Hoc Networks* 1 (1): 13–64.
5. Bai, F., and Helmy A. 2004. Wireless ad hoc and sensor networks. In *Chapter 1: a survey of mobility models*. http://nile.usc.edu/helmy/important/Modified-Chapter1-5-30-04.pdf.
6. Camp, T., J. Boleng, and V. Davies. 2002. A survey of mobility models for ad hoc network research. *Wireless Communications and Mobile Computing* 2 (5): 483–502.
7. Ariakhajorn, J., P. Wannawilai, and C. Sathiyatwiriyawong. 2006. A comparative study of random waypoint and gauss-markov mobility models in the performance evaluation of MANET. In *International Symposium on Communications and Information Technologies*, 894–899.
8. Bettstetter, C. 2001. Mobility modeling in wireless networks: categorization, smooth movement, and border effects. *ACM SIGMOBILE Mobile Computing and Communications Review* 5 (3): 55–66.
9. Bettstetter, C. 2001. Smooth is better than sharp: a random mobility model for simulation of wireless networks. In *Proceedings 4th International Symposium Modeling, Analysis and Simulation of Wireless and Mobile Systems MSWIM*, 19–27. Rome, Italy.
10. Jayakumar, G. and G. Ganapati. 2008. Reference point group mobility and random waypoint models in performance evaluation of MANET routing protocols. *Journal of Computer Systems, Networks, and Communications* 2008: 10. Article ID 860364.
11. Blakely, K., and B. Lowekamp. 2004. A structured group mobility model for the simulation of mobile ad hoc networks. In *International Conference of Mobile Computing Network, Proceedings of the 2nd International Workshop Mobile Management & Wireless Access Protocol*, 111–118. Philadelphia, USA.
12. Musolesi, M., S. Hailes, and C. Mascolo. 2004. An ad hoc mobility model founded on social network theory. In *Proceedings of the 7th ACM Int. Symposium on Modeling, Analysis and Simulation of Wireless and Mobile Systems*, 20–24. Venice, Italy.
13. Musolesi, M., and C. Mascolo. A community based mobility model for ad hoc network research. In *Proceedings of the 2nd ACM/SIGMOBILE International Workshop on Multi-hop Ad Hoc Networks: from Theory to Reality REALMAN'06, Colocated with MobiHoc2006*, 31–38. Florence, Italy.
14. Bittner, S., W.-U. Raffel, and M. Scholz. 2005. The area graph-based mobility model and its impact on data dissemination. In *Proceedings IEEE PerCom*, 268–272. Kuaai Island, Hawaii, USA.
15. Hsu, W.-J., K. Merchant, H.-W. Shu, C.-H. Hsu, and A. Helmy. 2005. Weighted waypoint mobility model and its impact on ad hoc net-works. *ACM SIGMOBILE Mobile Computing and Communications Review* 9 (1): 59–63.
16. Jardosh, A., E.M. Belding-Royer, K.C. Almeroth, and S. Suri. 2003. Towards realistic mobility models for mobile ad hoc networks. In *Proceedings IEEE MobiCom*, 217–229. San Diego, USA.
17. Jardosh, A.P., E.M. Belding-Royer, K.C. Almeroth, and S. Suri. 2005. Real-world environment models for mobile network evaluation. *IEEE Journal on Selected Areas in Communications* 23(3): 622–632.
18. Bettstetter, C., and C. Wagner. 2002. The spatial node distribution of the random waypoint mobility model. In *Proceedings of the 1st German Workshop Mobile Ad-Hoc Network WMAN'02*, 41–58. Ulm, Germany.

19. Gunes, M., and J. Siekermann. 2005. CosMos—communication scenario and mobility scenario generator for mobile Ad-Hoc networks. In *Proceedings of the 2nd International Workshop MANETs Interoper. Iss. MANETII'05*. Las Vegas, USA.
20. Kraaier, J., and U. Killat. 2005. The random waypoint city model—user distribution in a street-based mobility model for wireless network simulations. In *Proceedings of the 3rd ACM International Workshop on Wireless Mobile Applications and Services on WLAN hotspots*, 100–103. Cologne, Germany.
21. Chones, D.R., and F.E. Bustamante. 2005. An integrated mobility and traffic model for vehicular wireless networks. In *International Conference on Mobile Computing Network, Proceedings of the 2nd ACM International Workshop on Vehicular Ad Hoc Networks*, 69–78. Cologne, Germany.
22. Kumar, C., B. Bhushan, and S. Gupta. 2012. Evaluation of MANET performance in presence of obstacles. *International Journal of Ad hoc, Sensor & Ubiquitous Computing (IJASUC)* 3(3).
23. Amjad, K., M. Ali, S. Jabbar, M. Hussain, S. Rho, and M. Kim. 2015. Impact of dynamic path loss models in an urban obstacle aware ad hoc network environment. *Journal of Sensors* 2015: 8. Article ID 286270.
24. Perkins, C., and P. Bhagwat. 1994. Highly dynamic destination-sequenced distance-vector routing (DSDV) for mobile computers. In *Proceedings of Sigcomm Conference on Communications Architectures, Protocols and Applications*, 234–244. London, England, UK.
25. Johnson, D.B., and D.A. Maltz. 1996. Dynamic source routing in ad hoc wireless networks. In *Mobile computing*, vol. 5, ed. T. Imielinski and H. Korth, 153–181. Kluwer Academic Publishers.
26. Perkins, C.E., and E.M. Royer. 1997. Ad-Hoc on-demand distance vector routing. In *Proceedings of the 2nd IEEE Workshop on Mobile Computing Systems and Applications*, 1–11.
27. Alotaibi, E., and B. Mukherjee. 2011. A survey on routing algorithms for wireless Ad-Hoc and mesh networks. *Computer Networks the International Journal of Computer and Telecommunications Networking* 56 (2): 940–965.
28. Divecha, B., A. Abraham, C. Grosan, and S. Sanyal. 2007. Analysis of dynamic source routing and destination-sequenced distance-vector protocols for different mobility models. In *Proceedings of First Asia International Conference on Modelling & Simulation*, 224–229. Phuket, Thailand.
29. Morshed, M., H. Rahman, R.R. Mazumder, and K.A.M. Lutfullah. 2009. Simulation and analysis of Ad-Hoc on-demand distance vector routing protocol. In *Proceedings of ICIS*, 610–614. Seoul, Korea.
30. Huang, J., X. Fan, X. Xiang, M. Wan, Z. Zhuo, and Y. Yang. 2016. A clustering routing protocol for mobile ad hoc networks. *Mathematical Problems in Engineering* 2016: 10. Article ID 5395894.
31. Perkins, C.E., and E.M. Royer. 2001. Multicast operation of the Ad-Hoc on-demand distance vector routing protocol. In *Proceedings of 5th Annual ACM/IEEE International Conference on MOBILE Computing and Networking*, 207–218. Seattle, Washington, USA.
32. Mobaideen, W.A., H.M. Mimi, F.A. Masoud, and E. Qaddoura. 2007. Performance evaluation of multicast Ad-Hoc on-demand distance vector protocol. *Computer Communications* 30 (9): 1931–1941.
33. Malarkodi, B., P. Gopal, and B. Venkataramani. 2009. Performance evaluation of Ad-Hoc networks with different multicast routing protocols and mobility models. In *Proceedings of 2009 International Conference on Advances in Recent Technologies in Communication and Computing IEEE*. 81–84. India.
34. Rashid, U., O. Waqar, and A.K. Kiani. 2017. Mobility and energy aware routing algorithm for mobile Ad-Hoc networks. In *International Conference on Electrical Engineering (ICEE)*, 1–5.
35. Sarkar, S., and R. Dutta. 2017. Mobility-aware route selection technique for mobile *ad hoc* networks. *IET Wireless Sensor Systems* 3: 55–64.

36. Belghith, Abdelfettah, and M. Belhassen. 2012. CE-OLSR: a cartography and stability enhanced OLSR for dynamic MANETs with obstacles. *KSII Transactions on Internet and Information Systems* 6(1).
37. Belhassen, M., A. Dhraief, A. Belghith, and H. Mathkour. 2018. ALOE: autonomic locating of obstructing entities in MANETs. *Journal of Ambient Intelligence and Humanized Computing* 1(13).

Author Index

Printed in the United States
By Bookmasters